Anticancer Agents

ACS SYMPOSIUM SERIES **796**

Anticancer Agents

Frontiers in Cancer Chemotherapy

Iwao Ojima, Editor
State University of New York at Stony Brook

Gregory D. Vite, Editor
Bristol-Myers Squibb Pharmaceutical Research Institute

Karl-Heinz Altmann, Editor
Novartis Pharmaceuticals

American Chemical Society, Washington, DC

Library of Congress Cataloging-in-Publication Data

Anticancer agents : frontiers in cancer chemotherapy / Iwao Ojima, Gregory D. Vite, Karl-Heinz Altmann, editors.

    p. cm.—(ACS symposium series; 796)

    Includes bibliographical references and index.

    ISBN 0–8412–3745–X

    1. Antineoplastic agents—Congresses.

    I. Ojima. Iwao, 1945- II. Vite, Gregory D. III. Altmann, Karl-Heinz. IV. American Chemical Society. Division of Organic Chemistry. V. American Chemical Society. Division of Medicinal Chemistry. VI. American Chemical Society. Meeting (219th : 2000 : San Francisco, Calif.) VII. Series.

    [DNLM: 1. Antineoplastic Agents—therapeutic use—Congresses. 2. Neoplasms—drug therapy—Congresses. QZ 267 A6285 2001]

RS431.A64 A56   2001
616.99′4061—dc21   2001022569

The paper used in this publication meets the minimum requirements of American National Standard for Information Sciences—Permanence of Paper for Printed Library Materials, ANSI Z39.48–1984.

Copyright © 2001 American Chemical Society

Distributed by Oxford University Press

All Rights Reserved. Reprographic copying beyond that permitted by Sections 107 or 108 of the U.S. Copyright Act is allowed for internal use only, provided that a per-chapter fee of $20.50 plus $0.75 per page is paid to the Copyright Clearance Center, Inc., 222 Rosewood Drive, Danvers, MA 01923, USA. Republication or reproduction for sale of pages in this book is permitted only under license from ACS. Direct these and other permission requests to ACS Copyright Office, Publications Division, 1155 16th St., N.W., Washington, DC 20036.

The citation of trade names and/or names of manufacturers in this publication is not to be construed as an endorsement or as approval by ACS of the commercial products or services referenced herein; nor should the mere reference herein to any drawing, specification, chemical process, or other data be regarded as a license or as a conveyance of any right or permission to the holder, reader, or any other person or corporation, to manufacture, reproduce, use, or sell any patented invention or copyrighted work that may in any way be related thereto. Registered names, trademarks, etc., used in this publication, even without specific indication thereof, are not to be considered unprotected by law.

PRINTED IN THE UNITED STATES OF AMERICA

Foreword

The ACS Symposium Series was first published in 1974 to provide a mechanism for publishing symposia quickly in book form. The purpose of the series is to publish timely, comprehensive books developed from ACS sponsored symposia based on current scientific research. Occasionally, books are developed from symposia sponsored by other organizations when the topic is of keen interest to the chemistry audience.

Before agreeing to publish a book, the proposed table of contents is reviewed for appropriate and comprehensive coverage and for interest to the audience. Some papers may be excluded to better focus the book; others may be added to provide comprehensiveness. When appropriate, overview or introductory chapters are added. Drafts of chapters are peer-reviewed prior to final acceptance or rejection, and manuscripts are prepared in camera-ready format.

As a rule, only original research papers and original review papers are included in the volumes. Verbatim reproductions of previously published papers are not accepted.

ACS Books Department

Contents

Preface...xi

1. Cancer Drug Discovery and Development: New Paradigms for a New Millennium...1
   E. A. Sausville, J. I. Johnson, G. M. Cragg, and S. Decker

2. Evolutionary Biosynthesis of Anticancer Drugs.........................16
   George R. Pettit

3. Some Recent Developments in the Synthesis and Structure
   –Activity Relationship of Novel Taxanes.....................................43
   John F. Kadow, Thomas Altstadt, Shu-Hui Chen, Pierre Dextraze,
   Karen Du, Craig Fairchild, Jerzy Golik, Steven Hansel,
   Kathy A. Johnston, Robert A. Kramer, David R. Langley,
   Frank Lee, Byron Long, Harold Mastalerz, Carl Ouellet,
   Robert Perrone, William C. Rose, Paul Scola, Gene Schulze,
   Andrew Staab, Quifen May Xue, Michael Walker, Jen-Mei Wei,
   Mark Wittman, J. J. Kim Wright, Mary Zoeckler, and
   Dolatrai M. Yvas

4. New Generation Taxoids and Hybrids of Microtuble-Stabilizing
   Anticancer Agents..59
   Iwao Ojima, Scott D. Kuduk, Subrata Chakravarty, Songnian Lin,
   Tao Wang, Xudong Geng, Michael L. Miller, Pierre-Yves Bounaud,
   Evelyne Michaud, Young Hoon Park, Chung-Ming Sun, John C. Slater,
   Tadashi Inoue, Christopher P. Borella, John J. Walsh,
   Ralph J. Bernacki, Paula Pera, Jean M. Veith, Ezio Bombardelli,
   Antonella Riva, Srinivasa Rao, Lifeng He, George A. Orr,
   Susan B. Horwitz, Samuel J. Danishefsky, Giovanni Scambia,
   and Cristiano Ferlini

5. Discodermolide and Taxol: A Synergistic Drug Combination
   in Human Carcinoma Cell Lines..81
   Susan Band Horwitz, Laura A. Martello, Chia-Ping H. Yang,
   Amos B. Smith, III, and Hayley M. McDaid

6. **Highly Efficient Semisynthesis of Biologically Active Epothilone Derivatives**..................97
   Gregory D. Vite, Robert M. Borzilleri, Soong-Hoon Kim, Alicia Regueiro-Ren, W. Griffith Humphreys, and Francis Y. F. Lee

7. **Synthetic and Semisynthetic Analogs of Epothilones: Chemistry and Biological Activity**..................112
   Karl-Heinz Altmann, Marcel J. J. Blommers, Giorgio Caravatti, Andreas Flörsheimer, Kyriacos C. Nicolaou, Terrence O'Reilly, Alfred Schmidt, Dieter Schinzer, and Markus Wartmann

8. **Synthesis and Biological Activity of Epothilones**..................131
   Ulrich Klar, Werner Skuballa, Bernd Buchmann, Wolfgang Schwede, Thomas Bunte, Jens Hoffmann, and Rosemarie B. Lichtner

9. **Epothilones and Sarcodictyins: From Combinatorial Libraries to Designed Analogs**..................148
   Nicolas Winssinger and K. C. Nicolaou

10. **Synthesis and Structure–Activity Relationship Studies of Cryptophycins: A Novel Class of Potent Antimitotic Antitumor Depsipeptides**..................171
    Chuan Shih, Rima S. Al-Awar, Andrew H. Fray, Michael J. Martinelli, Eric D. Moher, Bryan H. Norman, Vinod F. Patel, Richard M. Schultz, John E. Toth, David L. Varie, Thomas H. Corbett, and Richard E. Moore

11. **Farnesyltransferase Inhibitors as Potential Anticancer Agents**..................190
    J. B. Gibbs, N. J. Anthony, I. Bell, C. A. Buser, J. P. Davide, S. J. deSolms, C. Dinsmore, S. L. Graham, G. D. Hartman, D. C. Heimbrook, H. Huber, K. S. Koblan, N. E. Kohl, R. B. Lobell, and T. M. Williams

12. **Farnesyltransferase Inhibitors: From Squalene Synthase Inhibitors to the Clinical Agent BMS-214662**..................199
    John T. Hunt

13. Inhibiting Farnesyl Protein Transferase with Sch-66336: Potentially a Selective Noncytotoxic Therapy for Human Cancer...............214
   A. G. Taveras, F. G. Njoroge, R. J. Doll, J. Kelly, S. Remiszewski, A. K. Mallams, C. S. Alvarez, J. del Rosario, R. R. Rossman, B. Vibulbhan, P. Pinto, J. Deskus, M. Connolly, J. Wang, J. Desai, R. Wolin, A. Afonso, A. B. Cooper, D. F. Rane, Y.-T. Liu, C. J. Aki, J. Chao, C. Strickland, P. Weber, M. Liu, M. S. Bryant, A. A. Nomeir, R. Patton, L. Wang, L. James, D. Carr, P. Kirschmeier, W. R. Bishop, V. Girijavallabhan, and A. K. Ganguly

14. Pyrrolo[2,3-*d*]pyrimidine and Pyrazolo[3,4-*d*]pyrimidine Derivatives as Selective Inhibitors of the EGF Receptor Tyrosine Kinase...............231
   G. Caravatti, J. Brüggen, E. Buchdunger, R. Cozens, P. Furet, N. Lydon, T. O'Reilly, and P. Traxler

15. STI571: A New Treatment Modality for CML?...............245
   Jürg Zimmermann, Pascal Furet, and Elisabeth Buchdunger

16. The Discovery and Development of Second-Generation Matrix Metalloproteinase Inhibitors for the Treatment of Cancer...............260
   Andy Baxter and John Montana

17. Prospects for Antiangiogenic Therapies Based upon VEGF Inhibition...............282
   Pascal Furet and Paul W. Manley

18. Carbohydrate-Based Tumor Antigens as Antitumor Vaccine Agents...............299
   Jennifer R. Allen and Samuel J. Danishefsky

19. Drugs to Enhance the Therapeutic Potency of Anticancer Antibodies: Antibody–Drug Conjugates as Tumor-Activated Prodrugs...............317
   Walter A. Blättler and Ravi V. J. Chari

Author Index...............339

Subject Index...............342

Preface

Cancer is a major health concern worldwide with an estimated six million new cases per year. In the United States alone, it is second only to heart disease as a cause of death, with lung cancer and breast cancer as the most devastating malignancies. The threat posed by cancer was recognized in 1971, with the legislation of the National Cancer Act, which increased research funding and which helped develop several national research centers. Considerable progress has been made since then to understand the mechanism of the disease and to control it, both with prevention and therapy. However, at the beginning of the 21st century, cancer is still a very difficult disease to treat and it is far from conquered. Accordingly, cancer chemotherapy will continue to be an extremely important field of medical research in the new century. The purpose of this American Chemical Society (ACS) Symposium Series book is to summarize the recent advances in the development of new anticancer agents after the introduction into clinical practice of paclitaxel (Taxol®), which captured tremendous attention in the 1990s and which has become arguably the most important drug in cancer chemotherapy presently. However, all anticancer drugs developed to date have substantial side effects and weakness against drug-resistant tumors. Therefore, at the beginning of the 21st century, it is very important for us to consolidate various ongoing approaches and to explore all possible ways to combat cancer.

It is clearly anticipated that chemistry will continue to play a key role in the development of next generation anticancer agents in the new century. In fact, the ACS held a symposium on paclitaxel in 1992, just before the drug was approved by the U.S. Food and Drug Administration (FDA), and three ACS divisions (Organic Chemistry, Medicinal Chemistry, and Chemical Health and Safety) held symposia on taxane anticancer agents in 1994. The ACS Symposium Series 583, *Taxane Anticancer Agents: Basic Science and Current Status* was published in 1995 based on the latter symposia and additional chapters. At the end of the 20th Century, Ojima, the managing editor of this book, organized the ACS Symposium on the "New Prospects in Anticancer Agents for the 21st Century" sponsored by the ACS Divisions of Organic Chemistry and Medicinal Chemistry as a part of the Society's "Chemistry in the 21st Century Celebration" program at its National Meeting in San Francisco, California, March 2000. Ojima gratefully acknowledges generous support from Bristol-Myers Squibb, Novartis Pharma AG, Serale/Monsanto, Merck Company, Lily Research Lab., Indena, SpA, Schering-Plough Research Institute, and ACS Chemistry in the 21st Century Celebration fund.

Although the symposium contributions covered many emerging cutting-edge approaches in cancer chemotherapy, the editors have expanded the scope of this book by incorporating several chapters from research laboratories that were not represented at the symposium. This expanded scope, the editors believe, makes this book even more attractive and informative. This book contains 19 chapters, covering a wide range of topics including the National

Cancer Institute's new paradigms for cancer drug discovery and development; various "statins" from marine natural products; new generation taxoids and other microtubule-interacting anticancer agents (e.g., epothilones, discodermolide, eleutherobin, sarcodictyins, and cryptophycins); the potentially tumor specific mechanism-based agents (e.g., farnesyl transferase inhibitors); inhibitors of signaling pathway (e.g., EGF receptor tyrosine kinase inhibitors); anti-angiogenesis agents including matrix metalloproteinase (MMP) inhibitors; VEGF inhibitors; Abl tyrosine kinase inhibitors that exhibit promise against chronic myeloid leukemia (CML); antitumor vaccines; and tumor activated prodrugs (TAPs) combining monoclonal antibody and cytotoxic agents.

Accordingly, this book describes various innovative approaches to the development of new generation anticancer drugs, which are aiming to not only increase potency against drug-resistant tumors, but also are trying to gain tumor specificity or using non-cytotoxic agents to reduce undesirable side effects. Tumor specificity and reduction of undesirable side effects are essential to improve the quality of life for cancer patients. The research on anticancer agents requires quite an interdisciplinary endeavor that brings together various disciplines such as synthetic organic, bioorganic, medicinal, and biological chemistry, chemical biology, pharmacology, biochemistry, and oncology. All these research communities will greatly benefit from this book. With the publication of this book, the editors and authors wish to further encourage and stimulate ongoing and new interdisciplinary collaborative research activities on the development of highly effective anticancer agents and therapies to combat this very difficult and deadly disease.

We dedicate this book to people who live with and fight cancer.

Iwao Ojima
Department of Chemistry
State University of New York at Stony Brook
Stony Brook, NY 11794–3400

Gregory D. Vite
Oncology Chemistry Department
Bristol-Myers Squibb Pharmaceutical Research Institute
P.O. Box 4000
Princeton, NJ 08543–4000

Karl-Heinz Altmann
Novartis Pharma AG
TA Oncology,
Chemistry Research, K–136.4.21
CH4002 Basel
SWITZERLAND

Chapter 1

Cancer Drug Discovery and Development: New Paradigms for a New Millennium

E. A. Sausville, J. I. Johnson, G. M. Cragg, and S. Decker

Developmental Therapeutics Program, Division of Cancer Treatment and Diagnosis, National Cancer Institute, 6116 Executive Boulevard, Suite 5000, Rockville, MD 20892

A revolution of information on the causes of the cancer phenotype is resulting in the former "empirical" drug development paradigm being replaced by a "rational" paradigm which emphasizes the effect of a test agent on a molecular drug target rather than on maximal tolerated dose. The new paradigm utilizes tools such as combinatorial chemistry, *in vitro* target-based testing, and data mining algorithms. Potential new anticancer therapeutics developed using the rational paradigm are now reaching human clinical trials. The optimal path for cancer drug discovery will likely result from a marriage of "informed empiricism" in generating new lead structures and rational design of eventual drug candidates, building on biological and structural data.

In the past twenty years a revolution has come to pass in our understanding of how cancer cells differ from normal cells. Accordingly, a shift is occurring in the methods used to discover anti-cancer drugs. Figure 1 illustrates the historically used "empirical" paradigm in which initial enthusiasm for a

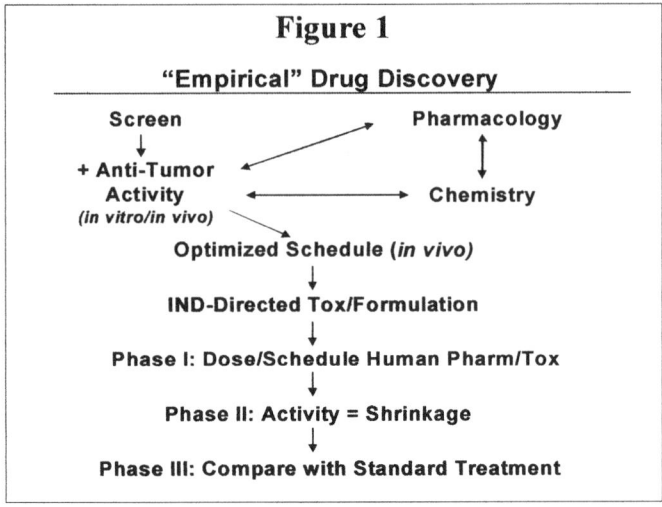

compound as an anti-neoplastic agent was based on evidence of toxicity for tumor cells either growing in tissue culture or as tumors in animals. This development path did not necessarily consider the mechanism of action of the agent as an important feature. The relation of the pharmacologic features of the agent to pharmacodynamic measures of drug action at the tumor site was also not considered.

In contrast, Figure 2 illustrates the paradigm to which the cancer drug development community is shifting. Initial selection of candidate drug molecules is made with a particular molecular target in mind. The strength of a target's importance is conveyed by an understanding of its role in the molecular "economy" of a tumor. Chemical optimization relies on pharmacology of drug with respect to its effect on the target in animals rather than on maximal tolerated dosing schemes. Then, at a relatively late stage in the molecule's development, there can occur a very focused *in vivo* evaluation in specialized models, measuring the effect on the biochemical target, as well as evidence of a potentially useful effect on tumor growth or progression.

Presented here is an introduction to this evolving knowledge, and laboratory and informatics tools available to researchers involved in the discovery and development of potential cancer therapies.

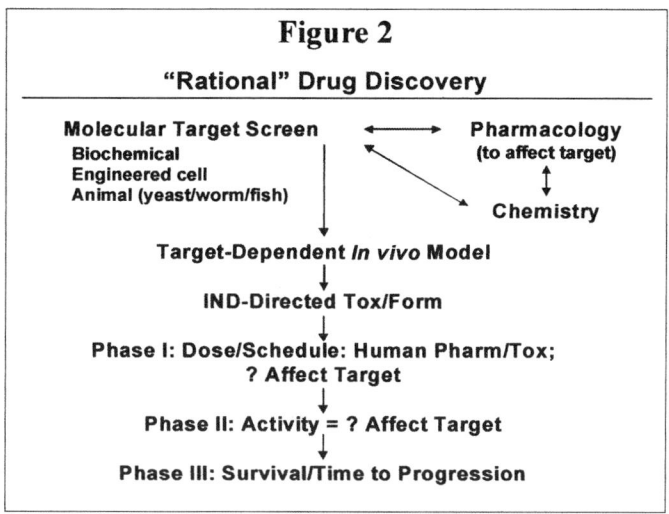

Diversity Generation

Whether the initial screen leading to a candidate drug is a mechanistically driven biochemical screen or an empirical antiproliferative screen, the more types of molecules examined, the more likely a successful novel structure will be identified. Examples such as reverse transcriptase and protease inhibitors directed against HIV in particular have emphasized the power of combining a knowledge of an intended target's structure at the atomic level with utilization of chemical diversity, and is the paradigm according to which generation of cancer therapeutics will increasingly be directed. Thus, methods which maximize the diversity of structures available to target-directed screens are of great interest.

Synthetic Compounds and Combinatorial Chemistry

The method of purposely synthesizing more than one compound as a result of a single reaction began about 15 years ago with the synthesis (*1*), of multiple numbers of peptides on polyethylene rods. Accordingly, the result represented "combinations" of products deriving from a set of reactants, and hence the term "combinatorial" chemistry. This technique of arranging solid supports in 96-well plates to allow facile addition of sequential reactants was dubbed "pin technology."

In contrast, the "split-pool" method, (wherein each individual reactant is added to equally divided solid support material for coupling, then solid supports pooled for washing, followed by division of supports for the next reactant coupling) was an improvement on the "pin" technique as the resultant products existed in theoretically equimolar amounts. The next technical advance was the enclosure of the solid support resin within porous polypropylene bags (the "tea-bag" method) (2), which also utilized split pool synthesis. This method allowed for the synthesis of mixtures of millions of peptides in solution phase which could be utilized for testing. A third variation of the split-pool approach involved the use of tagged polymeric beads as the solid support (3). The beads could either be used directly in assays, or the peptides could be cleaved from the bead into solution.

The synthesis of small peptide libraries, although an interesting chemical approach, was of limited use to cancer drug discovery, as small peptides do not enter cells and therefore are, in general, not suitable drug candidates. A variation of the chemistry was discovered which could produce peptide-like molecules (4) or "peptoids," which, lacking amide bonds, are less susceptible to enzymatic cleavage. The report (5) of the synthesis of a low-molecular weight non-peptidic benzodiazepine library demonstrated the utility of the combinatorial library approach for the first time as a drug discovery tool. Subsequent advances in the field include more efficient deconvolution schemes such as the positional scanning approach (6), a greater number of scaffolds on which to construct libraries, and a wider variety of reagents and amenable reaction conditions. Virtually all pharmaceutical companies now utilize combinatorial chemistry methods in their drug discovery and development programs.

Cancer drug discovery programs, however, also make use of existing sources of diversity. The National Cancer Institute's Developmental Therapeutics Program (DTP) has operated for more than 40 years a repository of synthetic and pure natural products which were evaluated as potential anticancer agents. These materials are from a variety of sources world-wide, and were selected to represent unique structural diversity. The collection contains both synthetic compounds and fully characterized pure natural products. DTP has recently catalogued the structural diversity of the non-proprietary portion of this collection with the goal of creating diverse sets of compounds to be utilized in cancer-relevant drug discovery research programs.

Natural Products

Plants have a long history of use in the treatment of cancer (7). Of the plant-derived anticancer drugs in clinical use, the longest known are the so-called vinca alkaloids, vinblastine and vincristine, isolated from the Madagascar periwinkle,

Catharanthus roseus. More recent additions to the armamentarium of naturally-derived chemotherapeutic agents are the taxanes and camptothecins. Paclitaxel initially was isolated from the bark of *Taxus brevifolia*, collected in Washington State as part of a random collection program by the U.S. Department of Agriculture for the NCI (*8*). Likewise, the clinically-active agents, topotecan (hycamptamine), irinotecan (CPT-11) and 9-aminocamptothecin, are semi-synthetically derived from camptothecin, isolated from the Chinese ornamental tree, *Camptotheca acuminata* (*9*). Camptothecin (as its sodium salt) was advanced to clinical trials by NCI in the 1970s, but was dropped because of severe bladder toxicity. The flavone, flavopiridol, currently in Phase I clinical trials, is scheduled to be advanced to Phase II trials against a broad range of tumors (*10*). While flavopiridol is totally synthetic, the basis for its novel structure is a natural product isolated from *Dysoxylum binectariferum*.

Despite more intensive investigation of terrestrial flora, it is estimated that only 5-15% of the approximately 250,000 species of higher plants have been systematically investigated, chemically and pharmacologically (*11*). The potential of large areas of tropical rainforests remains virtually untapped and may be studied through collaborative programs with source country organizations, such as those established by the NCI.

The marine environment is a rich source of bioactive compounds, many of which belong to totally novel chemical classes not found in terrestrial sources (*12*). As yet, no compound isolated from a marine source has advanced to commercial use as a chemotherapeutic agent, though several are in various phases of clinical development as potential anticancer agents. The most prominent of these is bryostatin 1, isolated from the bryozoan, *Bugula neritina* (*13*). This agent exerts a range of biological effects, thought to occur through modulation of protein kinase C. Phase II trials are either in progress or are planned against a variety of tumors, including ovarian carcinoma and non-Hodgkin's lymphoma (*10*).

The first marine-derived compound to enter clinical trials was didemnin B, isolated from the tunicate, *Trididemnum solidum* (*13*). Unfortunately, it has failed to show reproducible activity against a range of tumors in Phase II clinical trials, while demonstrating significant toxicity. Ecteinascidin 743, a metabolite produced by another tunicate *Ecteinascidia turbinata*, has significant *in vivo* activity against the murine B16 melanoma and human MX-1 breast carcinoma models, and is currently undergoing clinical evaluation in Europe and the United States (*12*).

Sponges are traditionally a rich source of bioactive compounds in a variety of pharmacological screens (*12*). In the cancer area, halichondrin B, a macrocyclic polyether initially isolated from the sponge, *Halichondria okadai* in 1985, was accepted for preclinical development by the NCI in 1992. Analogs derived from the total synthesis of halichondrin B have shown superior activity to the natural product (*14*), and are now in advanced preclinical development by the NCI in collaboration with Eisai Research Institute.

Exciting untapped resources are the deep-sea vents occurring along ocean ridges, such as the East Pacific Rise and the Galapagos Rift. Exploration of these regions is being performed by several organizations, and their rich biological resources of macro and micro organisms are being catalogued (*15,16*). Samples are being evaluated by the NCI in collaboration with chemists at Research Triangle Institute.

Antitumor antibiotics are amongst the most important of the cancer chemotherapeutic agents, which include members of the anthracycline, bleomycin, actinomycin, mitomycin and aureolic acid families (*17*). Clinically useful agents from these families are the daunomycin-related agents, daunomycin itself, doxorubicin, idarubicin and epirubicin; the glycopeptidic bleomycins A_2 and B_2 (blenoxane); the peptolides exemplified by dactinomycin; the mitosanes such as mitomycin C; and the glycosylated anthracenone, mithramycin. All were isolated from various *Streptomyces* species. Other clinically active agents isolated from *Streptomyces* include streptozocin and deoxycoformycin.

The large number of microbial agents reflects the major role played by the pharmaceutical industry in this area of drug discovery and development. Generally, industry has focused on the *Actinomycetales*, but expansion of research efforts, often supported by government funding, to the study of organisms from diverse environments, such as shallow and deep marine ecosystems and deep terrestrial subsurface layers, has demonstrated their potential as a source of novel bioactive metabolites (*18*).

The NCI 60 Cell Line Screen

Historically, pure chemical compounds and natural product extracts have been initially studied *in vivo* or *in vitro*. Both *in vitro* and *in vivo* approaches were primarily empirical, in that clear knowledge of a compound's target was not available. As described in the introduction to this chapter, these empirical approaches are being supplanted by approaches where definition of a compound's biochemical target exists before any efficacy experiment (*in vitro* or *in vivo*) is undertaken. Ultimately, whether a compound is created with a target in mind, or is simply a novel chemotype, assessment of antiproliferative potential is conventionally done both *in vitro* and *in vivo*.

In 1985, the NCI initiated feasibility studies to develop an *in vitro* screening model based on the use of established human tumor cell lines (*19*). This screen initially focused on lung cancer, but by 1990 when it became fully operational, the screen included cell lines from most of the major human malignancies. The current model utilizes approximately 60 human tumor cell lines representing hematologic malignancies, melanoma, and cancers of the lung, colon, breast, brain, kidney and prostate. The rationale for this approach was that human cell lines, particularly from solid malignancies, might yield an improved model for human cancer and might allow for selection of agents with activity towards specific histological types of cancer (20). The operation of the screen began with evaluation of toxicity in multiple cell lines using tetrazolium-based assays (21), but due to the variable behavior of tetrazolium dyes, the core assay was changed to a protein-based measurement which has been previously described (22). The selection of cell lines and assay parameters has been reviewed formerly (20).

The method employs continuous drug incubation for 48 hours with a relatively high starting cell inoculum (5,000-20,000 cells/well in a 96-well plate). Such an approach allows for calculation of net cell loss as a result of drug treatment and allows cell kill to be distinguished from growth inhibition (22). Furthermore, this approach minimizes the effects of the variable doubling times of cell lines and the masking of nonspecific antiproliferative effects. Using this method, the NCI *in vitro* anti-cancer drug screen has been operational since April of 1990. As of August, 2000, more than 79,000 pure compounds have been studied in 60 cell lines, 7,881 compounds have shown evidence of *in vitro* activity, and 4,384 have been selected for *in vivo* studies. A large fraction of the data is available to the public through a Web-based search capacity (http://dtp.nci.nih.gov).

A major issue in interpreting the results emerging from the 60 cell line screen was the orderly display of data and comparison of different patterns of dose-response curves. Paull and colleagues (23) developed the "mean graph" representation of screening data in which the mean concentration affecting all cells at three different levels (GI50 = concentration causing 50% inhibition of growth; TGI = concentration causing total growth inhibition; and LC50 = concentration causing 50% cell kill) is plotted in the midline of the graph, and the behavior of individual cell lines represented as a deflection to the left, by convention, for cells more resistant than the mean, and to the right for cells more sensitive than the mean. An unexpected and important application of this manner of presentation is the richly informative patterns of activity which emerge. Correlations of activity patterns could be quantified using a computer algorithm named COMPARE (24). COMPARE quantitatively estimates the similarity of a test compound's pattern of activity in the screen to the pattern of known molecules. Use of COMPARE led to the recognition that compounds with the same or similar mechanisms of action despite different chemical structure, often resulted in mean graph patterns which were closely correlated (24). This approach resulted in identification of new structures with tubulin-binding activity (25), topoisomerase I (26) and II activity (27), and antimetabolites (28). In another example, the use of COMPARE linked the effects of cucurbitacin (29) and jasplakinolide (*30*) to the actin cytoskeleton. Compounds with unique profiles of growth inhibition, suggesting modes of action not shared with known clinically active classes of chemotherapeutic agents, have also been identified (*31*). Alternatively, the COMPARE algorithm can suggest which in a series of analogs has a mechanism likely to be distinct from the lead.

The value of characterizing the expression of molecular targets in the NCI's human tumor cell line panel emerged from the observation that when function and expression of the *mdr-1* drug resistance efflux pump was examined, and the data analyzed by COMPARE, one could define in the database compounds with a similar pattern of activity which were subsequently shown to interact with the *mdr-1* gene product (*32*). This remarkable finding suggested that quantitation of targets in the cell lines could discover previously unreported activities of unknown compounds which interact with the target. This has been borne out particularly in experience with the epidermal growth factor receptor (*33*). Not all molecular target estimates yield meaningful correlations with compound activity though. For example, expression of *mrp* or *lrp* has not yielded correlations with compound activity in a way pointing to novel compounds affecting those targets.

The insight provided by COMPARE into mechanism of drug action continues to yield useful information, suggesting a continuing role for the 60 cell line screen in the new drug discovery paradigm. Development after discovery of a "novel" or "unique" COMPARE pattern will be highly coupled to elucidation of a compound's molecular target(s).

Examples of Rationally Improved Anti-cancer Agents

Early examples of the applications of molecular biology to cancer drug discovery include discovery of agents which interfere with the function of growth factor receptors as well as the *ras* oncogene. Armed with the knowledge of intracellular signaling proteins which act "downstream" from those receptors, emphasis is now on the search for small molecules which inhibit the intracellular signaling proteins, a few classes of which are described below.

Hsp90 Interactors

Heat shock protein 90 (Hsp90) is an abundant molecular chaperone that is involved in the folding of a defined set of signaling molecules including steroid-hormone receptors and kinases. Structural studies have identified the ATP-binding site in the N-terminal domain of Hsp90, which can be blocked by high-affinity inhibitors (*34*). Specifically, Hsp90 interacts with and stabilizes several newly synthesized oncogenic protein kinases (e.g., p185(erbB2), p60(v-src), and Raf-1) and is required for the stability and dominant-negative function of mutated p53 protein. Benzoquinone ansamycins are antibiotics with anticancer potential. First proposed to be tyrosine kinase inhibitors, they are now known to affect the stability of the target kinase by directly affecting HSP90 chaperone function. While herbimycin A and geldanamycin (GA) have been widely used in preclinical studies, both drugs are poor candidates for clinical trials owing to their *in vivo* toxicity and lack of stability. 17-Allylamino-17-demethoxygeldanamycin (17-AAG), is an ansamycin derivative which binds specifically to HSP90 in a manner similar to GA itself. 17-AAG also leads to degradation of the receptor tyrosine kinase p185erbB2, the serine/threonine kinase Raf-1, and mutant p53. Even though HSP90 binding by 17-AAG is weaker than by GA, 17-AAG and GA caused biologic effects in tumor cells at similar doses. Since 17-AAG has a better toxicity profile than GA, it is clearly a more engaging clinical candidate than benzoquinone ansamycin (*35*), and is in Phase I trials.

Recently a dimer of geldanamycin has been synthesized (*36*) which had selective activity against HER-kinases. Selectivity was a function of linker length and required two intact GM moieties. This geldanamycin dimer is reported to be a potent inducer of G1 block and apoptosis of breast cancer cell lines that overexpress HER2, but does not appreciably inhibit the growth of 32D cells that lack HER-kinases. This dimer could be useful in the treatment of carcinomas dependent on HER-kinases.

Kinase Antagonists

Cyclin-dependent kinases (CDKs) have been recognized as key regulators of cell cycle progression. Alteration and deregulation of CDK activity are pathogenic hallmarks of neoplasia. Therefore, inhibitors or modulators are of interest to explore as novel therapeutic agents in cancer, as well as other hyperproliferative disorders. In the last decade, the discovery and cloning of the CDKs have led to the identification of novel modulators of CDK activity. Initial experimental results demonstrated that these CDK modulators are able to block cell cycle progression, induce apoptotic cell death, promote differentiation, inhibit angiogenesis, and modulate transcription. Alteration of CDK activity may occur indirectly by affecting upstream pathways that regulate CDK activity or directly by targeting the CDK holoenzyme. The development of three CDK modulators, UCN-01, flavopiridol, and 9-nitropaullone, are discussed below.

UCN-01 (7-hydroxystaurosporine) is a potent inhibitor of tumor cell growth both in cell culture and has activity in *in vivo* xenograft models (*37,38*). It was found to be as potent an inhibitor of PKC as staurosporine but with a greater degree of specificity for PKC than other protein kinases. The ability of UCN-01 to inhibit the kinase activity of recombinant protein kinase C (PKC) isozymes was characterized using an *in vitro* kinase assay. Two distinct groups of isozymes could be defined on the basis of relative potency of kinase inhibition. UCN-01 was 15- to 20-fold more potent for inhibition of the Ca(2+)-dependent isozymes, compared with the Ca(2+)-independent isozymes. In contrast, UCN-02 (the diastereomer of UCN-01) and staurosporine exhibited less ability to discriminate between Ca(2+)-dependent and -independent isozymes. An analysis of the inhibition by UCN-01 and staurosporine of the kinase activity of PKC-alpha and -delta indicated mixed inhibition kinetics.

Cellular pharmacology studies revealed that the capacity of UCN-01 to inhibit PKC could be dissociated from growth inhibition (*39*), and instead was associated with inappropriate activation of CDK1. The Chk1 regulatory kinase (with effects on cdc25 phosphatase) has recently been identified as a target kinase for UCN-01 (*40*), and therefore refinement of the UCN-01 structure to address this target is now possible.

Flavopiridol is a semisynthetic flavonoid that emerged from an empirical screening program as a potent antiproliferative agent. Mechanistic studies went on to demonstrate this agent was a direct CDK 1, 2, and 4 competitive ATP site antagonist (*41*). Moreover, preclinical studies demonstrated the capacity of flavopiridol to induce programmed cell death, promote differentiation, inhibit angiogenic processes and modulate transcriptional events. Initial clinical trials have shown that concentrations that inhibit cell proliferation and CDK activity *in vitro* can be safely achieved in humans. Testing in early clinical human trials with infusional flavopiridol showed evidence of biologic effect in some patients with non-Hodgkin's lymphoma, renal, prostate, colon and gastric carcinomas. Main side effects were secretory diarrhea and a pro-inflammatory syndrome associated with hypotension. Biologically active plasma concentrations of flavopiridol are easily achievable in patients receiving infusional flavopiridol. Phase II trials with infusional flavopiridol in several tumor types, with other schedules, and in combination with standard chemotherapies are being assessed. In conclusion, flavopiridol is the first CDK inhibitor to be tested in clinical trials. Although important questions remain to be answered, this positive experience will stimulate the development of novel CDK modulators for cancer therapy (*42*).

To address the need for additional chemotypes that may serve as lead structures for drugs that would not have the toxicities associated with flavopiridol, compounds with a similar pattern of cell growth inhibitory activity in the NCI's *in vitro* anticancer drug screen have been recognized by COMPARE and then screened for anti-CDK activity in a biochemical screen. The benzodiazepine derivative, now known as kenpaullone, was revealed by that approach as a moderately potent (IC50 0.4 µM) inhibitor of CDK2, and led to the development of the class of compounds referred to as paullones (*43*). To investigate structure-activity relationships and to develop paullones with antitumor activity, derivatives of now kenpaullone were synthesized. The 9-nitro derivative, named alsterpaullone, showed a high CDK1/cyclin B inhibitory activity (IC50 = 0.035 µM) and exceeded the *in vitro* antitumor potency of the other paullones by an order of magnitude.

Although modulation of CDK activity is a well-grounded concept and now CDK modulators are being assessed for clinical testing, important scientific questions remain to be addressed (*44*). These questions include whether one or more CDKs should be inhibited, how CDK inhibitors should be combined with other chemotherapy agents, and which CDK substrates should be used to assess the biologic effects of these drugs in patients.

Farnesyltransferase Inhibitors

The *ras* oncogene and its 21 kD protein product, Ras, emerged during the last decade as a potentially exploitable target for anticancer drug development (*45*). The high frequency of *ras* mutation in human cancers prompted an intensive study to find ways to control its oncogenic function. After stimulation by various growth factors and cytokines, Ras activates several downstream effectors. Ras undergoes several posttranslational modifications that facilitate its attachment to the inner surface of the plasma membrane. The first and most critical modification is the addition of a farnesyl isoprenoid moiety in a reaction catalyzed by the enzyme protein farnesyltransferase (FTase). Different classes of FTase inhibitors have been identified that block farnesylation of Ras, reverse Ras-mediated cell transformation in human cell lines, and inhibit the growth of human tumor cells in nude mice. In transgenic mice with established tumors, FTase inhibitors cause regression in some tumors, which appears to be mediated through both apoptosis and cell cycle regulation. FTase inhibitors have been well tolerated in animal studies and do not produce the generalized cytotoxic effects in normal tissues that are a major limitation of most conventional anti-cancer agents.

The knowledge that Ras was readily prenylated by protein farnesyl transferase (PFTase) and that inhibition of this prenylation had functional consequences for the transformed phenotype that expressed oncogenic *ras* provided the rationale for the development of PFTase inhibitors. The initial enthusiasm for this approach seemed justified by the early identification of PFTase inhibitors that were potently and specifically able to block Ras processing, signaling and transformation in transformed and tumor cell lines *in vitro* and in certain selected animal models. Development efforts, however, have encountered numerous stumbling blocks that were not foreseen. Current research is pointing to the idea that Ras contributes to tumor development via multiple pathways, and efforts should be directed at determining the critical pathways in this development (46).

Iterations of Classical Targets

A common dictum in the development of small molecules for non-cancer indications is that molecules with high affinity for their particular target but which act in a biochemically reversible mechanism should be the goal of drug discovery paradigms. Affinity constants in the nM range are to be desired. Molecules which cause covalent modification of target proteins, e.g., Michael acceptors or molecules with "alkylating" potential, should be avoided. Since cancer treatments must, as described above, operate in a cellular environment relevant to several molecular targets, we must consider that some rethinking of these precepts may be in order to address the special case of cancer therapy-directed agents. Specifically, a very tight binding inhibitor that is functionally irreversible or indeed covalent may not be such a bad idea, provided that the binding occurs in an appropriately specific context. In fact, a covalent or kinetically irreversible interaction with a specific target may be yet one more way of augmenting drug selectivity to the range of targets expressed in the cancer cell milieu.

The development of irreversible or ultra-tight binding inhibitors also recognizes that an ultimate goal of cancer treatment for the major fraction of afflicted patients is reduction of cell mass. As the greatest fraction of cells of the average solid tumor may not be in an actively proliferating phase of the cell cycle at the time of treatment, such tight binding inhibitors may occupy their binding sites and therefore be available to afford a useful therapeutic effect at the time the cells next attempts cycle re-entry. While it could be argued that this paradigm is suspiciously similar to that which underlies the action of DNA-directed alkylating agents, we must remember that the latter drugs actually do work, albeit in the minority of diseases. One would profitably consider the characteristics which afforded these drugs their success, and attempt to emulate their properties, but with a range of novel targets differentially expressed in tumor cells.

Conclusion

It is apparent that the rich variety and diversity of targets emerging from biologic studies of cancer cells will provide opportunities for drug discovery unparalleled in the past. A key aspect leading to the success of this endeavor will be partnerships between chemists and biologists. Examples where such approaches have led to elucidation of lead structures by direct association with target (*47*) or by "smart" screening strategies (*48*) will become more frequent. The challenge will be rapidly turning these discoveries into useful therapeutics for patients.

References

1. Geysen, H. M.; Meleon, R. H.; Barteling, S. J. *Proc. Nat'l. Acad. Sci.* **1984**, *81*, 3998-4002.
2. Houghten, R. A.; Pinella, C.; Blondelle, S. E.; Appel, J. R.; Dooley, C. T.; Cuervo, J. H. *Nature* **1991**, *354*, 84-86.
3. Lam, K. S.; Salmon, S. E.; Hersh, E. M.; Hruby, V. .J.; Kazmierski, W. M.; Knapp, R. J. *Nature* **1991**, *354*, 82-84.
4. Zuckermann, R. N.; Martin, E. J.; Spellmeyer, D. C. *J. Med.. Chem.* **1994**, *37*, 2678-2685.
5. Bunin, B. A.; Ellman, J. *J. Am. Chem. Soc.* **1992**, *114*, 10997-10998.
6. Dooley, C. T.; Houghten, R. A. *Life Sci.* **1993**, *52*, 1509-1517.
7. Cragg, G. M.; Boyd, M. R.; Cardellina, J. H. II; Newman, D. J.; Snader, K .S.; McCloud, T. G. In *Ciba Foundation Symposium*; Chadwick, D. J.; Marsh, J. Eds.; Wiley & Sons, Chichester, U. K., 1994; *185*, pp 178-196.
8. Cragg, G. M.; Schepartz, S. A.; Suffness, M.; Grever, M. R. *J. Nat. Prod.*, **1993**, *56*, 1657-1668.
9. Wall, M. E.; Wani, M.C. *Cancer Res.* **1995**, *55*, 753-760.
10. Christian, M. C.; Pluda, J. M.; Ho, T. C.; Arbuck, S. G.; Murgo, A. J.; Sausville, E. A. *Sem. Oncol.*, **1997**, *24*, 219-240.
11. Balandrin, M. F.; Kinghorn, A. D.; Farnsworth, N. R. In *Human Medicinal Agents from Plants.* , Kinghorn, A. D.; Balandrin, M. F. Eds.; ACS Symposium Series 534; American Chemical Society: Washington, DC, 1993; pp 2-12.
12. Carté, B. K. *BioScience*, **1996**, *46*, 271-286.
13. McConnell, O.; Longley, R. E.; Koehn, F. E. In *The Discovery of Natural Products with Therapeutic Potential*, Gullo, V. P. Ed. Butterworth-Heinemann: Boston, MA, 1994; pp 109-174.
14. Towle, M. J.; Salvato, K. A.; Budrow, J.; Wels, B.; Kuznetsov, G.; Aalfs, K. K.; Welsh, S.; Zheng, W.; Seletsky, B. M.; Palme, M. H.; Habgood, G. J.; Singer, L. A.; DiPietro L. V.; Wang, Y.; Chen, J. J.; Quincy, D. A.; Yoshimatsu, K.; Yu, M. J.; Littlefield, B. A. *91st Annual Meeting Proceedings* AACR, **2000**, *41*, p 215.
15. Lutz, R. A.; Shank, T. M.; Fornari, D. J. *Nature*, **1994**, *371*, 663-664.
16. Lutz, R. A.; Kennish, M. J. *Rev. Geophys*, **1993**, *31*, 211-242.

17. *Cancer Chemotherapeutic Agents*, Foye, W. O., Ed.; ACS Professional Reference Book; American Chemical Society: Washington, DC, 1995.
18. Newman, D. J. *SIM News*, **1994**, *44*, 277-283.
19. Boyd, M. R. In *Accomplishments in Oncology;* Frei, E. II.; Freireich, E., Eds.; Lippincott: Philadelphia, PA, 1986; pp 68-76.
20. Boyd, M. R. *In Anticancer Drug Development Guide: Preclinical Screening, Clinical Trials, and Approval*; Teicher, B., Ed; Humana Press: Totowa, NJ, 1996; pp 23-42.
21. Scudiero, D. A.; Shoemaker, R. H.; Paull, K. D.; Monks, A.; Tierney, S.; Nofziger, T. H.; Currens, M. J.; Seniff, D.; Boyd, M. R. *Cancer Res.* **1988**, *48*, 4827-4833.
22. Monks, A.; Scudiero, D.; Skehan, P.; Shoemaker, R.; Paull, K.; Vistica, D.; Hose, C.; Langley, J.; Cronise, P.; Vaigro-Wolff, A. *J. Nat'l. Cancer Inst.* **1991**, *83*, 757-766.
23. Paull, K. D.; Shoemaker, R. H.; Hodes, L.; Monks, A.; Scudiero, D. A.; Rubinstein, L.; Plowman, J.; Boyd, M. R. *J. Nat'l. Cancer Inst.* **1989**, *81*, 1088-1092.
24. Paull, K. D.; Hamel, E.; Malspeis, L. In *Cancer Chemotherapeutic Agents;* Foye, W., Ed.; ACS Professional Reference Book 9; American Chemical Society: Washington, D.C., 1995, pp 9-45.
25. Paull, K. D.; Lin, C. M.; Malspeis, L.; Hamel, E. *Cancer Res.* **1992**, *52*, 3892-3900.
26. Kohlhagen, G.; Paull, K. D.; Cushman, M.; Nagafuji, P.; Pommier, Y. *Mol. Pharmacol* **1998**, *54*, 50-58.
27. Leteurtre, F.; Kohlhagen, G.; Paull, K. D.; Pommier, Y. *J. Nat'l. Cancer Inst.* **1994**, *86*, 239-244.
28. Cleaveland, E. S.; Monks, A.; Vaigro-Wolff, A.; Zaharevitz, D. W.; Paull, K.; Arbalan, K.; Conney, D. A.; Ford, H. Jr. *Biochem. Pharmacol.* **1995**, *49*, 947-954.
29. Duncan, K. K.; Duncan, M. D.; Alley, M. C.; Sausville, E. A. *Biochem. Pharmacol.* **1996**, *52*, 1553-1560.
30. Bubb, M. R.; Senderowicz, A. M.; Sausville, E. A.; Duncan, K. K.; Korn, E. D. *J. Biol. Chem.* **1994**, *269*, 14869-14871.
31. Bradshaw, T. D.; Wrigley, S.; Shi, D. F.; Schultz, R. J.; Paull, K. D.; Stevens, M. F. *Brit. J. Cancer* **1998**, *77*, 745-752.
32. Alvarez, M.; Paull, K.; Monks, A.; Hose, C.; Lee, J. S.; Weinstein, J.; Grever, M.; Bates, S.; Fojo, T. *J. Clin. Invest.* **1995**, *95*, 2205-2214.
33. Wosikowski, K.; Schuurhuis, D.; Johnson, K.; Paull, K. D.; Myers, T. G.; Weinstein, J. N.; Bates, S. E. *J. Nat'l. Cancer Inst.* **1997**, *89*, 1505-1515.
34. Buchner, J. *Trends Biochem. Sci.* **1999**, *4*, 136-141.

35. Schulte, T. W.; Neckers, L. M. *Cancer Chemother. Pharmacol.* **1998**, *42*, 273-279.
36. Zheng, F. F.; Kuduk, S. D.; Chiosis, G.; Münster, P. N.; Sepp-Lorenzino, L.; Danishefsky, S. J.; Rosen, N. *Cancer Res.* **2000**, *60*, 2090-2094.
37. Seynaeve, C. M.; Stetler-Stevenson, M.; Sebers, S.; Kaur, G.; Sausville, E. A.; Worland, P. J. *Cancer Res.* **1993**, *53*, 2081-2086.
38. Seynaeve, C. M.; Kazanietz, M. G.; Blumberg, P. M.; Sausville, E. A.; Worland, P. J. *Mol. Pharmacol.* **1994**, *45*, 1207-1214.
39. Wang, Q.; Worland, P .J.; Clark, J. L.; Carlson, B. A.; Sausville, E. A. *Cell Growth Differ.* **1995**, *6*, 927-936 [published erratum appears in *Cell Growth Differ.* **1995**, *6*, 1339].
40. Graves, P. R.; Yu, L.; Schwarz, J. K.; Gales, J.; Sausville, E. A.; O'Connor, P. M.; Piwnica-Worms, H. *J. Biol. Chem.* **2000**, *275*, 5600-5605.
41. Sausville, E. A.; Zaharevitz, D.; Gussio, R.; Meijer, L.; Louarn-Leost, M.; Kunick, C.; Schultz, R.; Lahusen, T.; Headlee, D.; Stinson, S.; Arbuck, S. G.; Senderowicz, A. *Pharmacol. Ther.* **1999**, *82*, 285-292.
42. Senderowicz, A. M. *Invest. New Drugs* **1999**, *17*, 313-320.
43. Schultz, C.; Link, A.; Leost, M.; Zaharevitz, D. W.; Gussio, R.; Sausville, E. A.; Meijer, L.; Kunick, C. *J. Med. Chem.* **1999**, *42*, 2909-2919.
44. Senderowicz, A. M.; Sausville, E. A. *J. Nat'l. Cancer Inst.* **2000**, *92*, 376-387.
45. Rowinsky, E. K.; Windle, J. J.; Von Hoff, D. D. *J. Clin. Oncol.* **1999**, *11*, 3631-3652.
46. McCormick, F. *Trends Cell. Biol.* **1999**, *9*, M53-56.
47. Liu, S.; Widom, J.; Kemp, C. W.; Crews, C. M.; Clardy, J. *Science* **1998**, *282*, 1324-1327.
48. Haggarty, S. J.; Mayer, T. U.; Miyamoto, D. T.; Fathi, R.; King, R. W.; Mitchison, T. J.; Schreiber, S. L. *Chem. Biol.* **2000**, *4*, 275-286.

Chapter 2

Evolutionary Biosynthesis of Anticancer Drugs

George R. Pettit

Cancer Research Institute and Department of Chemistry and Biochemistry, Arizona State University, P.O. Box 872404, Tempe, AZ 85287–2404

Natural biosynthetic products are the result of 3.8 billion years of evolutionary biosynthetic organic reactions directed at ever more specific molecular design and targeting. Presently, there may be at least 35 million different species of living organisms. That number may represent over 100 billion readily extractable compounds for biological evaluation. Most importantly, among such an obvious vast array there are surely many millions of substances useful for the broad spectrum of human medical problems and certainly for improving human cancer treatment. The plant drugs combretastatin, phyllantostatin/phyllanthoside and pancratistatin as well as the marine animal drugs bryostatin, spongistatin, dolastatin and cephalostatin provide important illustrations.

Overview

The U. S. National Cancer Institute (NCI) research programs directed at discovery of new and clinically useful animal, plant and microorganism anticancer constituents were implemented in 1957 and very soon demonstrated that 2-4% of plant specimens produce a great variety of anticancer agents (1-9). The dramatic discoveries arising from the early NCI research, such as camptothecin (10-13) and taxol (14-19) have stimulated considerable worldwide interest and initiation of analogous programs. Because of this vitally important NCI endeavor, new antineoplastic and/or cytotoxic biosynthetic products are being discovered now worldwide at an increasing rate. Certainly, the potential for discovering new animal (*20-22*), plant and microorganism (*23-24*) biosynthetic products for treatment of human cancer is great and offers promise of many curative approaches to the cancer problem. The world's flora may number up to 800,000 and the more conspicuous members of our terrestrial vegetation, the angiosperms, may number from 300,000 to some 500,000 (*25*). Furthermore, enormous numbers of marine animal (over 2,000,000) and microorganism species (over 30 million) are available for investigation (*25,26*). Even now, fewer than 10% of the higher plants and fewer than 0.5% of the marine animals have received even a cursory effort to detect antineoplastic constituents. So, the most important animal (*27*), microorganism and plant (*28*) cancer chemotherapeutic drugs still await discovery (*29*). Because of our long and very productive experience in discovering and developing new anticancer drugs to clinical trials based on the overall in vitro and in vivo antineoplastic activity combined with substantial evidence of important molecular targeting, the current clinical trial results have been very reassuring. That is especially true for advances involving apoptosis and bcl-2 modulation (*30,31*), down-regulation of protein kinase (*32,33*), tubulin inhibition (*34,35*) and cancer antiangiogenesis/vascular targeting (*36,37*). Those molecular targeting approaches, combined with our increasing efforts to utilize cyclin-dependent kinases (*38-41*) and tyrosine kinases (*42,43*), along with other mechanism-based approaches (*44*), should allow even better selection of future leads from the planet's exciting repository of marine animals, microorganisms, and plants.

The following examples resulting from our group's research efforts provide ample support for these conclusions (*45*).

The Combretastatins

Tropical and subtropical trees of the family Combretaceae represent a reservoir of constituents with potentially useful biological properties. The genus *Combretum* contains 25 species used in the traditional medical practices of Africa and India. Samples of the southern African tree *Combretum caffrum* (Eckl. Zehy.) Kuntze collected in 1973 and recollected in 1979 afforded extracts which showed activity in the U.S. National Cancer Institute's astrocyte reversal (9ASK) system and the P388 lymphocytic leukemia cell line. Historically, this extract was the first to be successfully fractionated by means of the 9ASK system. We reported the isolation in 1982 of the first combretastatin along with its structure and later synthesis (*46*). Subsequently, we isolated a number of active constituents designated combretastatins from this African bush willow used by the Zulu as a charm to ward off enemies and in traditional medical practices.

In 1987, we reported (47) the isolation and synthesis of combretastatin A-1 from the South African *Combretum caffrum* and this constituent showed significant cancer cell growth inhibition and antimitotic activity. That was followed by the isolation, structural elucidation and synthesis (48,49) of combretastatins A-2 and A-3 and the isolation and structure of combretastatins D-1 (50) and D-2 (51) as well as initial (52) structural modification. A comparison of the diphenol combretastatin A-1 with the monophenol counterpart combretastatin A-4 (53), the most active anticancer member of the combretastatin family, revealed a very similar antimitotic activity (IC_{50} 2-3 μM), but much greater cancer cell growth inhibition (ED_{50} ~0.0009 $\mu g/ml$, P388 cell line).

Development of combretastatin A-4 to the current Phase I human cancer clinical trials was accelerated following synthesis of the phosphate prodrug (54) and the uncovering of its very promising antiangiogenesis effects (55-57). Once administered, the phosphate prodrug was presumed to be converted into the parent drug via non-specific phosphatases and then transported intracellularly (54). The phosphate showed similar cancer cell line inhibition when compared to the parent compound (GI_{50} 0.0004 $\mu g/ml$, P388 cell line), while greatly increasing its aqueous solubility to 20 mg/ml (48,54)

Combretastatin A-4 and its phosphate prodrug have been shown to selectively damage tumor neovasculature with induction of extensive blood flow shutdown in the metastatic tumor compared to normal tissues. For example, six hours following treatment using the murine CaNT colon adenocarcinoma and a single i.p. injection of combretastatin A-4 prodrug (100 mg/kg), vascular function shutdown in the tumor was rapid, irreversible and extensive (56,59). In November, 1998 four Phase I human cancer clinical trials were initiated, two in the United States and two in England. Current clinical results (60) have been very encouraging and Phase II human cancer clinical trials are soon to be initiated (Spring, 2001).

Phyllanthoside/Phyllanthostatin 1

The genus *Phyllanthus* (Euphorbiaceae), which has more than 600 species in at least 10 subgenera, is found on all continents except for Europe and Antarctica. Many plants of this genus have been widely used in traditional medicine, and they are therefore of interest as potential sources for new drugs. *Phyllanthus acuminatus* Vahl (subgenus *Conami*, section *Nothoclema*) is widespread in Latin America, from southern Mexico through the West Indies and Central America to Peru and northern Argentina. Throughout this area,

Combretastatin D-2

Combretastatin A-4

Combretastatin A-1

Combretastatin A-4 prodrug

Phyllanthoside, R = COCH₃, R₁ = H
Phyllanthostatin, R = H, R₁ = COCH₃

Pancratistatin

Pancratistatin prodrug

concoctions of the plant (usually of the leaves) have been used to remedy headaches, asthma, and inflammations, especially of the breast, among other ailments (*61*).

As part of the exploratory plant evaluation program of the NCI, the roots of a Costa Rican tree believed to be *P. brasiliensis* Muell but later identified as *P. acuminatus* were collected in the early 1970s, in collaboration with the US Department of Agriculture. An ethanol extract of this collection was found to inhibit growth of P388 lymphocytic leukemia. Later collections of these roots were examined by the Kupchan group and by us. Following the 1978 recollection, we isolated and elucidated the structures of a series of biologically active glycosides from the dichloromethane extract of the roots (*61-64*). The structures of these compounds were determined by spectroscopic analyses and by X-ray crystallographic analyses of degradation products.

One of these glycosides was named phyllanthoside, of which the structure of the aglycone was proved by X-ray analysis. Of the other compounds isolated by us, phyllanthostatin 1 was shown to be the C-4 *O*-acetyl isomer of phyllanthoside. Both compounds markedly retarded growth of the P388 lymphocytic leukemia (52% life extension at 6.68 mg/kg; 62-90% life extension at 4-16 mg/kg) and the murine B16 melanoma (12% curative level at 8 mg/kg and 62-90% life extension at 4-16 mg/kg; 52-90% life extension at 6-48 mg/kg), and they significantly inhibited growth of a human myeloma cell line (private communication) with similar degrees of activity. Later it became clear that in solution the glycosides are interconvertible via an orthoacid rearrangement (*65*).

Phyllanthoside exhibited potentially useful antineoplastic activity against i.p. implanted murine B16 melanoma and P388 leukemia, as noted above, and it was an extremely potent inhibitor of protein synthesis, with an IC_{50} of about 0.22 mM. Concurrent pharmacology studies showed that the mouse is a poor model in which to evaluate the antitumor activity of this glycoside. The degradation in mouse plasma occurred much more rapidly (in less than 30 seconds after administration of 16 mg/kg) than in dog, monkey, and human plasma (*66*). In a human tumor colony forming assay, it elicited an overall response of 50% at approximately one microgram per milliliter, for a one-hour exposure, and against fresh human tumor specimens particular activity was seen with ovarian cancers. The strong activity against human neoplastic cell lines representing breast, CNS (TE671), colon (Colo 205), lung, ovary, and melanoma (Lox) cancers led to phyllanthoside being chosen for preclinical development by the NCI. Animal studies showed that the major toxicities at high doses were coagulopathy, hepatotoxicity, and lymph node necrosis, whereas neurotoxicity and hematotoxicity were not observed (private communication from the NCI).

In 1990, phyllanthoside was accepted for Phase 1 human trial by the Cancer

Research Campaign in Britain. The dose-limiting toxic effect in the trial candidates was peripheral neuropathy, suggesting that the drug may bind to tubulin proteins to block microtubule polymerization, as does vincristine. No objective responses were seen in this single Phase 1 study, which was as expected. Phase I trial patients are by definition generally refractory to all other therapies and usually have far-advanced disease. However, disease stabilization in two patients was noted. A continuous infusion schedule could overcome the short half-life and enhance the drug's overall antitumor activity and this needs to be done in the next clinical trials.

During the time in which the antitumor activity of phyllanthoside was being studied, we were also interested in the antiviral activity of our *Phyllanthus* isolates. *Phyllanthus acuminatus* has been used in folk medicine to treat problems of the genito-urinary tract, among other diseases, and initial testing of phyllanthoside against *Herpes* viruses indicated that the compound had good activity against HSV-1 and -2, with an 80% cure rate in the mouse vaginal model; it was also active in vitro against Epstein-Barr virus (EBV), although it was toxic in stationary and growing cells (private communication, C.K.-H. Tseng, NIAID).

Pancratistatin

One of the plant families we investigated in parallel with the Combretaceae was the Amaryllidaceae. Hippocrates, about 200 B.C., used extracts from plants of the genus *Narcissus* for treating breast cancer patients. The most active substance we located so far in this plant family was pancratistatin (67) (from *Pancratium littoralis*, later reidentified as *Hymenocallis littoralis*) (68). We were able to determine the structure by X-ray crystallography and later converted it to a phosphate prodrug. Synthesis of the phenol phosphate proved to be quite challenging and eventually we resorted to starting with a trivalent phosphite and oxidizing later in the synthesis (69). The prodrug again proved to have a very nice spectrum of anticancer activity. In addition to important antineoplastic activity, pancratistatin also has a promising spectrum of antiviral activity. So far, it is the first substance to cure Japanese encephalitis in an animal model (70). The pancratistatin prodrug is a very soluble (>230 mg/ml water) anticancer drug that can be prepared in high yield from pancratistatin and maintains the level of activity of the parent drug. Pancratistatin itself was chosen by the NCI for preclinical development because of its excellent anticancer activity, but it proved too difficult to formulate owing to its low solubility in water. Pancratistatin prodrug (NSC D-682901R) has a mean GI_{50} of $30.0 \pm 17.3 \times 10^{-8}$ M in the NCI 60-cell-line human tumor screen. Relative to the parent, the GI_{50} COMPARE correlation coefficient is 0.87-0.90, which suggests that loss of the phosphate ester occurs after contact with the cancer cells. The parent drug, pancratistatin, provided a 38-100% life extension at 0.38-3.0 mg/kg against the murine M5076 ovary sarcoma. Very importantly, in an antiangiogenesis study using the CaNT murine colon tumor, the pancratistatin prodrug was more effective at reducing tumor size than was combretastatin A-4 prodrug. The pancratistatin prodrug at 20 mg/kg reduced the tumor by a factor of a hundred, and at 100 mg/kg it effectively destroyed the tumor. Remarkable antiangiogenesis activity also was demonstrated with murine colon adenocarcinoma MAC 29, in which pancratistatin prodrug at 12.5 mg/kg caused a 72% decrease in functional vasculature in one hour and complete tumor vasculature shutdown in two hours.

Considerable research efforts (71) have been devoted to developing a very practical synthesis of pancratistatin. The first synthesis (Danishefsky) (72) provided racemic pancratistatin in 26 steps from bromobenzene (overall yield of 0.13%). The first enantioselective synthesis in 14 steps from bromobenzene (2% overall yield) of (+)-pancratistatin was reported by Hudlicky (73) in 1995. The same year, Trost (74) summarized a synthesis with an impressive 11% overall yield. More recently, Haseltine (75) and Magnus (76) (22 steps, 1.2% yield) have presented new syntheses of (+)-pancratistatin. In addition, a new synthesis of 7-deoxypancratistatin has been completed (21 steps, 7% overall yield) (77) and other new approaches are in progress (78). However, for clinical development we need a very efficient and practical total synthesis amenable to large scale-up requirements.

From the beginning of our synthetic approaches to pancratistatin, narciclasine (79) has remained an attractive precursor, as it is available in practical quantities from the bulbs of certain Amaryllidaceae species. Recently, it has been synthesized in twelve steps by the Hudlicky group starting from an enzymatic dihydrozylation of *m*-dibromo benzene (80). Earlier, we attempted to develop a practical synthesis of pancratistatin, but the last step, hydrogenolysis of the benzyl alcohol, did not lead to (+)-pancratistatin. Now we intend to develop practical methods of producing the clinical supply of pancratistatin prodrug through semisynthesis starting with narciclasine.

The Bryostatins

In 1965-66, we began the first systematic study of marine invertebrates and vertebrates as potential sources of new and potentially useful cancer chemotherapeutic drugs. By 1969, we found that 9-10% of marine animals yielded extracts with high and reproducible (confirmed active) antineoplastic

Bryostatin 3

Bryostatin 1

A =
B =
C =
D =
E =

R	R_1	
B	OH	Bryostatin 2
D	C	Bryostatin 4
A	C	Bryostatin 5
A	D	Bryostatin 6
A	A	Bryostatin 7
D	D	Bryostatin 8
D	A	Bryostatin 9
H	C	Bryostatin 10
H	A	Bryostatin 11
B	D	Bryostatin 12
H	D	Bryostatin 13
OH	C	Bryostatin 14
E	A	Bryostatin 15

Bryostatin 18

Bryostatin 17

Bryostatin 16

activity in the National Cancer Institute's P388 in vivo lymphocytic leukemia screening system (27,6). From this very high level of confirmed activity, it was abundantly clear that such natural products present an unusually good opportunity for discovering clinically useful anticancer drugs. To date, we have isolated a large number of new cytotoxic and/or antineoplastic agents from marine animals. Furthermore, these natural products are the result of <u>3.8 billion years (28) of evolutionary biosynthetic organic reactions aimed at even more specific molecular design and targeting</u>. The net result of what must have been many trillions of biologically directed organic reactions (biosynthetic combinatorial processes) is a great number of candidates for use as anticancer drugs and as drugs necessary across the medical spectrum. However, they need to be discovered and developed to the clinic. The bryostatins provide very compelling evidence for the extraordinary utility of this approach.

In 1968, we collected (upper Gulf of Mexico) a Bryozoa specimen that subsequently was found to be *Bugula neritina* and which gave extracts that more than doubled the life-span of animals with the P388 lymphocytic leukemia. Meanwhile, we were conducting expeditions in the Gulf of California and collected another bryozoan, again powerfully active, that turned out to be *Bugula neritina*. Research was continued with that specimen and later we found it again off the coast of California. Extensive bioassay (P388 leukemia) directed separation of the California *Bugula neritina* extracts led to isolation of the first milligram of bryostatin 1 in 1981 (81). We were able to crystallize this substance and complete the X-ray crystallographic structural determination. Bryostatin 1 was found to be a remarkable macrocyclic lactone (45,82).

Meanwhile, we have discovered new bryostatins and now have twenty of this series in hand from *Bugula neritina* collections that range from the Gulf of Mexico, Gulf of California, and coast of California, to Japan (Gulf of Sagami) (83). More recently, we explored *B. neritina* from two more remote areas in the Gulf of Japan (84) and from one of these found bryostatin 10 in fairly abundant amounts (about $10^{-3}\%$). Bryostatin 1 was found in $10^{-6}\%$ yields and some of the more rare bryostatins in yields of $10^{-8}\%$. By comparison, the yields of the spongistatins (see below) range from 10^{-6} to $10^{-8}\%$, and dolastatin 10 (see below) was found in some $10^{-6}\%$ yield. Some of the other dolastatins were isolated in 10^{-7} yields. Hence, the natural production of bryostatin 10 in $10^{-3}\%$ yields is very important for the future development of this compound to clinical trials.

Bryostatin 1 provided curative levels of activity against a variety of murine experimental cancer systems, including M5076 sarcoma and L10AB-cell lymphoma. Against human cancer models, in the nude or SCID models, bryostatin 1 evidenced excellent activity against myeloid and lymphoid leukemias, as well as against lymphoma cancer xenografts. Very importantly, bryostatin 1 has been shown to exhibit immunostimulatory properties, including stimulation of cytokine release, enhancement of T- and B-cell activation and lymphokine-activated killer cell activity. In addition, neutrophil phagocytic activity and degranulation have been demonstrated. Also very important have been observations that bryostatin 1 down-regulates *mdr*-1. Further details and over 450 bryostatin references will be included in a reference book now in preparation (*85*).

Clinical development (*86-92*) of bryostatin 1 has been ongoing from 1990. In addition to over 50 human cancer clinical trials of bryostatin 1 (either completed (20) or now ongoing in the United States, England, and Canada under the auspices of the U.S. National Cancer Institute, the British Cancer Research Campaign and the Cancer Institute of Canada, involving 178 principal investigators in 148 different institutions), the clinical development of bryostatin 1 is still expanding at the Phase I/II levels. To date, well over 500 patients have been treated with bryostatin 1 in phase I or phase II cancer clinical trials, and of these patients a good number have received promising benefits ranging from complete responses and partial responses to stable disease. Initial reports from clinical trials combining bryostatin 1 with well-known anticancer drugs such as taxol, vincristine, fludarabine (*88*), cisplatin, 2-CdA (*86*), and ARA-C suggest that bryostatin 1 will become an important anticancer drug and in combination with other anticancer drugs such as just noted should lead to useful improvements in human cancer treatment. Current research concerned with improving the clinical supply includes the first total synthesis of bryostatin 2 (and thereby bryostatin 1) (*93*) and attempts at eventual microbiological production from a bacterial symbiont of *Bugula neritina* (*94*).

The Spongistatins

In 1973 we found that a collection of the South African marine sponge *Spirastrella spinispirulifera* (*95*) gave extracts that would double the life span of animals given the P388 lymphocytic leukemia: a very important lead. Once we started the actual separations, some of these fractions were found to give a curative level of activity: a very exciting lead. From 1973 to 1979 we completed several scale-up operations and extensive separations aimed at the active constituents. Before we could isolate the active components on each attack, we ran out of active material. In 1979-80, we went to the next scale-up entailing about three tons of the red sponge. At one point in this endeavor, it was necessary to use HPLC columns that were nearly 3 m tall and 15 cm in diameter. By 1981, thanks to efforts on that scale, we isolated the first

Spongistatin 5, R = Cl, R₁ = H
Spongistatin 7, R = H, R₁ = H
Spongistatin 8, R = H, R₁ = COCH₃
Spongistatin 9, R = Cl, R₁ = COCH₃

Spongistatin 1, R = Cl, R₁ = R₂ = COCH₃
Spongistatin 2, R = H, R₁ = R₂ = COCH₃
Spongistatin 3, R = Cl, R₁ = H, R₂ = COCH₃
Spongistatin 4, R = Cl, R₁ = COCH₃, R₂ = H
Spongistatin 6, R = H, R₁ = COCH₃, R₂ = H

anticancer substance (800 μg) from the red sponge (*95*) and later designated it spongistatin 4.

Another investigation led to the final solutions of the red sponge anticancer components' structural challenges. In 1986, we collected a black sponge in the Republic of Maldives that was initially identified as a *Spongia* sp. but recently reidentified as a *Hyrtios* sp. We returned to the Maldives in 1988 for a 400-kg wet weight recollection. After launching a determined effort to separate the anticancer substances, we found some of the most active constituents to be the red sponge spongistatin series that we had been working with from 1973. We finally isolated the first of the Maldive spongistatins, albeit only a few milligrams, but by 1988 that was enough to solve the structure problem. The first of the Maldives series was named spongistatin 1. The structure posed a lot of difficulties with the NMR interpretations and required our solving it three times, in three different solvents, over the course of a year (*95*). The chiral centers of spongistatins 1 and 2 have now been established by total syntheses (*96-98*), and other syntheses are in progress (*99-105*).

The structural elucidation of spongistatin 1 quickly led to structural assignments for spongistatin 2 followed by spongistatin 3 (*106*). Both were found to involve some variations in the acetylation pattern of the parent macrocyclic lactone. Spongistatin 4 (*107*), as noted above, we discovered in 1981 in the African red sponge. Once we solved the structures of spongistatins 1-3, we were able to elucidate the structure of spongistatin 4 and continue in the same fashion with the other red sponge constituents (very small amounts). At present, we have determined the structures up to spongistatin 9 (*108,109*). With the red sponge components, there is a departure in the structure at number 5 due to an additional tetrahydrofuran ring. Spongistatin 5 (*107*), 7 (*108*), 8 (*109*), and 9 (*109*) all carry that ring system. However, the cancer cell growth inhibitory activity of spongistatin 5 was not diminished compared to that of spongistatin 1. Furthermore, spongistatin 6 (*108*) from the red sponge represents a return to the original spongipyran structure minus the tetrahydrofuran ring.

The profile of spongistatin 1 against the NCI human cancer cell line panel is probably the best to date in the NCI's evaluation programs. For example, with small cell lung cancer, the amount of spongistatin 1 required for 50% growth inhibition is 10^{-10} M, and that activity continues in lines from the colon cancer, renal cancer, ovarian cancer, and breast cancer sets. The latter are the most strongly inhibited, at roughly 10^{-12} M. Spongistatin 1 was found to show extraordinary activity against six major types of human cancer, namely: lung cancer (NCI-H522), leukemia (HL-60 and CCRF-CEM), colon (KM12, HT29, COLO 205, HCC-2998), ovarian (OVCAR-3), renal (A498) and breast (MDA-MB-435, MDA-N). As a result of such activity and in several corresponding human cancer xenografts, spongistatin 1 is considered a superior candidate for clinical development. For example, spongistatin 1, used against the OVCAR-3 xenograft at 25 µg/kg, led to better than 70% long term survivors. In terms of mechanism of action, spongistatin 1 inhibits mitosis and microtubule assembly (110). Both the anticancer in vivo and mechanistic studies are in progress while spongistatin 1 is being further advanced in preclinical development. Clearly, an efficient total synthesis of spongistatin 1 is a most urgent objective and a number of research groups world-wide are pressing forward. The first total synthesis of spongistatin 1 was summarized in 1998 by Kishi (111) and of spongistatin 2 by Evans in 1999 (112).

The Dolastatins

In 1972, specimens of the marine sea hare *Dolabella auricularia* were collected off the island of Mauritius in the Indian Ocean. We found extracts of this sea hare would more than double the life-span of animals with the P388 leukemia. Again, this was a very high-priority lead that we pursued very intensively, but it was not until we obtained an almost 2-ton collection that we were finally able to solve the extremely challenging isolation problem in 1984. The simplest way we found to isolate the key anticancer substance, dolastatin 10 (113) involved about 20,000 fractions and some 23 separate chromatographic steps using various techniques. About a year was needed to solve the structure problem with the first milligram, employing high-field NMR and high-resolution mass spectrometry, and finally total synthesis (114). The next challenging problem was the unknown nine chiral centers in that novel peptide. Furthermore, to prepare dolastatin 10 for eventual clinical trials, we would have needed about 700 tons of the sea hare. For ecological and many other reasons, that was not an option. Dolastatin 10 had to be synthesized, and that required determination of the chirality. Thus we relied on our knowledge of the high-field NMR characteristics of the components to direct the total synthetic approaches. Each of the total syntheses we completed required about 28 steps; they were not easy at the beginning and it took 15 total syntheses (114,115) to prepare the natural product, a definite improvement over the theoretical 512! Subsequently, we prepared dolastatin 10 for clinical trials by total synthesis, and NCI phase II cancer clinical trials are under way. Current results indicate, for example, a good future in the treatment of refractory AML. Meanwhile, one of a large number of dolastatin 10 structural modifications we synthesized and designated auristatin PE has been undergoing Phase I cancer clinical trials in Japan. Cancer Phase 2 clinical trials of auristatin PE are being readied for the U.S. and have already begun in Europe.

Meanwhile, we have been undertaking preclinical development of selected

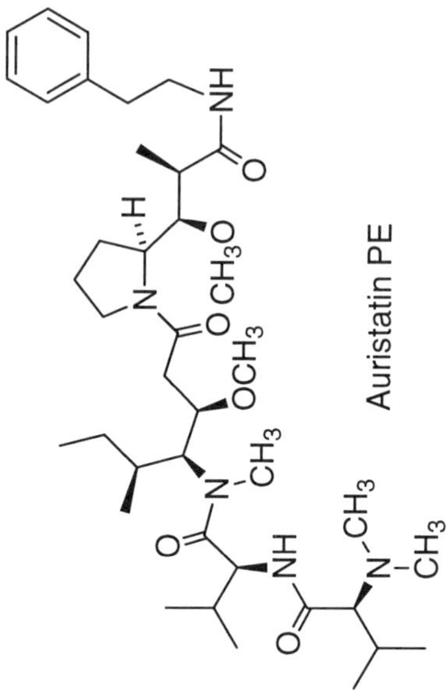

Auristatin PE

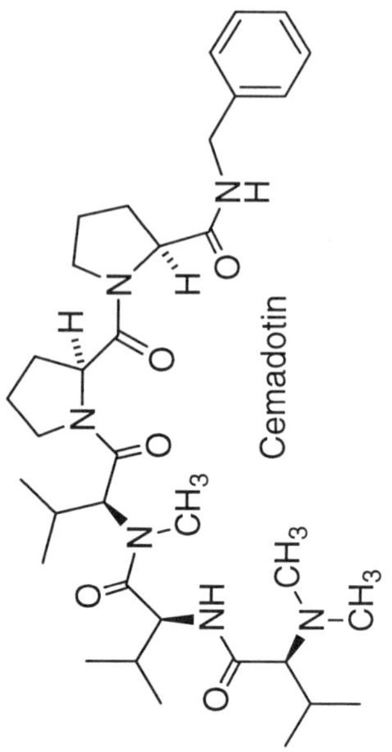

Cemadotin

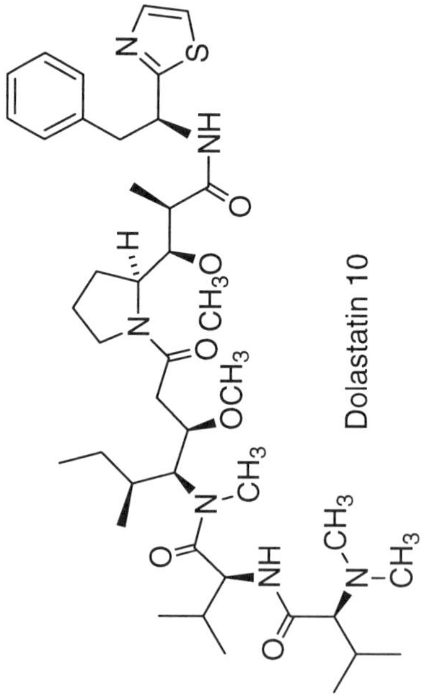

Dolastatin 10

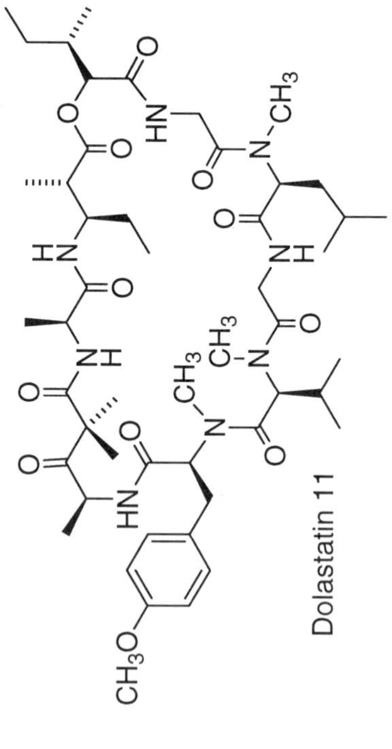

Dolastatin 11

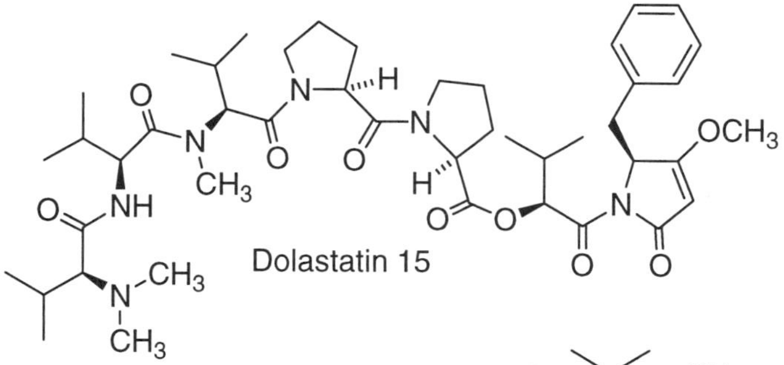

Dolastatin 15

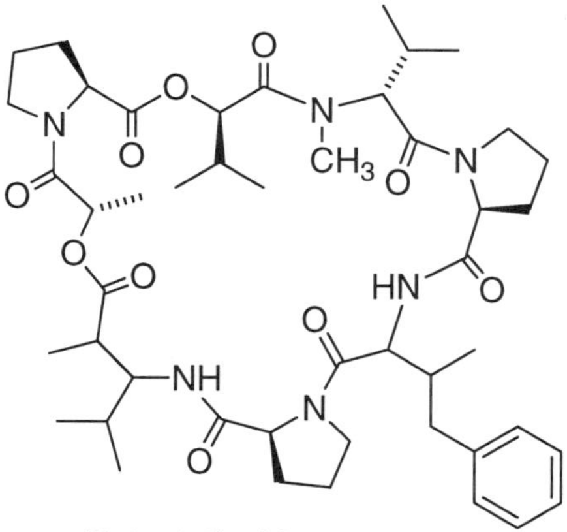

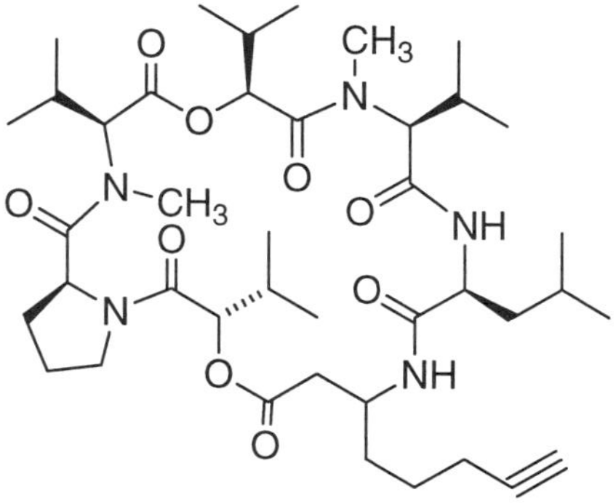

Dolastatin 17

Dolastatin 16

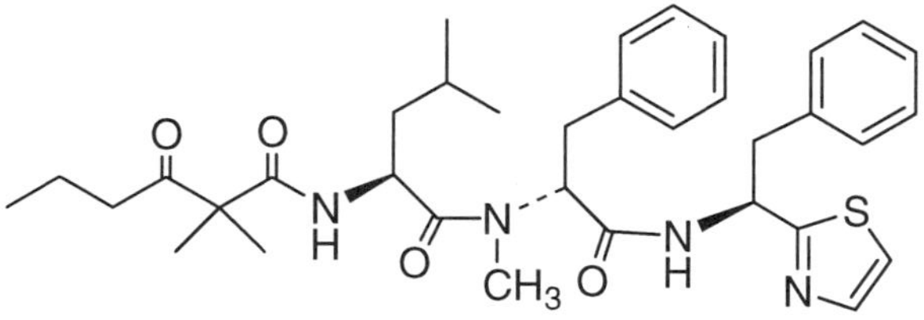

Dolastatin 18

members of the dolastatin 11-18 series (*116-120*). Currently dolastatin 11 (actin active) (*116*) and dolastatin 15 (*121,122*) are both very active substances with a different antineoplastic profile than that shown by dolastatins 10 and 11. In fact, the only unit common to dolastatin 10 and dolastatin 15 is the dolavaline group. The dolastatin 15 structural modification cemadotin (NSC D-669356) (*123*) has been undergoing Phase II human cancer clinical trials in Europe and the U.S. Present evidence suggests that the dolastatins represent an extraordinary resource for further discovery and development of potentially useful anticancer drugs (*124-132*).

The Cephalostatins

The discovery of cephalostatin 1 as a natural product with remarkable anticancer activity took many years and has been a continuing project from 1972 in our ASU Cancer Research Institute. The original collection of the natural source of cephalostatin 1 (*Cephalodiscus gilchristi*) took place in 1972. The genus *Cephalodiscus* is rare among marine worms and there are only a small number of species. Inhibitory activity in methanol and water extracts was found in the NCI's P388 murine lymphocytic leukemia system, with 32-41% life extension at 25-37.5 mg/kg. After 17 years of chemical investigation into the active constituents of *C. gilchristi*, we reported the structure of cephalostatin 1 (*133*). Cephalostatin 1 is a very potent cancer cell growth inhibitor (GI_{50} 10^{-9} - 10^{-10} M) and ranks among the most effective so far tested in the NCI's human cancer cell line panel.

Repeated recollections of *C. gilchristi* and further chemical separations eventually resulted in the discovery of a series of related compounds, now up to 19 (*134,135*). All have potent activity against human cancer cells. The mechanism of action of cephalostatin 1 is unknown but it demonstrates a unique and distinctive pattern of cancer cell inhibitory activity that sets it apart from other anticancer agents. Ganesan (*135*) has reviewed biological activities as well as other characteristics of these dimeric steroid alkaloids from marine animals and has noted that the cephalostatins do not contain functional groups commonly associated with cytotoxicity such as alkylation sites, Michael acceptors, intercalators, or redox-active quinones and further suggested that the steroidal dimeric compounds span the lipid bilayer of eukaryotic cells' membranes, thus perturbing membrane structure and function (*136*). Fuchs has proposed that cephalostatins may be enzyme inhibitors that form hydrogen bands with their target (which is presently unknown) (*137*) and later suggested that a set of reactions involving rearrangement of the right half of the dimeric steroid may take place and occur in vivo (*138*). The D-ring double bond would generate

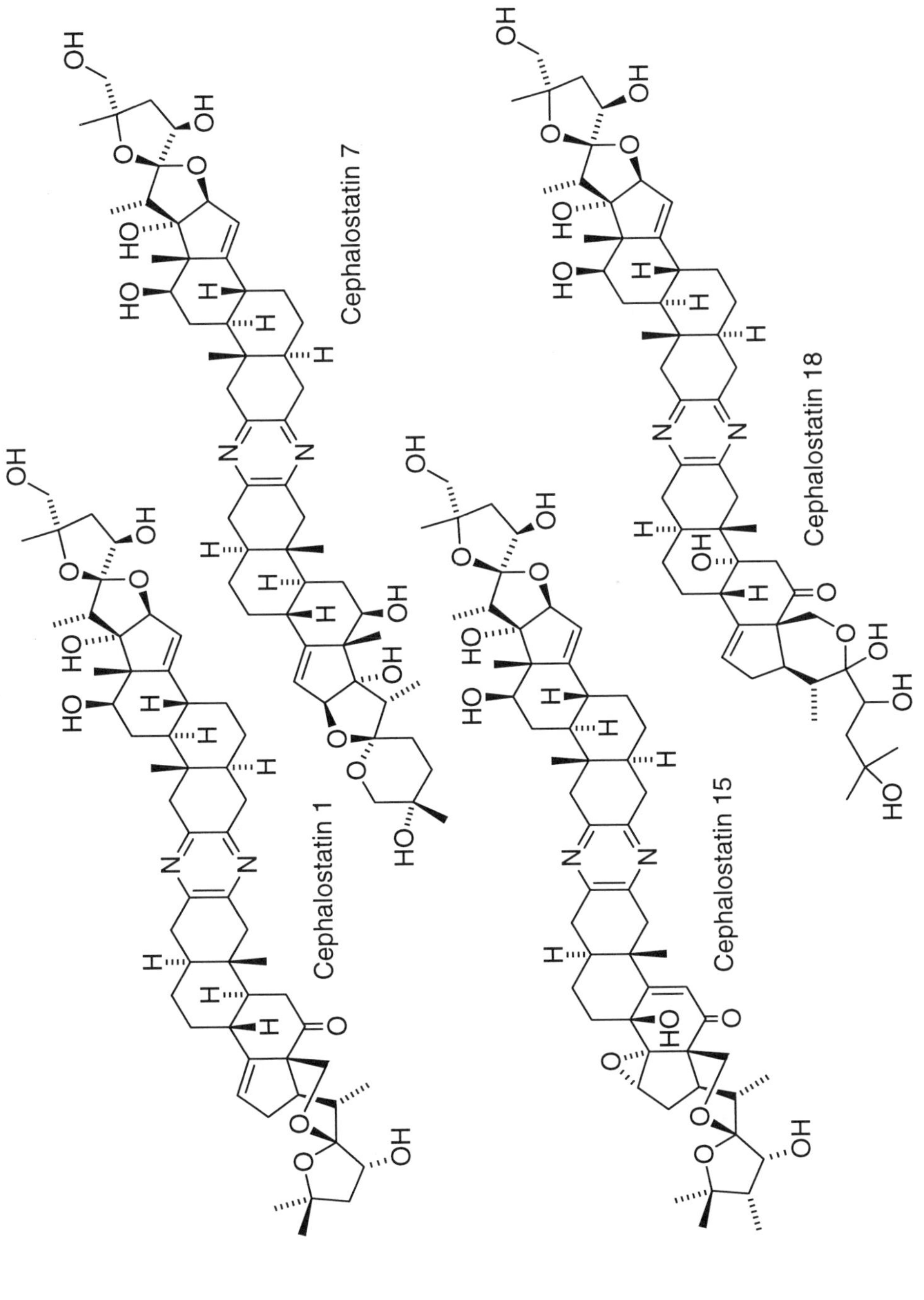

Cephalostatin 1

Cephalostatin 7

Cephalostatin 15

Cephalostatin 18

reactive electrophilic intermediates through protonation or epoxidation, the latter proving toxic to cell growth. The following possibilities for modes of action have been considered and eliminated: PKC, topoisomerase, mitotic spindle and tubulin. Eventual understanding of what should prove to be a novel mechanism will arise as we are able to increase available supplies through total synthesis. Cephalostatin 1 has been available in only limited supply from its natural source, despite extensive recollections and reisolations by our group. Recently a 65-step total synthesis was reported by Fuchs (*139*) but it is not practical for preclinical development and clinical investigation. The DCTD/NCI has decided to begin serious preclinical development of cephalostatin 1 and there is a most urgent need for several grams for the next stages of preclinical evaluation.

The preceding examples were selected from a considerable number of evolutionary biosynthesis-derived anticancer drugs currently under investigation in this research chemist's laboratory. To summarize, discovery of structurally novel and potentially very important anticancer drugs from animal, plant and microorganism sources followed by their synthesis and/or structural modification will continue to be a powerful route to improving human cancer treatment and greatly increasing the number of curative anticancer drugs.

Coda

For over two decades the ever increasing overly optimistic claims of certain molecular biology advocates for rapidly finding human cancer cures have instead seriously slowed the NCI anticancer drug discovery research. Hopefully, there will be increasing scientific and public awareness of this problem. The well proven path to discovery of improved anticancer drugs by focusing on the evolutionary biosynthetic anticancer constituents of marine animals, microorganisms and plants and their synthetic modifications will continue to be exceptionally productive.

Acknowledgment

I am very pleased to express my thanks and grateful appreciation to the very capable members of my research group such as Professor Cherry L. Herald, Professor Jean M. Schmidt and other colleagues over the past 43 years who contributed to advancing this anticancer drug discovery research. With pleasure I also want to acknowledge the expert contributions of Dr. Fiona Hogan, Mrs. Georgia Reimus and Mrs. Theresa Thornburgh to the final stages of manuscript preparation as well as the financial support provided by Outstanding Investigator Grant OIG CA44344-01A1-012 awarded by the U.S. National Cancer Institute,

DHHS, The Arizona Disease Control Research Commission, and the Robert B. Dalton Endowment Fund.

References

1. Cragg, G. M.; Newman, D. J. *Cancer Investigation* **1999**, *17*, 153-163.
2. Marshall, J. A.; Jiang, H. *J. Org. Chem.* **1999**, *64*, 971-975.
3. McKee, T. C.; Bokesch, H. R.; McCormick, J. L.; Rashid, M. A.; Spielvogel, D.; Gustafson, K. R.; Alavanja, M. M.; Cardellina, J. H., II; Boyd, M. R. *J. Nat. Prod.* **1997**, *60*, 431-438.
4. Pettit, G. R.; Hogan-Pierson, F.; Herald, C. L. *Anticancer Drugs from Animals, Plants, and Microorganisms*; John Wiley-Interscience: NY, 1993; Vol. 7.
5. Pettit, G. R.; Herald, C. L.; Smith, C. R. *Biosynthetic Products for Cancer Chemotherapy*; Elsevier Scientific Pub. Co.: NY, 1989; Vol.6.
6. Cassady, J. M.; Douros, J. D., Eds. *Anticancer Agents Based on Natural Product Models*; Academic Press: NY, 1980.
7. Suffness, M.; Douros, J. In *Methods in Cancer Research;* DeVita, A. V. T., Jr., Busch, H. Eds.; Drugs of Plant Origin; Academic Press: NY, 1979; Vol. XVI Chapter 3, p 73.
8. Kupchan, S. M. *Cancer Treatment Reports* **1976**, 1115-1126.
9. Carter, S. K.; Livingston, R. B. *Cancer Treatment Reports* **1976**, 1141-1156.
10. Combes, O.; Barré, J.; Duché, J-C.; Vernillet, L.; Archimbaud, Y.; Marietta, M. P.; Tillement, J-P.; Urien, S. *Investigational New Drugs* **2000**, *18*, 1-5.
11. Cao, Z.; Armstrong, K.; Shaw, M.; Petry, E.; Harris *Synthesis* **1998**, *12*, 1724-1730.
12. Lackey, K.; Sternbach, D. D.; Croom, D. K.; Emerson, D. L.; Evans, M. G.; Leitner, P. L.; Luzzio, M. J.; McIntyre, G.; Vuong, A.; Yates, J.; Besterman, J. M. *J. Med. Chem.* **1996**, *39*, 713-719.
13. Potmesil, M.; Pinedo, H. *Camptothecins: New Anticancer Agents*; CRC Press: FL, 1995.
14. Ojima, I. *Adv. Med. Chem.* **1999**, *4*, 69-124.
15. Gunatilaka, A. A.; Ramdayal, F. D.; Sarragiotto, M. H.; Kingston, D. G. I. Sackett, D. L.; Hamel, E. *J. Org. Chem.* **1999**, *64*, 2694-2703.
16. Jagtap, P. G.; Kingston, D. G. I. *Tetrahedron Lett.* **1999**, *40*, 189-192.
17. Morihira, K.; Hara, R.; Kawahara, S.; Mishimori, T.; Nakamura, N.; Kusama, H.; Kuwajima, I. *J. Am. Chem. Soc.* **1998**, *120*, 12980-12981.
18. Patel, R. M. *Annu. Rev. Microbiol.* **1998**, *98*, 361-395.

19. Suffness, M.; Wall, M. E. "Discovery and Development of Taxol," in *Taxol Science and Applications;* Suffness, M., Ed., CRC Press: NY, 1995; Chapter 1, pp 3-22.
20. Zewail-Foote, M.; Hurley, L. H. *J. Med. Chem.* **1999**, *42*, 2493-2497.
21. Forsyth, C. J.; Ahmed, F.; Cink, R. D.; Lee, C. S. *J. Am. Chem. Soc.* **1998**, *120*, 5597-5598.
22. Dhokte, U. P.; Khau, V. V.; Hutchison, D. R.; Martinelli, M. J. *Tetrahedron Lett.* **1998**, 8771-8774.
23. Nicolaou, K. C.; Ninkovic, S.; Sarabia, F.; Vourloumis, D.; He, Y.; Vallberg, H.; Finlay, M. R. V.; Yang, Z. *J. Am. Chem. Soc.* **1997**, *119*, 7974-7991.
24. Smith, A. L.; Nicolaou, K. C. *J. Med. Chem.* **1996**, *39*, 2103-2117.
25. Schultes, R. E.; Hoffman, A. *The Botany and Chemistry of Hallucinogens*; Charles C. Thomas Publisher: IL, 1973; p 13.
26. Bradner, W. T. In *Cancer and Chemotherapy*; Crooks, S. T.; Prestayko, A. W., Eds.; Academic Press: NY, 1980; Vol. 1, p 313.
27. Pettit, G. R.; Day, J. F.; Hartwell, J. L.; Wood, H. B. *Nature* **1970**, *227*, 962-963.
28. Sutherland, J. D.; Whitfield, J. N. *Tetrahedron* **1997**, *53*, 11493-11527.
29. Lawrence, R. N. *DDT* **1999**, *4*, 449-451.
30. Wall, N. R.; Mohammad, R. M.; Al-Katib, A. M. *Leukemia Res.* **1999**, *12*, 881-888.
31. Lundberg, A. S.; Weinberg, R. A. *Eur. J. Cancer* **1999**, *35*, 1886-1894.
32. Lorenzo, P. S.; Bögi, K.; Hughes, K. M.; Beheshti, M.; Bhattacharyya, D.; Garfield, S. H.; Pettit, G. R.; Blumberg, P. M. *Cancer Res.* **1999**, *59*, 5137-6144.
33. Varterasian, M. L; Mohammad, R. M.; Shurafa, M. S.; Hulburd, K.; Pemberton, P. A.; Rodriguez, D. H.; Spadoni, V.; Eilender, D. S.; Murgo, A.; Wall, N.; Dan, M.; Al-Katib, A. M. *Clin. Cancer Res.* **2000**, *6*, 825-828.
34. Pitot, H. C.; McElroy, E. A. Jr.; Reid, J. M.; Windebank, A. J.; Sloan, J. A.; Erlichman, C.; Bagniewski, P. G.; Walker, D. L.; Rubin, J.; Goldberg, R. M.; Adjei, A. A.; Ames, M. M. *Clin. Cancer Res.* **1999**, *5*, 525-531.
35. Villalona-Calero, M. A.; Baker, S. D.; Hammond, L.; Aylesworth, C.; Eckhardt, S. G.; Kroynak, M.; Fram, R.; Fischkoff, S.; Velagopudi, R.; Toppmeyer, D.; Razvillas, B.; Jakimowicz, K.; Von Hoff, D. D.; Rowinsky, E. *J. Clin. Oncol.* **1998**, *16*, 2770-2779.
36. Pettit, G. R.; Rhodes, M. R.; Herald, D. L.; Chaplin, D. J.; Stratford, M.; Pettit, R. K.; Chapuis, J-C.; Oliva, D. *Anti-Cancer Drug Des.* **1998**, *13*, 981-994.

37. Nihei, Y.; Suga, Y.; Morinaga, Y.; Ohishi, K.; Okano, A.; Ohsumi, K.; Hatanaka, T.; Nakagawa, R.; Tsuji, T.; Akiyama, Y.; Saito, S.; Hori, K.; Sato, Y.; Tsuruo, T. *Jpn. J. Cancer Res.* **1999**, *90*, 1016-1025.
38. Meijer, L.; Thunnissen, A-M.W.H.; White, A. W.; Garnier, M.; Nikolic, M.; Tsai, L-H.; Walter, J.; Cleverly, K. E.; Salinas, P. C.; Wu, Y. Z.; Biernat, J.; Mandelkow, E-M.; Kim, S-H.; Pettit, G. R. *Chem. and Biol.* **2000**, *7*, 51-63.
39. Schultz, C.; Link, A.; Leost, M.; Zaharevitz, D. W.; Gussio, R.; Sausville, E. A.; Meijer, L.; Kunick, C. *J. Med. Chem.* **1999**, *42*, 2909-2919.
40. Sausville, E. A.; Zaharevitz, D.; Gussio, R.; Meijer, L.; Louarn-Leost, M.; Kunick, C.; Schultz, R.; Lahusen, T.; Headlee, D.; Stinson, S.; Arbuck, S. G.; Senderowicz, A. *Pharmacol. Ther.* **1999**, *82*, 285-292.
41. Meijer, L.; Leclerc, S.; Leost, M. *Pharmacol. Ther.* **1999**, *82*, 279-284.
42. Garrett, M. D.; Workman, P. *Eur. J. Cancer* **1999**, *35*, 2010-2030.
43. Shokat, K. M. *Chem. Biol.* **1995**, *2*, 509-514.
44. Schreiber, S. L. *Science* **2000**, *287*, 1964-1973.
45. Pettit, G. R. *J. Nat. Prod.* **1996**, *59*, 812-821.
46. Pettit, G. R.; Cragg, G. M.; Herald, D. L.; Schmidt, J. M.; Lohavanijaya, P. *Can. J. Chem.* **1982**, *60*, 1374-1376.
47. Pettit, G. R.; Singh, S. B.; Niven, M. L.; Hamel, E.; Schmidt, J. M. *J. Nat. Prod.* **1987**, *50*, 119-131.
48. Pettit, G. R.; Singh, S. B. *Can. J. Chem.* **1987**, *65*, 2390-2396.
49. Lin, C. M.; Ho, H. H.; Pettit, G. R.; Hamel, E. *Biochem.* **1989**, 6984-6991.
50. Pettit, G. R.; Singh, S. B.; Niven, M. L. *J. Am. Chem. Soc.* **1988**, *110*, 8539-8540.
51. Singh, S. B.; Pettit, G. R. *J. Org. Chem.* **1990**, *55*, 2797-2800.
52. Singh, S. B.; Pettit, G. R. *Syn. Commun.* **1987**, *17*, 877-892.
53. Pettit, G. R.; Singh, S. B.; Hamel, E.; Lin, C. M.; Alberts, D. S.; Garcia-Kendall, D. *Experientia* **1989**, *45*, 209-211.
54. Pettit, G. R.; Rhodes, M. *Anti-Cancer Drug Des.* **1998**, *13*, 183-191.
55. Horsman, M. R.; Ehrnrooth, E.; Ladekarl, M.; Overgaard, J. *Int. J. Radiation Oncology Biol. Phys.* **1998**, *42*, 895-898.
56. Dark, G.; Hill, S. A.; Prise, V. E.; Tozer, G. M.; Pettit, G. R.; Chaplin, D. J. *Cancer Res.* **1997**, *57*, 1829-1834.
57. Chaplin, D. J.; Pettit, G. R.; Hill, S. A. *Anticancer Res.* **1999**, *19*, 189-195.
58. Stevenson, J. P.; Gallagher, M.; Sun, W.; Algazy, K.; Vaughn, D. J.; Haller, D. G.; Hiller, K.; Halloran, L.; O'Dwyer, P. J. *Proc. Amer. Assoc. Cancer Res.* **2000**, *41*, 544.

59. Grosios, K.; Loadman, P. M.; Swaine, D. J.; Pettit, G. R.; Bibby, M. C. *Anticancer Res.* **2000**, *20*, 229-234.
60. Schwikkard, S.; Zhou, B-N.; Glass, T. E.; Sharp, J. L.; Mattern, M. R.; Johnson, R. K.; Kingston, D. G. I. *J. Nat. Prod.* **2000**, *63*, 457-460.
61. Pettit, G. R.; Cragg, G. M.; Gust, D.; Brown, P.; Schmidt, J. M. *Can. J. Chem.* **1982**, *60*, 939-941.
62. Pettit, G. R.; Cragg, G. M.; Gust, D.; Brown, P. *Can. J. Chem.* **1982**, *60*, 544-546.
63. Pettit, G. R.; Cragg, G. M.; Niven, M. L.; Nassimbeni, L. R. *Can. J. Chem.* **1983**, *61*, 2630-2632.
64. Pettit, G. R.; Cragg, G. M.; Suffness, M. I.; Gust, D.; Boettner, F. E.; Williams, M.; Saenz-Renauld, J. A.; Brown, P.; Schmidt, J. M.; Ellis, P. D. *J. Org. Chem.* **1984**, *49*, 4258-4266.
65. Pettit, G. R.; Cragg, G. M.; Suffness, M. *J. Org. Chem.* **1985**, *50*, 5060-5063.
66. Powis, G.; Moore, D. J. *J. Chromatog.* **1985**, *342*, 129-134.
67. Pettit, G. R.; Gaddamidi, V.; Cragg, G. M.; Herald, D. L.; Sagawa, Y. *J. Chem. Soc. Chem. Commun.* **1984**, 1693-1694.
68. Pettit, G. R.; Pettit, G. R., III; Backhaus, R. A.; Boettner, F. E. *J. Nat. Prod.* **1995**, *58*, 37-43.
69. Pettit, G. R.; Freeman, S.; Simpson, M. J.; Thompson, M. A.; Boyd, M. R.; Williams, M. D.; Pettit, G. R. III; Doubek, D. L. *Anti-Cancer Drug Des.* **1995**, *10*, 243-250.
70. Gabrielsen, B.; Monath, T. P.; Huggins, J. W.; Kefauver, D. F.; Pettit, G. R.; Groszek, G.; Hollingshead, M.; Kirsi, J. J.; Shannon, W. M.; Schubert, E. M.; Dare, J.; Ugarkar, B.; Ussery, M. A.; Phelan, M. J. *J. Nat. Prod.* **1992**, *55*, 1569-1581.
71. For reviews, refer to, Hoshino, O. *The Alkaloids*; Cordell, G. A., Ed.; Academic Press: 1998; Vol. 51, p 323.
72. Danishefsky, S.; Lee, J. Y. *J. Am. Chem. Soc.* **1989**, *111*, 4829-4837.
73. Tian, X.; Hudlicky, T.; Königsberger, K. *J. Am. Chem. Soc.* **1995**, *117*, 3643-3644.
74. Trost, B. M.; Pulley, S. R. *Am. Chem. Soc.* **1995**, *117*, 10143-10144.
75. Doyle, T.; Hendrix, M. M.; Van der Veer, D.; Jaanmard, S.; Haseltine, J. *Tetrahedron* **1997**, *53*, 11153-11170.
76. Magnus, P.; Sebhat, I. K. *Tetrahedron* **1998**, *54*, 15509-15524.
77. Keck, G. E.; McHardy, S. F.; Murry, J. A. *J. Org. Chem.* **1999**, *64*, 4465-4476.
78. Grubb, L. M.; Dowdy, A. L.; Blanchette, H. S.; Friestad, G. K.; Branchaud, B. P. *Tetrahedron Lett.* **1999**, *40*, 2691-2694. Aceña, J. L.;

Arjona, O.; Iradier, F.; Plumet, J. *Tetrahedron Lett.* **1996**, *37*, 105-106.
Gauthier, D. R.; Bender, S. L. *Tetrahedron Lett.* **1996**, *37*, 13-16.
79. Ceriotti, G. *Nature* **1967**, *213*, 595-596.
80. Gonzalez, D.; Martinot, T.; Hudlicky, T. *Tetrahedron Lett.* **1999**, *40*, 3077-3080.
81. Pettit, G. R.; Herald, C. L.; Doubek, D. L.; Herald, D. L.; Arnold, E.; Clardy, J. *J. Am. Chem. Soc.* **1982**, *104*, 6846-6848.
82. Pettit, G. R. The Bryostatins. In *Progress in the Chemistry of Organic Natural Products*; Tamm, Ch., Eds. No. 57; Springer-Verlag: NY, 1991; pp 153-195.
83. Pettit, G. R.; Gao, F.; Blumberg, P. M.; Herald, C. L.; Coll, J. C.; Kamano, Y.; Lewin, N. E.; Schmidt, J. M.; Chapuis, J-C. *J. Nat. Prod.* **1996**, *59*, 286-289.
84. Kamano, Y.; Zhang, H-P.; Morita, H.; Itokawa, H.; Shirota, O.; Pettit, G. R.; Herald, D. L.; Herald, C. L. *Tetrahedron* **1996**, *52*, 2369-2376.
85. Pettit, G. R.; Herald, C. L.; Hogan, F. In *Anticancer Drug Design*, Academic Press, in preparation.
86. Ihmad, I.; Al-Katib, A. M.; Beck, F. W. J.; Wall, N. R.; Mohammad, R. M. *Proc. Amer. Assoc. Cancer Res.* **2000**, *41*, 756.
87. Mohammad, R. M.; Varterasian, M. L.; Almatchy, V. P.; Pettit, G. R.; Al-Katib, A. *Clin. Cancer Res.* **1998**, *4*, 1337-1343.
88. Grant, S.; Cragg, L.; Roberts, J.; Smith, M.; Feldman, E.; Winning, M.; Tombes, M. *Blood* **1999**, *94*, 96a-97a.
89. Varterasian, M. L.; Mohammad, R. M.; Shurafa, M. S.; Hulburd, K.; Pemberton, P. A.; Rodriguez, D. H.; Spadoni, V.; Eilender, D. S.; Murgo, A.; Wall, N.; Dan, M.; Al-Katib, A. M. *Clin. Cancer Res.* **2000**, *6*, 825-828.
90. Bangalore, N. S.; Baidas, S.; Bhargava, P.; Rizvi, N.; El-Ashry, D.; Ness, L.; Marshall, J. L. Phase I study of bryostatin 1 and cisplatin in patients with advanced cancer. Proceedings of ASCO; Volume 19, 2000.
91. Kaubisch, A.; Kelsen, D. P.; Saltz, L.; Kemeny, N.; O'Reilly, E.; Ilson, D.; Endres, S.; Barazzuol, J.; Piazza, A.; Schwartz, G. K.; Phase 1 trial of weekly sequential bryostatin-1 (bryo), cisplatin, and paclitaxel in patients with advanced solid tumors. Proceedings of ASCO; Volume 19, 2000.
92. Sumacz, T. A.; Dowlati, A.; Remick, S. C.; Whitacre, C. M. *Proc. Amer. Assoc. Cancer Res.* **2000**, *41*, 145.
93. Evans, D. A.; Carter, P. H.; Carreira, E. M.; Prunet, J. A.; Charette, A. B.; Lautens, M. *Angew Chem.,Int. Ed. Engl.* **1998**, *37*, 2354-2359.
94. Davidson, S. K.; Haygood, M. G. *Biol. Bull.* **1999**, *196*, 273-280.
95. Pettit, G. R.; Cichacz, Z. A.; Gao, F.; Herald, C. L.; Boyd, M. R.;

Schmidt, J. M.; Hooper, J. N. A. *J. Org. Chem.* **1993**, *58*, 1302-1304.
96. Guo, J.; Duffy, K. J.; Stevens, K. L.; Dalko, P. I.; Roth, R. M.; Hayward, M. M.; Kishi, Y. *Angew. Chem., Int. Ed. Engl.* **1998**, *37*, 187-192.
97. Hayward, M. M.; Roth, R. M.; Duffy, K. J.; Dalko, P. I.; Stevens, K. L.; Guo, J.; Kishi, Y. *Angew. Chem. Int. Ed. Engl.* **1998**, *37*, 192-196.
98. Evans, D. A.; Trotter, B. W.; Coleman, P. J.; Côté, B.; Dias, L. C.; Rajapakse, H. A.; Tyler, A. N. *Tetrahedron* **1999**, *55*, 8671-8726.
99. Claffey, M. M.; Hayes, C. J.; Heathcock, C. H. *J. Org. Chem.* **1999**, *64*, 8267-8274.
100. Samadi, M.; Munoz-Letelier, C.; Poigny, S.; Guyot, M. *Tetrahedron Lett.* **2000**, *41*, 3349-3353.
101. Paterson, I.; Wallace, D. J.; Oballa, R. M. *Tetrahedron Lett.* **1998**, *39*, 8545-8548.
102. Lemaire-Audoire, S.; Vogel, P. *J. Org. Chem.* **2000**, *65*, 3346-3356.
103. Micalizio, G. C.; Roush, W. R. *Tetrahedron Lett.* **1999**, *40*, 3351-3354.
104. Terauchi, T.; Nakata, M. *Tetrahedron Lett.* **1998**, *39*, 3795-3798.
105. Smith, A. B., III; Lin, Q.; Nakayama, K.; Boldi, A. M.; Brook, C. S.; McBriar, M. D.; Moser, W. H.; Sobukawa, M.; Zhuang, L. *Tetrahedron Lett.* **1997**, *38*, 8675-8678.
106. Pettit, G. R.; Cichacz, Z. A.; Gao, F.; Herald, C. L.; Boyd, M. R. *J. Chem. Soc., Chem. Commun.* **1993**, *14*, 1166-1168.
107. Pettit, G. R.; Herald, C. L.; Cichacz, Z. A.; Gao, F.; Schmidt, J. M.; Boyd, M. R.; Christie, N. D.; Boettner, F. E. *J. Chem. Soc. Chem. Commun.* **1993**, 1805-1807.
108. Pettit, G. R.; Herald, C. L.; Cichacz, Z. A.; Gao, F.; Boyd, M. R.; Christie, N. D.; Schmidt, J. M. *Nat. Prod. Lett.* **1993**, *3*, 239-244.
109. Pettit, G. R.; Cichacz, Z. A.; Herald, C. L.; Gao, F.; Boyd, M. R.; Schmidt, J. M.; Hamel, E.; Bai, R. *J. Chem. Soc. Chem. Commun.* **1994**, 1605-1606.
110. Bai, R.; Taylor, G. F.; Cichacz, Z. A.; Herald, C. L.; Kepler, J. A.; Pettit, G. R.; Hamel, E. *Biochemistry* **1995**, *34*, 9714-9721.
111. Hayward, M. M.; Roth, R. M.; Duffy, K. J.; Dalko, P. I.; Stevens, K. L.; Guo, J.; Kishi, Y. *Angew. Chem. Int. Ed.* **1998**, *37*, 192-196.
112. Evans, D. A.; Trotter, B. W.; Coleman, P. J.; Côté, B.; Dias, L. C.; Rajapakse, H. A.; Tyler, A. N. *Tetrahedron* **1999**, *55*, 8671-8726.
113. Pettit, G. R.; Kamano, Y.; Herald, C. L.; Tuinman, A. A.; Boettner, F. E.; Kizu, H.; Schmidt, J. M.; Baczynskyj, L.; Tomer, K. B.; Bontems, R. J. *J. Am. Chem. Soc.* **1987**, *109*, 6883-6885.
114. Pettit, G. R.; Singh, S. B.; Hogan, F.; Lloyd-Williams, P.; Herald, D. L.; Burkett, D. D.; Clewlow, P. J. *J. Am. Chem. Soc.* **1989**, *111*, 5463-5465.

115. Pettit, G. R.; Srirangam, J. K.; Singh, S. B.; Williams, M. D.; Herald, D. L.; Barkóczy, J.; Kantoci, D.; Hogan, F. *J. Chem. Soc. Perkin I* **1996**, *8*, 859-863.
116. Bates, R.; Brusoe, K. G.; Burns, J. J.; Caldera, S.; Cui, W.; Gangwar, S.; Gramme, M. R.; McClure, K. J.; Rouen, G. P., Schadow, H.; Stessman, C. C.; Taylor S. R.; Vu, V. H.; Yarick, G. V.; Zhang, J.; Pettit, G. R.; Bontems, R. *J. Am. Chem. Soc.* **1997**, *119*, 2111-2113.
117. Pettit, G. R.; Kamano, Y.; Herald, C. L.; Fujii, Y.; Kizu, H.; Boyd, M. R.; Boettner, F. E.; Doubek, D. L.; Schmidt, J. M.; Chapuis, J-C.; Michel, C. *Tetrahedron* **1993**, *49*, 9151-9170.
118. Pettit, G. R.; Xu, J.-P.; Hogan, F.; Williams, M. D.; Doubek, D. L.; Schmidt, J. M.; Cerny, R. L.; Boyd, M. R. *J. Nat. Prod.* **1997**, *60*, 752-754.
119. Pettit, G. R.; Xu, J.-P.; Hogan, F.; Cerny, R. L. *Heterocycles* **1998**, *47*, 491-496.
120. Pettit, G. R.; Xu, J.-P.; Williams, M. D.; Hogan, F.; Schmidt, J. M.; Cerny, R. L. *BioMed. Chem. Lett.* **1997**, *7*, 827-832.
121. Pettit, G. R.; Kamano, Y.; Dufresne, C.; Cerny, R. L.; Herald, C. L.; Schmidt, J. M. *J. Org. Chem.* **1989**, *54*, 6005-6006.
122. Pettit, G. R.; Herald, D. L.; Singh, S. B.; Thornton, T. J.; Mullaney, J. T. *J. Am. Chem. Soc.* **1991**, *113*, 6692-6693.
123. de Arruda, M.; Cocchairo, C. A.; Nelson, C. M.; Grinnell, C. M.; Janssen, B.; Haupt, A.; Barlozzari, T. *Cancer Res.* **1995**, *55*, 3085-3092.
124. Pettit, G. R. The Dolastatins, In *Progress in the Chemistry of Organic Natural Products*; Herz, W.; Kirby, G. W.; Moore, R. E.; Steglich, W.; Tamm, Ch., Eds.; No. 70; Springer-Verlag: NY, 1997; pp 1-79.
125. Pettit, G. R.; Srirangam, J. K.; Barkóczy, J.; Williams, M. D.; Boyd, M. R.; Hamel, E.; Pettit, R. K.; Hogan, F.; Bai, R.; Chapuis, J-C.; McAllister, S. C.; Schmidt, J. M. *Anti-Cancer Drug Des.* **1998**, *13*, 243-277.
126. Pettit, G. R.; Srirangam, J. K.; Barkóczy, J.; Williams, M. D.; Durkin, K. P. M.; Boyd, M. R.; Bai, R.; Hamel, E.; Schmidt, J. M.; Chapuis, J-C. *Anti-Cancer Drug Des.* **1995**, *10*, 529-544.
127. Mirsalis, J. C.; Schindler-Horvat, J.; Hill, J. R.; Tomaszewski, J. E.; Donohue, S. J.; Tyson, C. A. *Cancer Chemother. Pharmacol.* **1999**, *44*, 395-402.
128. Pettit, R. K.; Pettit, G. R.; Hazen, K. C. *Antimicrobial Agents Chemother.* **1998**, *42*, 2961-2965.
129. Luesch, H.; Yoshida, W. Y.; Moore, R. E.; Paul, V. J. *J. Nat. Prod.* **1999**, *62*, 1702-1706.

130. Kigoshi, H.; Yamada, S. *Tetrahedron* **1999**, *55*, 12301-12308.
131. Harrigan, G. G.; Luesch, H.; Yoshida, W. Y.; Moore, R. E.; Nagle, D. G.; Paul, V. J. *J. Nat. Prod.* **1999**, *62*, 655-658.
132. Harrigan, G. G.; Luesch, H.; Yoshida, W. Y.; Moore, R. E.; Nagle, D. G.; Paul, V. J.; Mooberry, S. L.; Corbett, T. H.; Valeriote, F. A. *J. Nat. Prod.* **1998**, *61*, 1075-1077.
133. Pettit, G. R.; Inoue, M.; Kamano, Y.; Dufresne, C.; Christie, N.; Niven, M. L.; Herald, D. L. *J. Chem. Soc. Chem. Commun.* **1988**, 865-867.
134. Pettit, G. R.; Tan, R.; Xu, J-P.; Ichihara, Y.; Williams, M. D.; Boyd, M. R. *J. Nat. Prod.* **1998**, *61*, 955-958.
135. Ganesan, A. *Angew. Chem. Int. Ed. Engl.* **1996**, *35*, 611-615.
136. Ganesan, A.; Heathcock, C. H. *Chemtracts; Org. Chem.* **1988**, *1*, 311-312.
137. Pan, Y.; Merriman, R. L.; Tanzer, L. R.; Fuchs, P. L. *Bioorg. Med. Chem. Lett.* **1992**, *2*, 967-972.
138. Bhandaru, S.; Fuchs, P. L. *Tetrahedron Lett.* **1995**, *36*, 8351-8354.
139. LaCour, T. G.; Guo, C.; Bhandaru, S.; Boyd, M. R.; Fuchs, P. L. *J. Am. Chem. Soc.* **1998**, *120*, 692-707.

Chapter 3

Some Recent Developments in the Synthesis and Structure–Activity Relationship of Novel Taxanes

John F. Kadow[1], Thomas Alstadt[1], Shu-Hui Chen[1],
Pierre Dextraze[1], Karen Du[1], Craig Fairchild[2], Jerzy Golik[1],
Steven Hansel[1], Kathy A. Johnston[1], Robert A. Kramer[2],
David R. Langley[1], Frank Lee[2], Byron Long[2], Harold Mastalerz[1],
Carl Ouellet[1], Robert Perrone[1], William C. Rose[2], Paul Scola[1],
Gene Schulze[1], Andrew Staab[1], Quifen May Xue[1],
Michael Walker[1], Jen-Mei Wei[1], Mark Wittman[1],
J. J. Kim Wright[1], Mary Zoeckler[1], and Dolatrai M. Vyas[1]

[1]Bristol-Myers Squibb Pharmaceutical Research Institute, 5 Research Parkway, Wallingford, CT 06492
[2]Oncology Drug Discovery, Pharmaceutical Research Institute, Bristol-Myers Squibb Company, Route 206 and Province Line Road, Princeton, NJ 08543

The 7-methyl thioether **15** (BMS-184476) and the C-4 methylcarbonate **6** (BMS-188797) analogs of paclitaxel **1** were found to be more active than the parent in several preclinical *in-vivo* anticancer models. Both compounds are currently undergoing Phase I and II clinical trials. Chemistry for synthesizing C-7 sulfur analog of 15 and 6-alpha fluoro paclitaxel **27**, a compound which should be resistant to metabolic hydroxylation at C-6 is described. The synthesis of 7-deoxy-6-beta hydroxy paclitaxel **36** and 7-deoxy-6-thiomethyl paclitaxel **37** is also summarized.

First approved for clinical use in 1992, the taxane anticancer agent TAXOL® (paclitaxel **1**) is now a widely used drug, approved for a number of indications either in combination or as a single agent(1-3). Because TAXOL® displays activity against ovarian, breast, lung, Kaposi sarcoma, bladder, prostate, esophageal, head and neck, cervical, and endometrial cancers, considerable efforts are still being expended on clinical studies to maximize the utility of this drug(4). Although it will be difficult and challenging to better the clinical activity of TAXOL® with another taxane, we have been involved in an analog program aimed at finding taxanes with a superior antitumor profile against a variety of paclitaxel sensitive/resistant/partially resistant in-vivo preclinical tumor models. Such analogs will have the potential to demonstrate an expanded or improved spectrum of activity in cancer patients without displaying any additional toxicity. TAXOTERE®(docetaxel, **2**), a semisynthetic taxane marketed by Aventis has also been approved for clinical use.

Paclitaxel, is an antimitotic cytotoxic which alters equilibrium between assembled microtubules and disassembled tubulin(5-7). The *in vitro* SAR of paclitaxel is by now well developed and has been reviewed extensively (8-13). Our analog program, like others in the field, was directed at modifying both the Northern and Southern portion of the molecule.

Southern Region Modifications

The sidechain, the C-2 benzoate, the C-4 acetate, and the oxetane ring form the southern region of the molecule and are considered to be the portions of the molecule which directly interact with the microtubules formed by the tubulin dimers. Deletions of any of these functionalities leads to significant losses in potency in assays which measure tubulin polymerization or microtubule stabilization. We have continued our efforts to find an improved taxane by studying taxanes containing cores modified at the C-4 position. The general chemical strategy pursued was similar to that which is by now well known in the taxane literature. Analogs with modified baccatin cores were prepared and then after identification of promising modifications, numerous sidechain-core hybrids were synthesized.

C-4 Analogs with a Paclitaxel Sidechain. Our syntheses and selective methodology for the preparation of C-4 esters and carbonate analogs have already been reported (14, 15). To select compounds for *in vivo* evaluation we

Table 1 In vitro Profile of C-4 Ester and Carbonate Analogs

Compound	R	Tubulin polym. Ratio[a]	HCT 116 IC_{50} (nM)[b]	HCT 116 R/S Ratio[c]
1	-CH$_3$	1	4.0	114
3	-CH$_2$CH$_3$	0.6	2.0	63
4	-(CH$_2$)$_2$CH$_3$	0.4	1.1	31
5	-cyc-propyl	0.2	1.0	2.4
6	-OCH$_3$	0.4	2.0	68
7	-OCH$_2$CH$_3$	0.6	1.0	2.9
8	-O(CH$_2$)$_2$CH$_3$	0.8	2.6	10.7

[a]The ratio in the tubulin polymerization assay is the potency of the analog / the potency of paclitaxel. Ratios less than 1 reflect analogs that are more potent than paclitaxel. [b] Paclitaxel sensitive Human Colon Tumor 116. [c]Ratio of IC_{50} for HCT 116(VM)46 MDR resistant cell line /IC_{50} for HCT 116 sensitive cell line.

optimized in vitro properties, primarily cytoxicity against paclitaxel resistant cell lines, however potency in tubulin polymerization and sensitive cytotoxicity assays was also evaluated. As shown in Table 1, moderately increased size at C-4 correlated with an improved ability to overcome resistance in the HCT 116(VM)46 cell line(16). This cell line is more than 100 fold resistant to paclitaxel due to overexpression of P-glycoprotein (Pgp) resulting in decreased intracellular concentrations of **1** and leading to the multidrug resistance (MDR) phenotype (17, 18)) and the resulting decreased intracellular concentrations of **1** in the resistant cell line. Clinical resistance for paclitaxel is not completely understood but Pgp overexpression and altered tubulin have been suggested as two likely mechanisms (19-21). As can be seen in table 1, the cyclopropyl ester **5** and ethyl carbonate **7** analogs of paclitaxel were able to effectively overcome the effects of Pgp-mediated resistance to paclitaxel (R/S IC50 ratios near 1). Studies with the cyclopropyl analog **5** showed that the analog was overcoming MDR as actual measurements of intracellular concentrations in the resistant cell line were close to those in the sensitive cell line unlike the large decrease seen for paclitaxel(22). Analogs were initially selected for evaluation in *in vivo* models based on their ability to overcome resistance in the HCT 116(VM)46 cell line *in vitro*.

Table 2 C-4 analogs which overcome resistance vs HCT 116(VM)46

Compound[a]	R_2	R_1	R (C-4)	HCT 116 R/S ratio[b]	T/C,(dose,mg/kg) M109 [c]
1	Bz	Ph	-CH$_3$	125	159-228
9	t-Boc	2-Furyl	-(CH$_2$)$_2$CH$_3$	0.5	96 (1)
5	**Bz**	**Ph**	**-cyc-propyl**	**2.4**	**188 (6)**
10	t-Boc	Ph	-cyc-propyl	0.9	115 (1.6)
11	Bz	2-Furyl	-cyc-propyl	5.2	132 (3)
12	t-Boc	2-Furyl	-cyc-propyl	1.3	100 (4)
13	**t-Boc**	**2-Furyl**	**-OCH$_3$**	**3.6**	**183 (32)**
7	Bz	Ph	-OCH$_2$CH$_3$	2.9	142 (6)
14	t-Boc	2-Furyl	-OCH$_2$CH$_3$	1.4	144 (4)

[a]Tubulin data has been previously reported (14). [b]Ratio of IC$_{50}$ for HCT 116(VM46)MDR resistant cell line /IC$_{50}$ for HCT 116 sensitive cell line. [c] Madison Murine Lung Carcinoma. Drug is given ip on days 1-5 post ip implantation. Values are %T/C analog (maximum tolerated dose) % T/C for paclitaxel at a historically determined maximally efficacious dose. %T/C values of greater than 125 % are considered active.

C-4 Analogs with Modified Sidechains The most promising C-4 modified cores (the cyclopropyl ester and the ethyl carbonate) and close analogs (methyl carbonate for example) were combined with modified sidechains. Table 2 contains C-4 analogs with modified sidechains which were synthesized and which overcame resistance *in vitro*. The analogs were initially screened *in vivo*

Table 3 *In vitro* profiles of two active taxanes

6(BMS-188797)   15 (BMS-184476)

Compound	Tubulin Polym. Ratio[a]	Cytotoxicity IC50 values (nM)[a]			
		HCT 116[b]	HCT116(VM)46[c] (R/S)[d]	A2780[e]	A2780/PTX22[f] (R/S)[g]
Paclitaxel	1.0	3.6	375 (106[b])	4.5	62 (14)
6	0.4	2.3	156 (68)	6.1	35 (5.8)
15	1.1	2.1	25 (12)	3.0	16 (5.3)

[a]The ratio in the tubulin polymerization assay is the potency of the analog / potency of paclitaxel. Ratios less than 1 reflect analogs that are more potent than paclitaxel. [b] Paclitaxel sensitive Human Colon Tumor 116. [c]MDR resistant human colon tumor cell line. [d]Ratio of IC_{50} for HCT 116(VM)46 MDR resistant cell line /IC_{50} for HCT 116 sensitive cell line. [e]Paclitaxel sensitive human ovarian tumor cell line. [f]Paclitaxel resistant human ovarian tumor cell line known to be resistant due to a β-tubulin mutation (T. Fojo, NCI(23)). [g] Ratio of IC_{50} for human ovarian resistant cell line /IC_{50} for A2780 senstive cell line.

against an ip/ip M109 tumor model(16). This initial *in vivo* screen minimizes the impact of pharmokinetics since both the tumor and drug are placed in the intraperitoneal cavity and provides initial indications of activity and potency. Analogs which displayed a T/C value greater than 125 were considered active. Subsequently, we evaluated the majority of these active analogs in one or more secondary distal tumor models (iv administration of drug to a mouse bearing a subcutaneously implanted tumor) in an effort to identify compounds more efficacious than paclitaxel. Unfortunately, none of these compounds was found to have advantages over paclitaxel in *in vivo* studies. It became apparent that enhanced ability to overcome MDR resistance did not correlate with enhanced *in vivo* efficacy in the C-4 analog series; and in fact actually lowered the therapeutic index. Secondary tumor models with iv (intravenous) dosing experiments more closely mimic the likely clinical protocol. However the *in vivo* data from the initial ip/ip M109 screen in Table 2, despite its limitations(24), appeared to show that the most active analogs were **5**, the cyclopropyl analog with the paclitaxel sidechain, and **13**, the C-4 methyl carbonate with the furyl Boc sidechain. Interestingly, the furyl Boc sidechain analog was inactive in the cyclopropyl ester series, suggesting that the *in vivo* activity of the C-4 methyl carbonate might be more robust and better able to tolerate the incorporation of a more potent sidechain. To explore this hypothesis, the *in vivo* activity of **6**, the C-4 methyl carbonate analog with a paclitaxel sidechain, was examined, in several distal tumor models despite the compound's relatively modest ability to overcome resistance *in vitro*.

BMS 188797, the C-4 Methyl Carbonate Analog of Paclitaxel. As can be seen in Table 3, methyl carbonate **6**, was only minimally able to overcome MDR in the HCT 116(VM)46 cell line and also remained cross resistant to the A2780/PTX22 cell line, a cell line known to be resistant due to a β-tubulin mutation (T. Fojo, NCI(23)). However, in distal tumor models, the compound showed impressive activity (table 4):

a) Against the L2987 human lung cancer xenograft model the compound produced >5.2 LCK (log cell kill(**16**)) reduction in tumor growth and 3 out of 8 mice were cured. Paclitaxel at its optimum dose and schedule provided 3.6 LCK and only 1/7 cures. A difference of one LCK is considered significant.

b) The HCT/pk tumor is a model derived in the laboratory but in which resistance was developed by mimicking the levels of drug that would be present after clinical doses of paclitaxel. Against this xenograft, **6** produced 4.4 LCK and 3/7 cures while paclitaxel again at its optimum dose and schedule produced 1.7 LCK and no cures.

Overall, the analog was found to be more active than paclitaxel in four *in vivo* distal tumor models after iv administration. Two of these models were partially resistant human carcinomas (one clinically derived). A complete description of the *in vivo* models and biological evaluation of this compound has been submitted for publication(25). The compound was always at least as active as paclitaxel in other models and was progressed to development where it is now currently in Phase I and Phase II clinical trials.

Table 4. Distal Tumor activity of 6 (BMS-188797) and 15 (BMS-184476)

Compound	Human Tumor[a]	Schedule[b]	Optimal dose[c]	Analog LCK (c/t)[d]	Paclitaxel LCK (c/t)
6	L2987	Q2dx5; 23	24	>5.2 (3/8)	3.6 (1/7)
6	HCT/pk	Q2dx5; 12	24	4.6 (3/7)	1.7 (0/0)
15	A2780	Q2dx5; 8	30	(8/8)	6.1 (0/8)
15	HCT pk	Q2dx5; 10	24	(6/8)	2.8 (1/8)

[a] L2987 lung carcinoma, HCT/pk colon carcinoma, A2780 ovarian carcinoma. [b] q2dx5: 23 = test compound was administered every other day X 5, starting on day 23 post tumor implantation. [c] dose (mg/kg). The next highest dose was excessively toxic. [d] LCK, gross log cell kill, with cures/total mice (c/t) shown in parentheses.

Northern Region Modifications

The northern region, spanning positions 6, 7, 19, 9, and 10 has been found to be fairly tolerant of changes (within reasonable molecular weight ranges) and modifications in general, do not drastically diminish potency in tubulin polymerization or microtubule stabilization assays. However, these positions play an important role in the *in vivo* efficacy of the taxanes. They may modulate the solubility, pharmacokinetics, metabolism, and distribution of the compounds in the body.

Previously we and others have described 7-deoxy(26-28), 7-halo(29, 30), or 7,19 cyclopropane derivatives(29-32), many incorporating modified sidechains. During the course of our program and others, 7 ethers, acetal ethers, and methyl thioethyl ethers(33-35), were also prepared. Modifications at the C-7 position of paclitaxel have been shown to reduce the effect of p-glycoprotein mediated multiple drug resistance in cancer cells(18, 36, 37) and now we also describe some modifications which can produce this effect.

7-methyl thiomethyl (MTM) analogs. A three step synthesis of the 7-methyl thio ether analog **15** of paclitaxel from paclitaxel has previously been reported(38). As shown in Table 3, **15** was essentially equipotent to paclitaxel in a tubulin polymerization assay but showed an ability to overcome resistance (MDR) against the HCT 116(VM)46 cell line. The analog was also found to partially overcome resistance in the A2780/PTX22 ovarian cell line which is resistant due to a β-tubulin mutation. 7-MTM analogs with modified sidecahins were also synthesized and evaluated for SAR purposes. As can be seen from table 5, the ability to overcome Pgp-mediated MDR resistance was enhanced even further by incorporating modified sidechains. Reasoning that analogs with the greatest structural and biological differences from paclitaxel should offer the greatest chance for a compound with a broadened or altered spectrum of activity, the 7-methylthiomethyl analogs with the greatest ability to overcome resistance (the lowest R/S ratios) were evaluated in an initial ip/ip M109 screen. Compounds with T/C values greater than 125 were considered to be active and as in the previous series progressed to evaluation in *in vivo* secondary distal tumor models. As can be seen from the ip/ip M109 data, the analogs with the combination of either the isobutenyl or 2-furyl as the 3' substituent with the tBoc at the 3'N were inactive in the ip M109 *in vivo* model despite their impressive ability to overcome MDR. Unfortunately, we were unable to identify a 7-MTM analog with a modified sidechain which demonstrated a statistical improvement in activity over paclitaxel in distal tumor models (49).

Table 5. Biological activity of paclitaxel 7-methyl thio ethers

Analog	R_1	R_2	HCT 116[a] IC_{50} (nM)[b]	R/S ratio[b]	T/C (dose,mg/kg) M109 [c]
15	Phenyl	Phenyl	2.1	10.5	Not tested
16	Phenyl	*t*-Butoxy	0.6	1.7	143 (12)
17	Phenyl	*n*-Butoxy	1.2	4.3	177 (8)
18	Phenyl	*i*Propoxy	1.7	1	191 (8,16)
19	2-furyl	*t*-Butoxy	1.7	1	104 (3)
20	2-furyl	Phenyl	0.6	6.2	191 (6)
21	*i*-Butenyl	Phenyl	0.5	3.9	168 (3)
22	*i*-Butenyl	*t*-Butoxy	0.7	3.3	109 (4)

aIC_{50} Human colon tumor cell line sensitive to paclitaxel. [b]Ratio of IC_{50} for HCT 116(VM)46 MDR resistant cell line /IC_{50} for HCT 116 sensitive cell line. [c] Madison Murine Lung Carcinoma. Drug is given ip on days 1-5 post ip implantation. Values are %T/C analog (maximum tolerated dose) / % T/C for paclitaxel (200 to 259)at a historically determined efficacious dose but does not represent the MTD. %T/C values of > 125 % are considered active.

BMS 184476, The 7-MTM analog of paclitaxel. The parent MTM compound, **15**, which showed a significant but more moderate ability to overcome resistance, was also evaluated in distal tumor models. The compound

was found to be highly active and curative in some moderately staged tumor models.
a) Against a paclitaxel sensitive A2780 human ovarian xenograft model, **15**, provided 8/8 cures while paclitaxel at its optimum dose provided 6.1 LCK and no cures (table 4).
b) Against the moderately resistant, HCT/pk tumor model described above, provided 6/8 cures while paclitaxel provided 2.8 LCK and 1/8 cures

A manuscript describing the complete *in vivo* profile of this analog has been submitted for publication (25). Based on these results and its profile against a panel of other tumor models, the 7-MTM analog of paclitaxel was also advanced to clinical development where it is presently in Phase 1 and Phase 2 trials.

Scheme 1

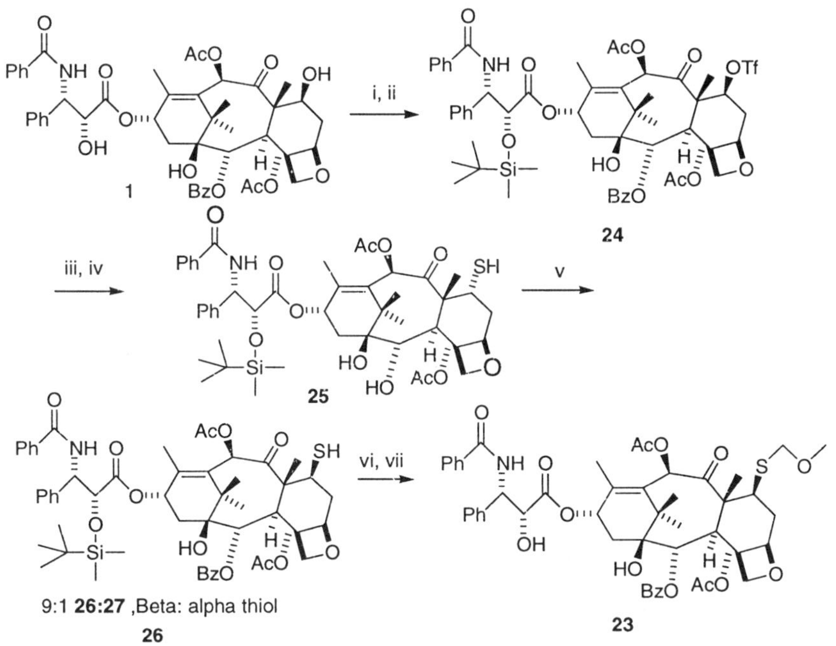

(i) tBuMe$_2$SiCl, DMF, imidazole, 99% (ii) Tf$_2$O, DMAP, CH$_2$Cl$_2$, 98%
(iii) KS(O)CH$_3$ EtOH, RT, 45h, (dark), 89% (iv) NH$_3$, EtOH, 61%
(v) Toluene, DBU, 90°, 89% (vi) BrCH$_2$OCH$_3$, DBU, 67% (vii) nBu$_4$NF, 77%

Our efforts to identify useful taxane analogs from the northern half of the molecule have continued. The syntheses and chemistry for the synthesis of a number of additional analog classes has been developed. Preliminary *in vitro* biology for these compounds is tabulated at the end of the chapter and SAR would suggest that these new compounds will have activity but since the paradigm and criteria for selecting new analogs must shift in order to find analogs with other unique advantages over our current leads **6** and **15**, further discussion is left for the future.

C-7 sulfur analogs. The enhanced efficacy of **15** led us to develop a synthesis of the isomeric C-7 thio methyl methyl ether **23** from a C-7 thiol(39). As can be seen in Scheme 1, displacement of the 7 triflate in **24** with potassium thioacetate followed by ammonolysis provided the 7-epi thiol **25**. Epimerization of the thiol moiety with DBU in toluene provided a 9:1 ratio of the desired 7-beta-mercapto paclitaxel **26** to starting epimer **25**. Unlike the parent 7-hydroxy system, where the 7-epi alcohol is favored, the thiol equilibrium appears to be controlled by steric factors. The poor hydrogen bonding ability of the thiol would be expected to make hydrogen bonding to the C-4 acetate inconsequential. Alkylation of thiol **26** with bromomethyl methyl ether and desilylation provided the desired thio methyl methyl ether **23** which, as shown in table 6 retained potency in the tubulin polymerization assay and against the sensitive HCT 116 cell line. The compound displayed a moderate ability to overcome resistance against the MDR cell line.

Synthesis of 6-alpha fluoro paclitaxel. The major human metabolites of paclitaxel arise from cytochrome P450 mediated C-6 hydroxylation from the alpha face. This tranformation, is essentially a metabolic deactivation, since it produces 6-alpha hydroxy paclitaxel, which has been reported to be approximately 30 fold less active in whole cell cytotoxicity asays(40). A synthesis of 6-alpha fluoro paclitaxel **27**, an analog which should be resistant to metabolic hydroxylation because the fluorine atom blocks the site of metabolism was developed. Olefin **28** was prepared from paclitaxel as reported earlier(32). While we and others showed that olefin can be epoxidized in high yield in a stereospecific manner, we were unable to use the epoxide as an intermediate for synthetically useful functionalizations of C-6(41-43). However, x-ray crystallography on a baccatin derivative, established that epoxidation proceeded from the more accessible alpha face(44). This tendency for approach of reagents from the bottom face at positions 6 and 7 proved general for a variety of chemical reactions. The alpha diol **29** was prepared stereospecifically as previously described (Scheme 2) (42, 45). Selective oxidation of the more accessible 6-hydroxy group using a TEMPO based oxidation provided the hydroxy ketone **30** in high yield. Fortuitously, the C-7 hydroxy group epimerized upon exposure to silica gel to exclusively produce the desired configuration at C-7. Reduction with triacetoxy borohydride provided only the beta diol **31** which was unavailable from direct hydroxylation of the olefin. Formation of the cyclic sulfite and subsequent oxidation provided the cyclic sulfate **32**. This intermediate could then be reacted with nucleophiles which predictably reacted at the more sterically accessible C-6 position. Reaction with tetrabutylammonium fluoride followed by acid catalyzed liberation of the C-7 hydroxy group provided a 66% yield of 6-alpha fluoropaclitaxel **27**. This compound was not metabolized in human liver S9 preparations but unfortunately neither was the paclitaxel in the same experiment so the enhanced metabolic stability of the fluoro analog can for now only only be postulated.

Scheme 2

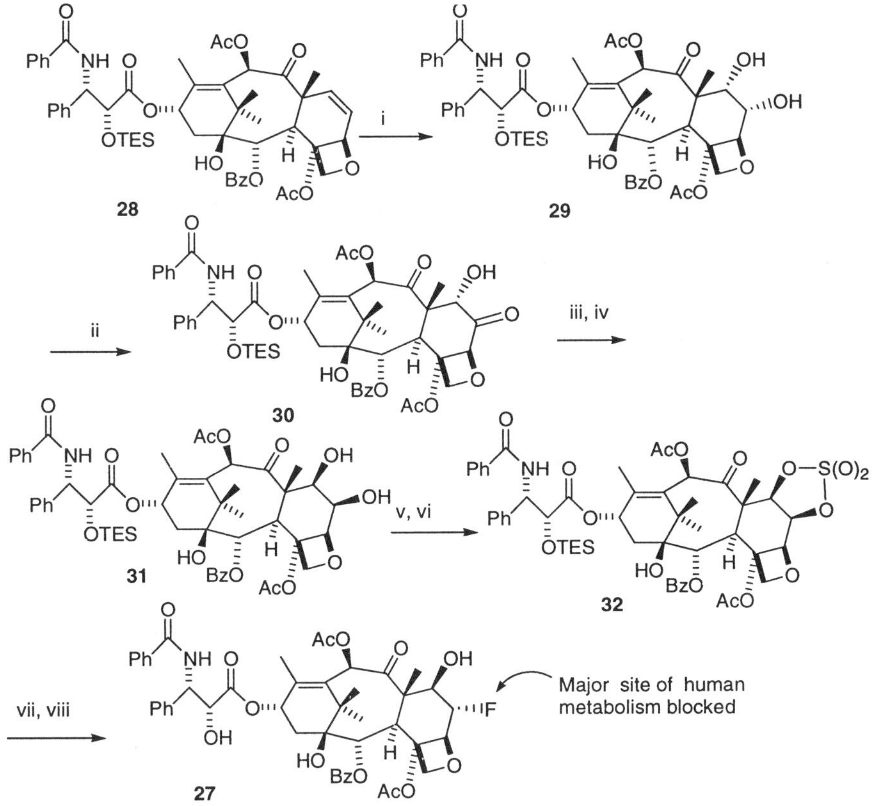

(i) OsO_4 NMO, Acetone/ water (8:1) 86% (ii) 4-BzO-TEMPO, CH_2Cl_2, H_2O, >90% (iii) Silica gel, CH_2Cl_2 (iv) $NaBH(OAc)_3$, CH_3CN 75% (v) $SOCl_2$, Et_3N, CH_2Cl_2 (vi) $NaIO_4$, $RuCl_3$, CCl_4, CH_3CN, H_2O 89% for v and vi. (vii) nBu_4NF, THF, 3h, RT (viii) H_2SO_4, H_2O, 1h, RT 66% for vii and viii.

In rodents, C-6 hydroxylation is not a major pathway of metabolism so this species difference makes preclinical evaluation of the effects of decreased metabolism non-trivial. As can be seen in table 6 the compound possessed an *in vitro* profile which was similar to paclitaxel. Analogous chemistry which have been described in the patent literature allowed introduction of alpha bromides, chlorides, azides or nitrates and this chemistry(46).

7-deoxy-6 substituted analogs. In a second approach, compounds were targeted in which the C-7 position was deoxygenated and the 6 position was hydroxylated on the beta face. This is essentially a hydroxy transposition which we reasoned would maintain the number of hydroxy groups present in the analog with the parent which should be beneficial for cellular or *in vivo*

Scheme 3

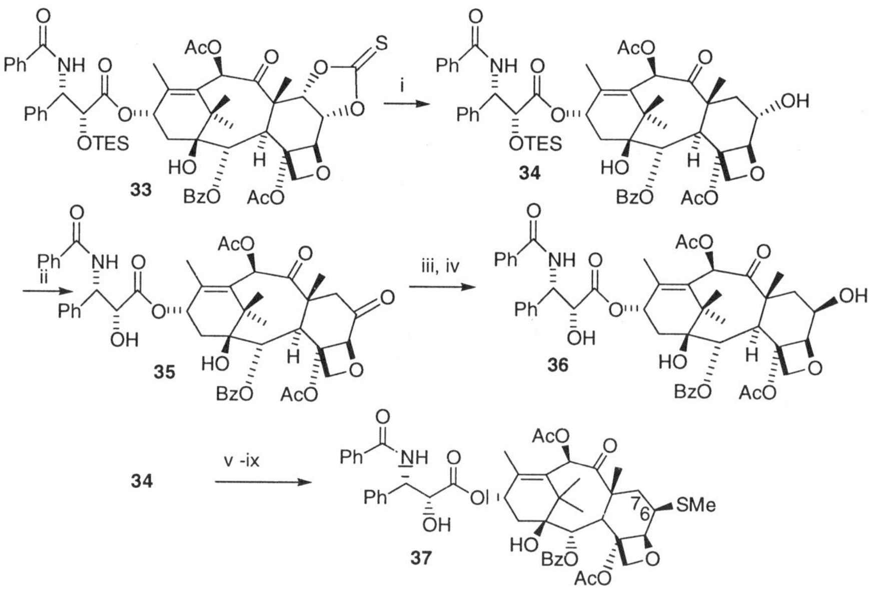

(i) Bu$_3$GeH, AIBN, THF/ toluene, 100 °C 71% (ii) TPAP, NMO, CH$_2$Cl$_2$, 95% (iii) NaBH$_4$, EtOH, 0°, 87% (iv) 1N HCl, CH$_3$CN, 0°, 95% (v) Tf$_2$O, DMAP 84% (vi) KSAc, DMF, 94% (vii) NH3, EtOH, 69% (viii) MeI, DBU, PhH, 94%(ix)1M HCl, CH3CN 76%

activity. Deoxygenation at the 7-position had previously been reported to improve potency(26) and we reasoned that this modification in combination with substitution on the beta face of C-6 might improve activity.

As previously reported the cyclic thiocarbonate **33** underwent extremely selective C-7 deoxygenation with tributyl germanium hydride to generate the 7-deoxy 6 alpha hydroxy compound **34** in good yield (45, 47). Oxidation of **34** using TPAP provided the 6-ketone **35** in high yield as the exclusive regioisomer. Reduction with sodium borohydride from the alpha face, followed by hydrolysis with 1N HCl in acetonitile provided the desired 7-deoxy 6-beta hydroxy taxane target **36** efficiently and as a single isomer. Alcohol **36** underwent metabolic oxidation in an *in vitro* human liver S9 assay to provide the corresponding C-6 ketone demonstrating that P450 mediated C-6 hydroxylation was still possible. The *in vitro* data for **36** showed the compound to be less potent than paclitaxel. However, the compound was active (T/C 158 at a dose of 100mg/kg on days 5 and 8) in the ip/ip M109 in vivo screen.

Alcohol **34** can be converted to the triflate and used as a handle for displacement chemistry. For example displacement with potassium thioacetate, ammonolysis, methylation with iodomethane /DBU, and desilylation provided the 7-deoxy-6-thio methyl ether **37**(48). As can be seen in table 6, compound **37**, retained potency in *in vitro* assays and displayed a modest ability to overcome MDR in the resistant cell line.

Table 6. In vitro Biological Activity for New Taxanes

Compound	Tubulin [a]	HCT 116 IC_{50} (nM)[b]	HCT 116(VM)46 IC_{50} (nM)[c]
23	1.9	0.5	17.8 (35.6)[d]
27	1.4	1.4	>114 (>81)
36	1.1	1.2	>117 (>97)
37	1.0	1.1	15.0 (13.6)

[a]The ratio in the tubulin polymerization assay is the potency of the analog over the potency of paclitaxel. Ratios less than 1 reflect analogs that are more potent than paclitaxel. [b] Paclitaxel sensitive Human Colon Tumor 116. [c]MDR resistant human colon tumor cell line. [d]Ratio of IC_{50} for HCT 116(VM)46 MDR resistant cell line /IC_{50} for HCT 116 sensitive cell line.

Conclusion

Exploring both the southern, tubulin binding region as well as the northern region of paclitaxel was important in the discovery of more efficacious, novel taxanes. Both BMS 184476 (**15**) and BMS 188797 (**6**) are presently in Phase I and II clinical trials. Clinical results will determine the relevance and predictive ability of our preclinical selection criteria.

Literature cited:

1. *Taxol Science and Applications;* Suffness M., Ed.; CRC Press: Boca Raton, FL, 1995.
2. Rowinsky, E. K. and Donehower, R. C. *New Engl. J. Med.* **1995,** *332,* 1004-1014.
3. Rowinsky, E. K. *Annu. Rev. Med.* **1997,** *48,* 353-374.
4. Rowinsky, E. K. *Semin. Oncol.* **1997,** *24,* 1-12.
5. Wilson, L. and Jordan, M. A. *Chem. Biol.* **1995,** *2,* 569-73.
6. Jordan, M. A., Wendell, K., Gardiner, S., Derry, W. B., Copp, H. and Wilson, L. *Cancer Res.* **1996,** *56,* 816-25.
7. Rao, S., He, L., Chakravarty, S., Ojima, I., Orr, G. A. and Horwitz, S. B. *J. Biol. Chem.* **1999,** *274,* 37990-37994.
8. Kingston, D. G. I. *J. Nat. Prod.* **2000,** *63,* 726-734.
9. Kingston, D. G. I. in *Taxane Anticancer Agents* Georg, G. I., Chen, T. T., Ojima, I., Vyas, D.M., Eds.; ACS Symp. Ser. Washington, D.C., 1995, Vol. 583, 203-16.
10. Georg, G. I., Boge, T. C., Cheruvallath, Z. S., Clowers, J. S., Harriman, G. C. B., Hepperle, M. and Park, H In *Taxol Science and Applications;* Suffness M., Ed.; CRC Press: Boca Raton, FL, 1995, 317-75.
11. Vyas, D. M. and Kadow, J. F. In *Prog. Med. Chem.*Ellis, G.P., Luscombe, D. K.Eds.; 1995, 32, 289-337.
12. Blagosklonny, M. V. and Fojo, T. *Int. J. Cancer* **1999,** *83,* 151-156.
13. Ojima, I., Kuduk, S. D. and Chakravarty, S. *Adv. Med. Chem.* **1999,** *4,* 69-124.
14. Chen, S.-H., Wei, J.-M., Long, B. H., Fairchild, C. A., Carboni, J., Mamber, S. W., Rose, W. C., Johnston, K., Casazza, A. M. and et al. *Bioorg. Med. Chem. Lett.* **1995,** *5,*2741-6.
15. Chen, S.-H., Kadow, J. F., Farina, V., Fairchild, C. R. and Johnston, K. A. *J. Org. Chem.* **1994,** *59,*6156-8.
16. Rose, W. C. In *Taxol Science and Applications;* Suffness M., Ed.; CRC Press: Boca Raton, FL, 1995; 209-35.
17. Ecker, G. and Chiba, P. *Expert Opin. Ther. Pat.* **1997,** *7,* 589-599.
18. Wu, Q., Bounaud, P.-Y., Kuduk, S. D., Yang, C.-P. H., Ojima, I., Horwitz, S. B. and Orr, G. A. *Biochemistry* **1998,** *37,*11272-11279.
19. Gonzalez-Garay, M. L., Chang, L., Blade, K., Menick, D. R. and Cabral, F. *J. Biol. Chem.* **1999,** *274,* 23875-23882.
20. Monzo, M., Rosell, R., Sanchez, J. J., Lee, J. S., O'Brate, A., Gonzalez-Larriba, J. L., Alberola, V., Lorenzo, J. C., Nunez, L., Ro, J. Y. and Martin, C. *J. Clin. Oncol.* **1999,** *17,*1786-1793.
21. Casazza, A. M. and Fairchild, C. R. *Drug Resistance* **1996,***Vol 87,*149-171.
22. Fairchild, C. R., Bristol-Myers Squibb Co. Unpublished Results.

23 Giannakakou, P., Sackett, D. L., Kang, Y.-K., Zhan, Z., Buters, J. T. M., Fojo, T. and Poruchynsky, M. S. *J. Biol. Chem.* **1997,**Vol *272,*17118-17125.
24 Since this is a screen, the analogs are not tested concomitantly. Comparison of historical data for activity can sometimes be misleading.
25 Rose, W. C. *Canc. Chemotherap. Pharmacol.* **2000,** In Press.
26 Chaudhary, A. G., Rimoldi, J. M. and Kingston, D. G. I. *J. Org. Chem.* **1993,** *58,* 3798-9.
27 Chen, S. H., Huang, S., Kant, J., Fairchild, C., Wei, J. and Farina, V. *J. Org. Chem.* **1993,** *58,* 5028-9.
28 Chen, S.-H., Kant, J., Mamber, S. W., Roth, G. P., Wei, J.-M., Marshall, D., Vyas, D. M. and Farina, V. *Bioorg. Med. Chem. Lett.* **1994,** *4,* 2223-8.
29 Chen, S. H., Huang, S. and Farina, V. *Tetrahedron Lett.* **1994,** *35,* 41-4.
30 Hester, J. B., Jr., Johnson, R. A., Kelly, R. C., Midy, E. G. and Skulnick, H. I. PCT Int. Patent Appl. WO 9413655, 1994.
31 Chen, S. H. and Farina, V. Eur. Patent Appl. EP 577083, 1994.
32 Johnson, R. A., Nidy, E. G., Dobrowolski, P. J., Gebhard, I., Qualls, S. J., Wicnienski, N. A. and Kelly, R. C. *Tetrahedron Lett.* **1994,** *35,* 7893-6.
33 Kelly, R. C. and Gebhard, I. PCT Int. Patent Appl. WO 9600724, 1996.
34 Golik, J., Kadow, J. F., Kaplan, M. A., Li, W.-S., Perrone, R. K., Thottathil, J. K., Vyas, D., Wittman, M. D., Wong, H. and Wright, J. J. Eur. Patent Appl. EP 639577, 1995.
35 Wittman, M. D. and Wong, H. Eur. Patent Appl. EP 694539,
36 Ojima, I., Bounaud, P.-Y., Takeuchi, C., Pera, P. and Bernacki, R. J. *Bioorg. Med. Chem. Lett.* **1998,** *8,*189-194.
37 Chakravarty, S., Oderda, C. L. F., Bounaud, P.-Y., Pera, P., Bernacki, R. J., Ojima, I *Book of Abstracts, 217th ACS National Meeting,* **1999,** MEDI 039.
38 Golik, J., Wong, H. S. L., Chen, S. H., Doyle, T. W., Wright, J. J., Knipe, J., Rose, W. C., Casazza, A. M. and Vyas, D. M. *Bioorg. Med. Chem. Lett.* **1996,** *6,* 1837-1842.
39 Mastalerz, H. and Kadow, J. F. U.S. Patent 6,017,935, 2000.
40 Harris, J. W., Katki, A., Anderson, L. W., Chmurny, G. N., Paukstelis, J. V. and Collins, J. M. *J. Med. Chem.* **1994,** *37,* 706-9.
41 Roth, G. U.S. Patent 5,395,850, 1995.
42 Liang, X., Kingston, D. G. I., Lin, C. M. and Hamel, E. *Tetrahedron Lett.* **1995,** *36,* 2901-4.
43 Menichincheri, M., Botta, M., Ceccarelli, W., Ciomei, M., Corelli, F., D'Anello, M., Fusar-Bassini, D., Mongelli, N., Pesenti, E. and et al. *Med. Chem. Res.* **1995,** *5,* 534-55.
44 Altstadt, T. J., Gao, Q., Wittman, M. D., Kadow, J. F. and Vyas, D. M. *Tetrahedron Lett.* **1998,** *39,* 4965-4966.
45 Wittman, M. D., Alstadt, T. J., Kadow, J. F., Vyas, D. M., Johnson, K., Fairchild, C. and Long, B. *Tetrahedron Lett.* **1999,** *40,* 4943-4946.

46 Wittman, M. D. and Kadow, J. F. U.S. Patent 5,912,264, 1999.
47 Wittman, M. D., Altstadt, T. J., Kadow, J. F., Kingston, D. G. I. and Liang, X. U.S. Patent 5,773,461, 1998.
48 Staab, A. J., Kadow, J. F., Vyas, D. M., Wittman, M. D. and Mastalerz, H. A. U.S. Patent 5,977,386, 1999.
49 Wittman, M. D. A manuscript containing a more detailed description of the chemistry and preclinical biology of these analogs (including tubulin polymerization data) has been submitted for publication.

Chapter 4

New Generation Taxoids and Hybrids of Microtuble-Stabilizing Anticancer Agents

Iwao Ojima[1], Scott D. Kuduk[1], Subrata Chakravarty[1],
Songnian Lin[1], Tao Wang[1], Xudong Geng[1], Michael L. Miller[1],
Pierre-Yves Bounaud[1], Evelyne Michaud[1], Young Hoon Park[1],
Chung-Ming Sun[1], John C. Slater[1], Tadashi Inoue[1],
Christopher P. Borella[1], John J. Walsh[1], Ralph J. Bernacki[2],
Paula Pera[2], Jean M. Veith[2], Ezio Bombardelli[3], Antonella Riva[3],
Srinivasa Rao[4], Lifeng He[4], George A. Orr[4], Susan B. Horwitz[4],
Samuel J. Danishefsky[5], Giovanni Scambia[6], and Cristiano Ferlini[6]

[1]Department of Chemistry, State University of New York at Stony Brook, Stony Brook, NY 11794–3400
[2]Department of Experimental Therapeutics, Roswell Park Cancer Institute, Elm and Carlton Streets, Buffalo, NY 14263
[3]Indena S.P.A., Viale Ortles, 12, 20139 Milan, Italy
[4]Department of Molecular Pharmacology, Albert Einstein College of Medicine, 1300 Morris Park Avenue, Bronx, NY 10461
[5]Laboratory for Bioorganic Chemistry, Memorial Sloan-Kettering Cancer Center, 1275 York Avenue, New York, NY 10021
[6]Department of Obstetrics and Gynecology, Catholic University of the Sacred Heart, Largo A. Gemelli 8, 00168 Rome, Italy

Structure-activity relationship (SAR) studies on taxoids derived from 14-OH-baccatin III led to the discovery of a highly potent anticancer agent IDN5109. IDN5109 exhibits much superior activity against a variety of drug-resistant cancers as compared to paclitaxel and docetaxel, and is orally active with excellent bioavailability. Human clinical trials began on the basis of highly promising preclinical study results. A series of other "Second Generation Taxoids" that possess 2-3 orders of magnitude higher potencies against drug-resistant human cancer cell lines as compared to paclitaxel have been developed as well. On the basis of SAR results, conformational analyses, and molecular modeling, a common pharmacophore for paclitaxel, epothilones, eleutherobin, and discodermolide, all of which stabilize microtubules, has been proposed. Based on this common pharmacophore, various hybrids that would become significant lead compounds for the next generation *de novo* microtubule-stabilizing anticancer agents have been synthesized and their activities assayed. Photoaffinity labeling of microtubules and P-glycoprotein using photoreactive radiolabeled taxoids has disclosed the drug binding domain of tubulin as well as Pgp. Together with the information for microtubule-bound conformation of fluorine-labeled paclitaxel/taxoid obtained by solid state ^{19}F NMR studies, the bioactive conformation of paclitaxel and taxoids appears to emerge, which is extremely important for the design and development of the next generation anticancer agents targeting microtubules.

Introduction

TAXOL® (paclitaxel), a naturally occurring diterpenoid isolated from the bark of the Pacific yew tree (*Taxus brevifolia*), is currently considered one of the most important drugs in cancer chemotherapy.*(1-4)* Paclitaxel was approved by the FDA for the treatment of advanced ovarian cancer in 1992, for metastatic breast cancer in 1994, and for AIDS related Kaposi's sarcoma in 1997. A semisynthetic analog of paclitaxel, Taxotère® (docetaxel)*(5)* was also approved by the FDA in 1996 for the treatment of advanced breast cancer. These two taxane anticancer drugs are currently undergoing clinical trials worldwide for the treatment of other cancers such as lung, head and neck, prostate, and cervical cancers. Effective chemotherapy in combination with other anticancer agents has also been developed.

These drugs bind to the β-subunit of the tubulin heterodimer, and accelerate the polymerization of tubulin and stabilizes the resultant microtubules, thereby inhibiting their depolymerization. The stabilization of microtubules results in the arrest of the cell division cycle mainly at the G2/M stage, leading to apoptosis of the cancer cells through cell signaling cascade. *(6-9)*

Although both paclitaxel and docetaxel possess potent antitumor activity, recent reports have shown that treatment with these drugs often encounters undesired side effects as well as various drug resistance.*(3,4)* Therefore, it is important to develop new taxoid anticancer agents with fewer side effects, superior pharmacological properties, and improved activity against various classes of tumors, especially against drug-resistant human cancers.

Paclitaxel: R^1 = Ph, R^2 = Ac
Docetaxel: R^1 = t-BuO, R^2 = H

10-DAB

It was recognized early in the development of paclitaxel as an anticancer drug that the drug supply was a serious problem. Paclitaxel is isolated from the bark of the Pacific yew tree (*Taxus brevifolia*), a nonrenewable resource, through a low-yielding extraction process. Accordingly, the long-term use of this drug could not be secured due to the limited natural abundance of the trees. In 1985, Potier et al. isolated 10-deacetylbaccatin III (DAB), a diterpenoid comprising the complex tetracyclic core of paclitaxel from the leaves of the European yew tree (*Taxus baccata*).*(10)* The isolation of DAB from a renewable resource has secured the long term supply of paclitaxel through semisynthesis and enabled extensive SAR studies. Among many semisynthetic methods studied, the use of the β-lactam synthon method (β-LSM) *(11)* with the Ojima-Holton coupling of *O*-metalated DAB derivatives has proven to be the most efficient and versatile process. *(11-13)* This method has opened facile routes to a diverse array of new taxoids, which made extensive SAR studies of the taxoid anticancer agents possible in addition to its application to the total synthesis*(14-17)* as well as commercial synthesis of paclitaxel.

Evolution of Taxoids: From Paclitaxel/Docetaxel to the Advanced Second Generation Taxoids

Extensive SAR studies of paclitaxel and its congeners in these laboratories have focused on the discovery of new taxoids exhibiting excellent cytotoxicity against different cancer types including those expressing the Pgp based multidrug resistance (MDR) phenotype. These new taxoids should also possess improved pharmacological properties, and reduced undesirable side effects. In fact, these efforts have led us to the development of the "second generation taxoids", which are remarkably potent against drug-resistant human cancer cell lines and human tumor xenograft models.

In our early SAR study of taxoids, it was found that the phenyl moieties of paclitaxel at the C-2, C-3', and C-3'N positions were not essential for the strong anti-mitotic activity and cytotoxicity because these phenyl moieties were effectively replaced by cyclohexyl groups (SB-T-1031, SB-T-1041, and SB-T-1051).*(18)* This intriguing finding prompted us to investigate the effects of nonaromatic substitutents at the C-3' position of docetaxel on cytotoxicity. Thus, a series of new docetaxel analogs possessing various alkyl and alkenyl substituents at the C-3' position were synthesized and their potency assayed. *(19)*

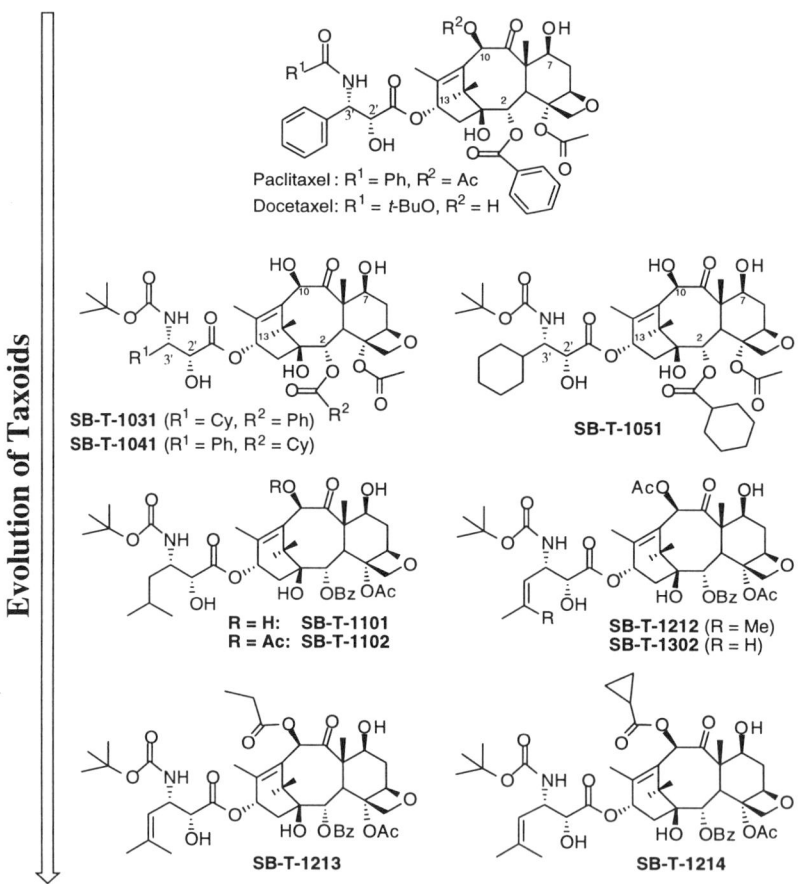

While a number of these taxoids exhibited excellent cytotoxicity against several human cancer cell lines, we were particularly encouraged by the substantial increase in activity against a drug-resistant breast cancer cell line, MCF7-R (paclitaxel, IC_{50} = 300 nM; SB-T-1102, SB-T-1212, and SB-T-1302, IC_{50} = 36, 12, and 14 nM, respectively). We also recognized that there was a significant difference in the activity against MCF7-R between 10-OH and 10-Ac taxoids, e.g., SB-T-1101 (IC_{50} = 107 nM) vs. SB-T-1102 (IC_{50} = 36 nM). Accordingly, we synthesized a series of taxoids with various substituents and functional groups at the C-10 position including ether, ester, carbamate, and carbonate to examine the effects of the C-10 modification on the activity against MCF7-R.*(20)* Then, it was found that a number of these taxoids exhibited one order of magnitude stronger activity (up to IC_{50} = 0.09 nM) than paclitaxel against drug-sensitive cancer cell lines, i.e., A121 (ovarian), A549 (non-small lung), HT-29 (colon), and MCF7 (breast). However, the most significant result for this series of taxoids is their remarkable activity (IC_{50} = 2.1-9.1 nM; paclitaxel and docetaxel, IC_{50} = 300 and 235 nM, respectively) against MCF7-R. Among these "second generation taxoids", SB-T-1213 and SB-T-1214 (IC_{50} = 2.2 and 2.1 nM, respectively) are exceptionally potent, possessing *two orders of magnitude* better activity than paclitaxel and docetaxel.*(20)* The observed marked effects of the C-10 modification forms a sharp contrast to other reports concluding that the C-10 deletion or acylation of paclitaxel does not modulate the activity based on the results obtained only for drug-sensitive cell lines.*(21-23)*

The Advanced Second-Generation Taxoids

As described above, an appropriate modification at the C-10 position and replacement of the phenyl moiety with an alkenyl or alkyl group at the C-3' position led us to develop the second generation taxoids. In the mean time, Kingston et al. reported that modifications at the 3-position of the C-2 benzoate with certain substituents (e.g., CN, N_3, MeO, and Cl, especially N_3) considerably increased the anticancer activity against P-388 cell line.*(24-26)* We anticipated that a combination of these two modifications should provide a series of highly active taxoids. Thus, a series of the second generation taxoids with modifications at the C-2 position was synthesized and their activities assayed.*(27)* In fact, some of these new taxoids were found to possess *2–3 orders of magnitude higher potency* against the drug-resistant cancer cell lines LCC6-MDR and MCF7-R as compared to paclitaxel and docetaxel. The most characteristic feature of these new second-generation taxoids is the very small difference in the IC_{50} values between drug-sensitive and drug-resistant cell lines. Although it has been shown that the 3-azidobenzoyl analog exhibits the highest potency in the SAR of paclitaxel,*(24,25)* in these "*advanced* second-generation taxoids", the ones with 3-methoxybenzoyl group possess superior activity, especially against MCF7 and MCF7-R. It is noteworthy that three of the newly developed *advanced* second generation taxoids (SB-T-110303, 121303, and 121304) show essentially no difference in activity against drug-resistant and drug-sensitive cell lines, virtually overcoming MDR, which is truly remarkable.

SB-T-121303 (X = MeO)
SB-T-121304 (X = N₃)

SB-T-110303

IC50 (nM)

Taxoid	LCC6-WT	LCC6-MDR	R/S	MCF7	MCF7-R	R/S
Paclitaxel	3.1	346	112	1.7	300	177
Docetaxel	1.0	120	120	1.0	235	235
SB-T-110303	0.9	0.8	0.9	0.36	0.33	0.92
SB-T-121303	1.0	0.9	0.8	0.36	0.43	1.19
SB-T-121304	0.9	1.2	1.3	0.9	1.1	1.2

Taxoids from 14β-OH-10-Deacetylbaccatin III: Discovery of IDN5109, a Highly Potent and Orally Active Taxoid

14β-Hydroxy-10-deacetylbaccatin III (14-OH-DAB) was first isolated from the needles of the Himalayan yew tree (*Taxus wallichiana Zucc.*) and its structure was determined by X-ray crystallographic analysis in 1992.*(28)* Due to the presence of an extra hydroxyl group at the C-14 position, 14-OH-DAB possesses substantially better water solubility than DAB. It was anticipated that new taxoids derived from 14-OH-DAB would help to improve water solubility, bioavailability, and also reduce the hydrophobicity-related drug resistance. If indeed it is the case, these improved pharmacological properties may have good effects on the modification of undesirable toxicity and the activity spectra against different cancer types.

Various taxoids were synthesized from 14-OH-DAB using β-LSM (*vide supra*) and their cytotoxicity assayed, which include 14-OH-docetaxel (**1**), *(29)* 1,14-carbonate analogs of paclitaxel and docetaxel (**2**: R^2 = Ph),*(30)* pseudo-taxoids (**3**),*(30)* and nor-seco-taxoids (**4**).*(31,32)* 14-OH-Docetaxel (**1**) showed almost the same potency as docetaxel with much better water solubility and the 1,14-carbonate analogs (**2**: R^2 = Ph) exhibited the same level of cytotoxicity as that of paclitaxel. Pseudo-taxoids and nor-seco-taxoids retained reduced yet substantial cytotoxicity and the latter implied a possible replacement of baccatin skeleton by surrogates.

Since docetaxel-1,14-carbonate (**2**: R^1 = t-BuO, R^2 = Ph, R^3 = H) exhibited excellent *in vivo* activity,*(30)* we examined the effects of the C-3' and C-10 modifications on the activity of this novel series of taxoids **2***(33)* in parallel to the development of the second generation taxoids derived from DAB. These second generation taxoids (**2**: $R^2 = i$-Bu, 2-methyl-1-propenyl, 1-propenyl, CF_3,*(34,35)* CF_2H*(35,36)*) derived from 1,14-OH-DAB-1,14-carbonate were found to possess excellent potency against drug-sensitive ($IC_{50} \geq 0.2$ nM) and drug-resistant ($IC_{50} = 10$ nM level) cell lines.*(36)* Among these potent second generation taxoids, SB-T-101131 (**2**: $R^1 = t$-Bu, $R^2 = i$-Bu, $R^3 = $ Ac), SB-T-101141 (**2**: $R^1 = t$-Bu, $R^2 = $ 2-methyl-1-propenyl, $R^3 = $ Ac), and SB-T-101144 (**2**: $R^1 = t$-Bu, $R^2 = $ 2-methyl-1-propenyl, $R^3 = Me_2NCO$) were chosen for candidates for extensive *in vivo* assay against human tumor xenografts in mice. Then, SB-T-101131 was selected as the best compound for preclinical studies (code name was changed to IDN5109) by Indena, S.p.A. and associated medical institutions in the U.S. and Italy. This selection was made mainly on the basis of the fact that SB-T-101131 (IDN5109) did not show any appreciable neurotoxicity nor cardiotoxicity and exhibited much less general toxicity while maintaining excellent cytotoxicity, which provides a wide therapeutic window.*(37-40)* IDN5109 has shown much superior activity as compared to paclitaxel across the board against the NCI's panel of drug-resistant cancer cell lines. IDN 5109 also exhibited 2-3 orders of magnitude higher potency than paclitaxel in inducing apoptosis.*(37)*. In addition to these excellent features and antitumor activity by the standard intravenous administration,*(38)* IDN5109 has been found to be highly orally active with excellent bioavailability (50.4% in mice based on pharmacokinetics analysis using Tween 80/EtOH formulation), providing the first highly promising orally active taxoid anticancer agent.*(39-41)* [Note: paclitaxel is not orally active at all, see *in vivo* antitumor activity assay results against MX-1 mammary carcinona xenograft in mice shown in Figure 1 (PTX = paclitaxel; p.o. = oral administration; i.v. = intravenous administration)].

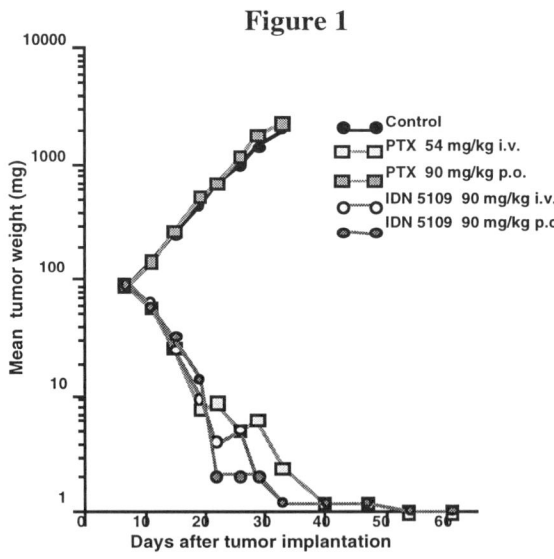

	Cell cycle / Apoptosis Analysis (CEM VBL-R)		
Taxane		G1/G2	DNA breakdown (%)
Paclitaxel			
	500 nM	3.4	13.6
	2500 nM	1.1	36.7
Docetaxel			
	100 nM	2.6	21.0
	500 nM	0.3	41.0
SB-T-101131			
	20 nM	3.5	24.5
	100 nM	0.2	46.1

IDN5109 (SB-T-101131)

IC_{50} (nM)

Taxane	A121 (ovarian)	A549 (NSCLC)	HT-29 (colon)	MCF7 (breast)	MCF7-R (breast)	CEM VBL-R (Leukemia)
Paclitaxel	6.3	3.6	3.6	1.7	1040	1200
Docetaxcel	1.2	1.0	1.2	1.0	280	580
IDN5109	1.2	0.7	1.5	1.1	35	50

Figure 1

IDN5109 also exhibits remarkable antitumor activity against drug-resistant tumors expressing P-glycoprotein (Pgp)*(37)* such as human colon carcinoma SW-620 (xenograft in mice) in sharp contrast with paclitaxel, which is not effective even by i.v. with its maximum tolerated dose (MTD) (Figure 2).*(41)* Moreover, it has been observed that the best antitumor efficacy of IDN5109 is often reached with doses lower than MTD, which indicates an excellent therapeutic index of this anticancer agent. *(38)*

Figure 2

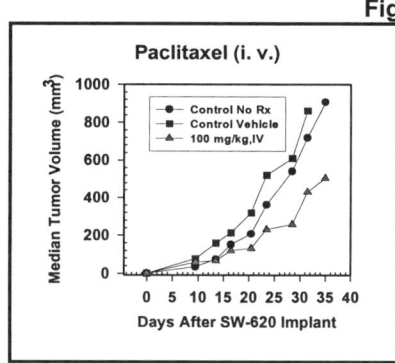
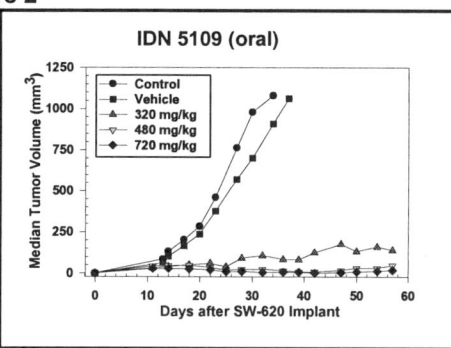

Animals treated IV on days 10,14,18 and 21 after implant

Animals treated PO on days 11,15,19 and 23 after implant

In light of these highly promising preclinical results, IDN5109 has recently entered human clinical trials and further clinical development will be performed by Bayer Corporation.

Unique Dual Function of Second Generation Taxoids

As discussed above, the second generation taxoids derived from DAB and 14-OH-DAB including IDN5109 exhibit exceptional activity against drug-resistant cancer cell lines and tumors expressing multidrug resistance (MDR). How do these taxoids overcome MDR?

Multidrug resistance (MDR), a common and serious problem in cancer chemotherapy, is a property by which tumor cells become resistant to various hydrophobic cytotoxic agents through the action of P-glycoprotein (Pgp), a transmembrane protein that is overexpressed in MDR cancer cells. Pgp acts as an efflux-pump that binds and eliminates cytotoxins from the cell membrane and cytosol, thereby keeping the intracellular drug concentration at innocuous levels.*(42-44)*

On the basis of the observation that the presence of an appropriate acyl group at the C-10 position plays a crucial role in the dramatic increase in potency against drug-resistant cancer cell lines expressing Pgp, we hypothesized, in the beginning, that these second generation do not bind to Pgp, thereby escaping from the Pgp efflux pump. To our surprise, however, our recent studies on the interaction of the second generation taxoids, e.g., SB-T-1213 and IDN5109, with Pgp using the rhodamine 123 fluorescence probe have revealed that these taxoids possess MDR reversal activity (!), i.e., these novel taxoids act as highly cytotoxic agents and, at the same time, inhibitors of the Pgp efflux pump.*(45)* Further studies on the unique dual function and its molecular basis are actively underway in these laboratories.

In order to understand the mode of interaction of paclitaxel and taxoids with Pgp in connection with MDR, we have developed efficient paclitaxel photoaffinity labels with tritium as a radioactive marker (*vide infra*).*(46,47)* Also, we have been actively developing novel and drug-specific taxane-based MDR reversal agents that efficiently modulate Pgp, restoring the potency of paclitaxel almost completely ($\leq 99.8\%$) against drug-resistant cancer cells.*(48-50)*

Development of the Common Pharmacophore Model for Microtubule-Stabilizing Anticancer Agents

Paclitaxel was the first among a new class of chemotherapeutic agents known as microtubule-stabilizing anticancer agents. Recently, four other natural products, structurally dissimilar to paclitaxel, have been found to share its mechanism of action.*(6,9)* Epothilones A and B*(51,52)* eleutherobin,*(53)* and discodermolide*(54)* (Figure 3) show activities comparable to or better than those of paclitaxel in various assays.*(52-56)* Moreover, these compounds were found to competitively inhibit the binding of {^{3}H}-paclitaxel to microtubules *(12,52,57,58)*, which strongly suggests the existence of a common or at least closely overlapping binding site for these agents.

Figure 3

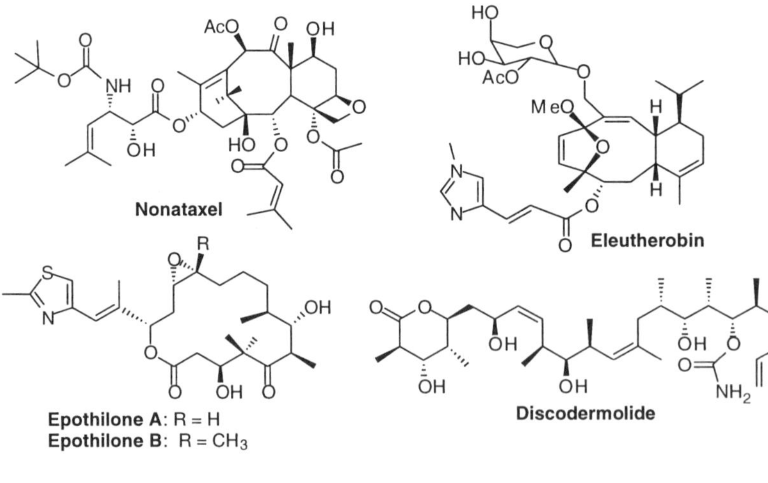

Nonataxel

Eleutherobin

Epothilone A: R = H
Epothilone B: R = CH$_3$

Discodermolide

The identification of a three-dimensional pharmacophore common to these structurally diverse agents, would enable us to develop the next generation cytotoxic agent, targeting microtubules. Thus, it is a very important endeavor to identify such a common pharmacophore. We have succeeded in identifying significant regions of structural homology through extensive molecular modeling studies (Sybyl and Biosym with Insight II programs) of these newly found natural products, paclitaxel, and nonataxel, *(59)* a highly potent fully non-aromatic second generation taxoid. *(60)* We selected a lower energy conformation of nonataxel arising from the restrained molecular dynamics (RMD) calculations keeping NOE-defined atom distances. Then, this structure was used as the template for searching a common pharmacophore with epothilone B, eleutherobin, and discodermolide. Resulting overlay structures are shown in Figure 4.

Figure 4

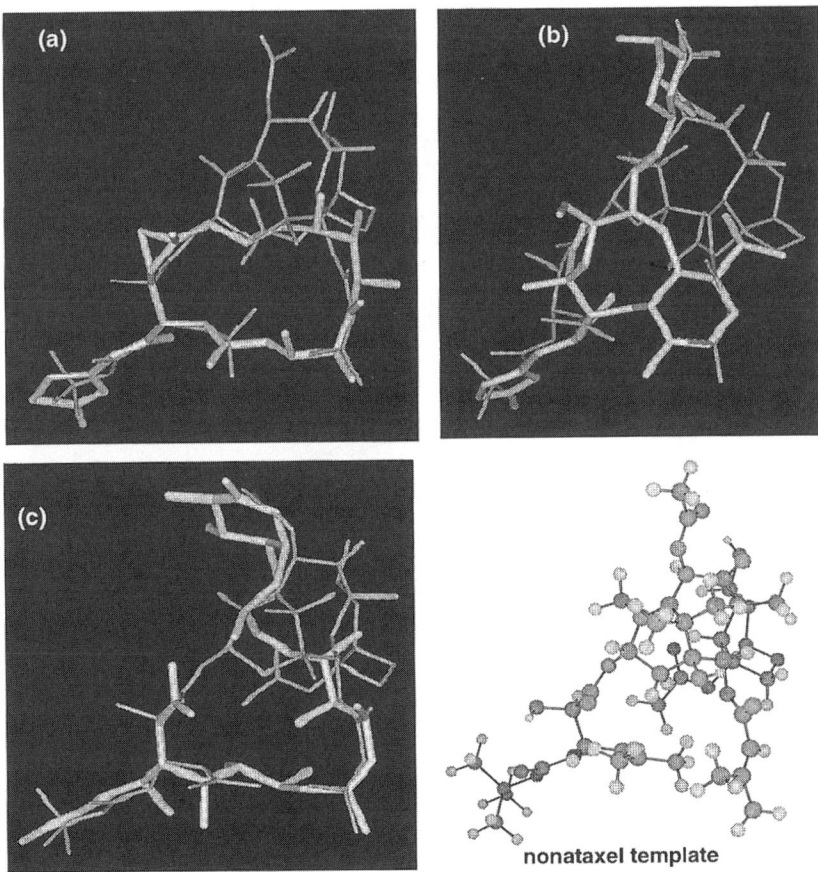

The observed structural homology between epothilone B and nonataxel nicely accommodates the vast SAR data that has been accumulated for the former (Figure 4a).*(60-65)* Another notable feature of the proposed common pharmacophore is that it accommodates the non-essential nature of the C12–C13 epoxide in the epothilones. In fact, the 12,13-deoxyepothilone B is showing a very promising profile of *in vivo* activity*(63,64)* and the 12,13-methanoepothilone B, i.e., cyclopropane analog, exhibits the same activity as epothilone B.*(66)* This common pharmacophore also accommodates the SAR data available for eleutherobin very well.*(65,67)* For example, the proposed pharmacophore indicates the essential nature of the C-8 side chain including an imidazole moiety and expendability of the C-3 sugar moiety, which is in complete agreement with the SAR results (Figure 4b)*(65,67)* Because of the highly flexible nature of discodermolide molecule, the search process was somewhat arbitrary. Nevertheless, the overlay (Figure 4c) suggests that discodermolide generally conforms to the common pharmacophore proposal.

The common pharmacophore model paves the way for designing *the third generation taxoids* that could be essentially baccatin-free, or "hybrids" integrating the structures of paclitaxel and the new antitumor agents. On the basis of the model, a series of novel cytotoxic hybrid constructs containing 16~18-membered macrocycles were designed and synthesized.

The first series of macrocyclic hybrids connected the C-2 and C-3' positions of the taxoid skeleton (C-linked hybrids).*(68)* Our retrosynthetic analysis indicated that macrocyclic taxoids such as SB-TE-1120 should be obtained by using a Ru-catalyzed ring-closing metathesis (RCM) reaction*(69-73)* in the key

Scheme 1

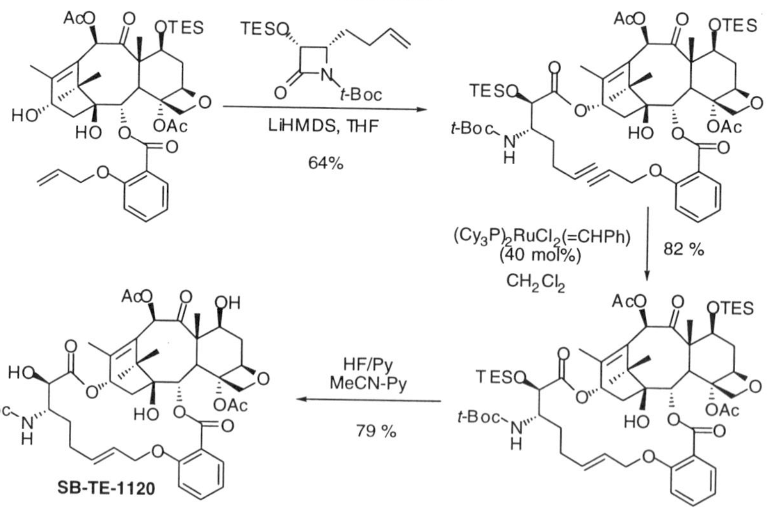

synthetic step and the key precursors should be readily accessible using the β-LSM (*vide supra*). The synthesis of SB-TE-1120 is outlined in Scheme 1 as an example of 15 macrocycles synthesized.*(68)* It is noteworthy that the Ru-catalyzed RCM is found to be applicable to the macrocyclization of highly complex and multifunctional systems such as these macrocyclic taxoids.*(68)* The C-linked hybrids exhibited good cytotoxicity against a human breast cancer cell line MDA-435/LCC6, e.g., SB-TE-1120: $IC_{50} = 0.39$ μM).

Next, a series of N-linked macrocyclic hybrids was designed and synthesized after we recognized the short distance (3.5 Å) between the methyl of the C-3'-*N-t*-Boc moiety and the C-2 benzoyl group in the X-ray crystal structure of docetaxel *(74)* and also this structure fits into our common pharmacophore model with small distortion. The new series of N-linked hybrids were synthesized by using Ru-catalyzed RCM as well as Pd-catalyzed Heck reaction.*(75)* The synthesis of SB-TE-14801 and SB-TE-14811 using the Heck reaction in the key step is exemplified in Scheme 2, wherein the key precursor was obtained from a C-2 modified baccatin using the β-LSM. The N-linked hybrids exhibited considerably better activity than the C-linked hybrids, e.g., SB-TE-14801, $IC_{50} = 0.071$ μM; SB-TE-14811, $IC_{50} = 0.060$ μM.

Scheme 2

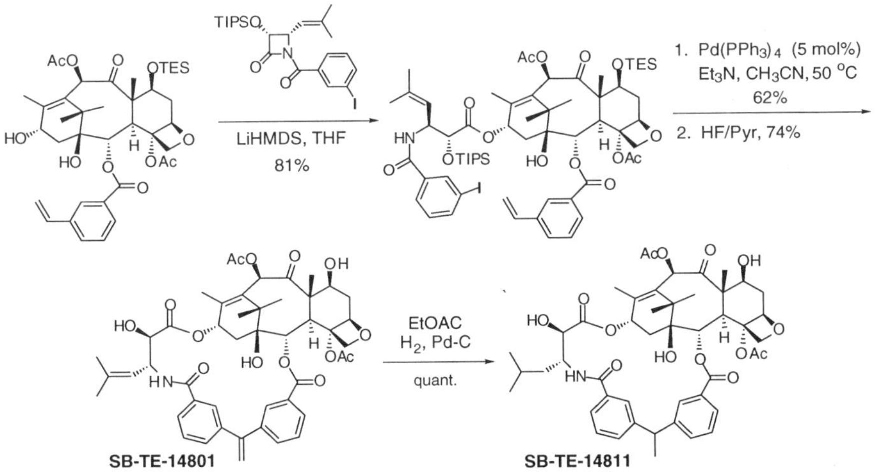

Further studies on the incorporation of more of the epothilone characteristics into these macrocyclic hybrids as well as the replacement of the baccatin skeleton with a simpler scaffold are actively underway.

Probing the Drug Binding Domains of Microtubules and P-Glycoprotein as well as Bioactive Conformation of Taxoids

Determination of the bioactive conformation of paclitaxel or its congeners is quite important since it constitutes a powerful approach to the design of the next generation anticancer agents targeting microtubules, complementary to the one based on a common pharmacophore discussed above.

Paclitaxel and its congeners bearing fluorine markers serve as excellent probes for identifying bioactive conformations of paclitaxel and taxoids.*(76)* Conformational analyses of taxoids bearing fluorines at the key positions revealed highly dynamic nature of their structures and suggested the conformations likely to be recognized by microtubules at the time of drug binding based on VT-^{1}H- and ^{19}F-NMR in combination with molecular dynamics analyses.*(77)*

Solid state ^{19}F NMR analyses, i.e., RFDR*(78)* and REDOR*(79)*, of microtubule-bound fluoro-taxoids could provide direct and crucial information about the bioactive conformation of paclitaxel and its congeners. Our preliminary study based on RFDR of F_2-docetaxel has indicated the ^{19}F-^{19}F distance of the two *para*-positions of the C-2 benzoate and C-3' phenyl to be 6.2 Å (Figure 5a).*(80)* A recent paper by Schaefer, Bane and collaborators using REDOR reports the two ^{19}F-^{13}C distances (see Figure 5b) between the *para*-position of the C-2 benzoate (^{19}F) and the C3' carbon as well as the carbonyl carbon of the C-3'N-benzoyl group (9.8 Å and 10.3 Å, respectively).*(81)* The combination of these two experimental data and molecular modeling would deduce the most likely microtubule-bound bioactive conformation of paclitaxel, which provides very useful information to the refinement of the common pharmacophore model discussed above. Study along this line is actively in progress in these laboratories.

Figure 5

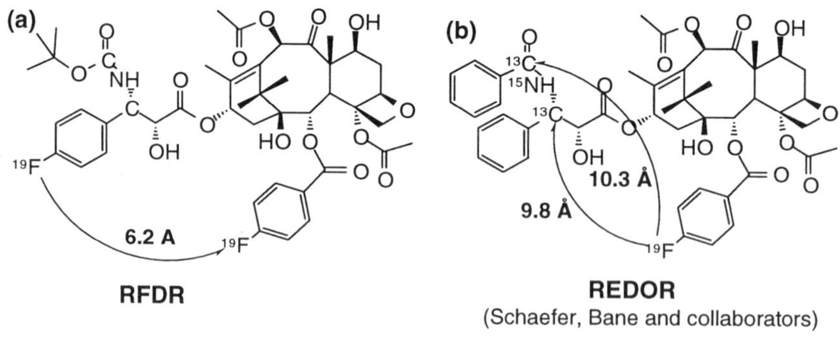

(a) RFDR

(b) REDOR
(Schaefer, Bane and collaborators)

In order to probe the paclitaxel binding domains of microtubules and Pgp, we designed and synthesized photoaffinity labeling taxoids. The use of these photoaffinity labels should provide structural information about the three-dimensional drug-binding site on the microtubules for paclitaxel, thus allowing for the rational design of anticancer agents targeting tubulin. Previous photoaffinity labeling studies using {^{3}H}-3'-(4-azidobenzamido)paclitaxel *(82,83)* and {^{3}H}-2-(3-azidobenzoyl)paclitaxel*(84)* identified the N-terminal 1-31 amino acid residues and the 217-233 amino acid residues of β-tubulin, respectively. To further investigate the drug-binding site on microtubules, new and highly efficient three photoreactive analogs of paclitaxel, {^{3}H}-SB-T-5101, {^{3}H}-SB-T-5111, and {^{3}H}-SB-T-5121, bearing benzophenone as photoreactive moiety, were synthesized (>99.9% radiochemical purity with 30-50 Ci/mmol) (Figure 6).*(46,47)*

Figure 6

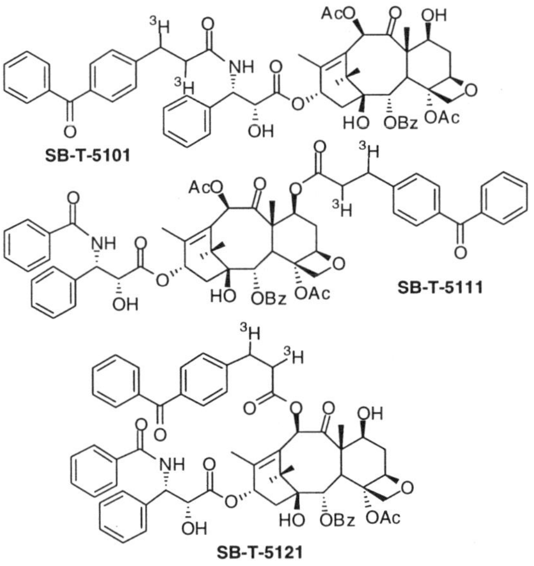

It is noteworthy that the photoaffinity labeling study using {^{3}H}-SB-T-5111 successfully identified the single amino acid residue (Arg 282) located in the β-tubulin subunit (Figure 7).*(85)* This exciting result in conjunction with the information from previous studies*(82-84)* as well as the crystal structure of the zinc-stabilized α,β-tubulin dimer sheet determined by electron crystallography *(86)* allowed for visualization of the likely structure of the paclitaxel-bound β-tubulin subunit in microtubules. It should be noted that the photoaffinity labeling studies were performed with biologically relevant microtubules while the electron crystallography was carried out with artificial paclitaxel-bound zinc-stabilized β-tubulin dimer sheet. This study has revealed that the position of the baccatin skeleton is largely in good agreement, but the positioning of the C-13 side chain amino acid in the proposed crystal structure is not consistent with the previous photoaffinity labeling results. The proposed model that can accommodate all three critical photoaffinity labeling study results is shown in Figure 7.*(85)*

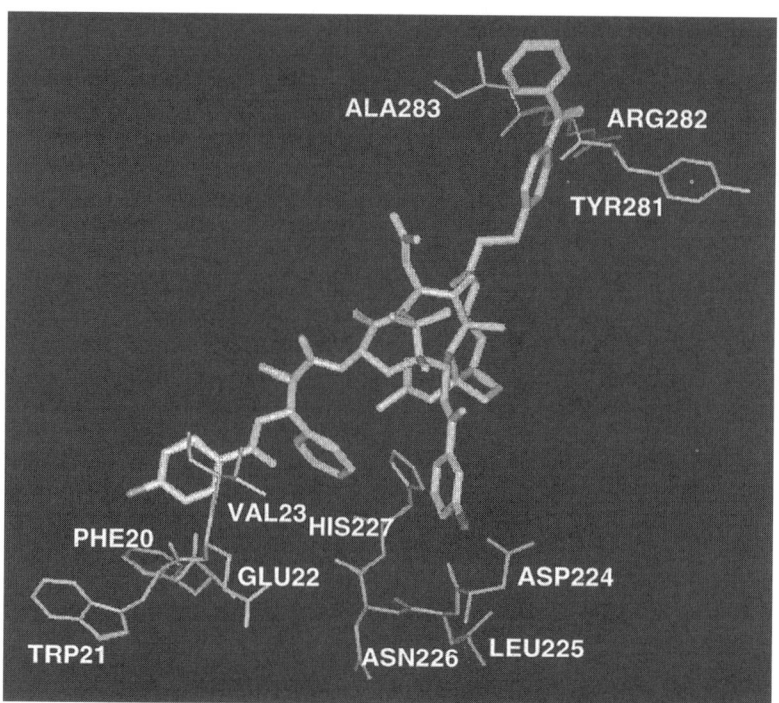

After our proposal of the first common pharmacophore of these microtubule-stabilizing agents, two other proposals have been made on the basis of microtubule mutations*(87)* and interesting activity of 2-(3-azidobenzoyl)baccatin III.*(88)* These proposals overlay the thiazole moiety of epothilone B with the C-2 benzoate moiety of paclitaxel (or 3-azidobenzoyl moiety of the baccatin derivative) as well as the epoxide oxygen of epothilone B with the oxetane oxygen of paclitaxel. However, recent disclosures of highly active cyclopropane analogs of epothilone B *(66)* and docetaxel*(89)* do not support the proposed superposition of these two oxygens. Also, the proposed model leaves the C-13 phenylisoserine moiety completely out of the overlapping region, which does not account for the vast SAR studies on taxoids.

These new photolabeling taxoids have also been successfully used for mapping Pgp. {^{3}H}-SB-T-5101 specifically photolabeled isoforms of murine Pgp present in drug-resistant cell lines. A 25-fold molar excess of paclitaxel effectively displaced {^{3}H}-SB-T-5101 from the binding site on Pgp. To the best of our knowledge, this is the first successful photoaffinity labeling of Pgp with a photoreactive analog of an anticancer agent.*(46)* In the same manner, the photolabeling of Pgp with {^{3}H}-SB-T-5111 was successfully carried out. The subsequent immuno-domain mapping studies of the labeled Pgp identified two separate binding regions, i.e., {^{3}H}-SB-T-5101 was incorporated into the peptide sequence 985-1088, while {^{3}H}-SB-T-5111 into the sequence 683-760.*(90)* The results strongly suggest that there is a specific binding site(s) for paclitaxel in Pgp. This information should eventually lead to identification of the binding pocket(s) in Pgp for paclitaxel, and may reveal the molecular architecture of the transporter domain.

Conclusion

As discussed above, our SAR studies on taxoids have led to the discovery and development of highly promising second generation taxoids including IDN5109, which is the first promising orally active taxane anticancer drug candidate, and advanced second generation taxoids that virtually overcome MDR by modulating Pgp efflux. These new anticancer agents possess dual functions, i.e., cytotoxicity and Pgp modulation, which appears to form a new class of anticancer agents. A common pharmacophore was proposed for taxoids, epothilones, eleutherobin, and discodermolide, which could provide the basis for development of the third generation taxoids or designing *de novo* anticancer agents targeting microtubules. Macrocyclic taxoid-epothilone hybrids have been designed and synthesized based on our common pharmacophore model. Some of these hybrids showed good cytotoxicity and tubulin polymerization ability. Solid state NMR analyses of a microtubule-bound fluoro-taxoid and photoaffinity labeling study have provided substantial information for the bioactive conformation of paclitaxel and taxoids. Photoaffinity labeling study of Pgp has revealed a specific binding site for paclitaxel in Pgp. This finding might have profound implications for the widely accepted mechanism of Pgp efflux causing MDR, which is based on the non-specific binding of substrates. Further studies based on chemistry and chemical biology of taxoids, hybrids, microtubules, and Pgp are actively underway in these laboratories, aiming at the discovery and development of new generation anticancer agents.

Acknowledgments: This research was supported by grants from National Institutes of Health (I.O., R.J.B., S.B.H., S.J.D). Generous support from Indena, SpA (I.O., R.J.B., G.S., C.F.) is also gratefully acknowledged.

References

1. Georg, G. I.; Chen, T. T.; Ojima, I.; Vyas, D. M. *Taxane Anticancer Agents: Basic Science and Current Status*; American Chemical Society: Washington D.C., 1995.
2. Suffness, M. *Taxol: Science and Applications*; CRC Press: New York, 1995.
3. Rowinsky, E. K.; Onetto, N.; Canetta, R. M.; Arbuck, S. G. *Seminars in Oncology* **1992**, *19*, 646-662.
4. Rowinsky, E. K. *Ann. Rev. Med.* **1997**, *48*, 353-374.
5. Guénard, D.; Guéritte-Vogelein, F.; Potier, P. *Acc. Chem. Res.* **1993**, *26*, 160-167.
6. (a) Schiff, P. B.; Fant, J.; Horwitz, S. B. *Nature* **1979**, *277*, 665-667. (b) Schiff, P. B.; Horwitz, S. B. *Proc. Natl. Acad. Sci., U. S. A.* **1980**, *77*, 1561-1565.
7. Schiff, P. B.; Horwitz, S. B. *Biochemistry* **1981**, *20*, 3247-3252
8. Jordan, M. A.; Toso, R. J.; Wilson, L. *Proc. Natl. Acad. Sci. USA* **1993**, *90*, 9552-9556.
9. Blagoskolnny, M. V.; Fojo, T. *Int. J. Cancer* **1999**, *83*, 11-156.
10. Denis, J.-N.; Greene, A. E.; Guénard, D.; Guéritte-Voegelein, F.; Mangatal, L.; Potier, P. A. *J. Am. Chem. Soc.* **1988**, *110*, 5917-5919.
11. Ojima, I. *Acc. Chem. Res.* **1995**, *28*, 383-389 and references cited therein.
12. Ojima, I.; Kuduk, S. D.; Chakravarty, S. In *Adv. Med. Chem.*; JAI Press: Greenwich, CT, 1998; pp 69-124.
13. Holton, R. A.; Biediger, R. J.; Boatman, P. D. Semisynthesis of Taxol and Taxotere. *Taxol®: Science and Applications*; CRC Press: New York, 1995; pp 97-121.
14. Holton, R. A.; Kim, H.-B.; Somoza, C.; Liang, F.; Biediger, R. J.; Boatman, P. D.; Shindo, M.; Smith, C. C.; Kim, S.; Nadizadeh, H.; Suzuki, Y.; Tao, C.; Vu, P.; Tang, S.; Zhang, P.; Murthi, K. K.; Gentile, L. N.; Liu, J. H. *J. Am. Chem. Soc.* **1994**, *116*, 1599-1600.
15. Nicolaou, K. C.; Yang, Z.; Liu, J. J.; Ueno, H.; Nantermet, P. G.; Guy, R. K.; Claiborne, C. F.; Renaud, J.; Couladouros, E. A.; Paulvannan, K.; Sorensen, E. J. *Nature* **1994**, *367*, 630-634.
16. Danishefsky, S.; Masters, J.; Young, W.; Link, J.; Snyder, L.; Magee, T.; Jung, D.; Isaacs, R.; Bornmann, W.; Alaimo, C.; Coburn, C.; Di Grandi, M. *J. Am. Chem. Soc.* **1996**, *118*, 2843-2859.
17. Wender, P. A.; Badham, N. F.; Conway, S. P.; Floreancig, P. E.; Glass, T. E.; Houze, J. B.; Krauss, N. E.; Lee, D.; Marquess, D. G.; McGrane, P. L.; Meng, W.; Natchus, M. G.; Shuker, A. J.; Sutton, J. C.; Taylor, R. E. *J. Am. Chem. Soc.* **1997**, *119*, 2757-2758.
18. Ojima, I.; Duclos, O.; Zucco, M.; Bissery, M.-C.; Combeau, C.; Vrignaud, P.; Riou, J. F.; Lavelle, F. *J. Med. Chem.* **37**, *37*, 2602-2608.
19. Ojima, I.; Sun, C. M.; Park, Y. H. *J. Org. Chem.* **1994**, *59*, 1249-1250.

20. Ojima, I.; Slater, J. C.; Michaud, E.; Kuduk, S. D.; Bounaud, P.-Y.; Vrignaud, P.; Bissery, M.-C.; Veith, J.; Pera, P.; Bernacki, R. J. *J. Med. Chem.* **1996**, *39*, 3889-3896.
21. Georg, G. I.; Boge, T. C.; Cheruvallath, Z. S.; Clowers, J. S.; Harriman, G. C. B.; Hepperle, M.; Park, H. The Medicinal Chemistry of Taxol. *Taxol®: Science and Applications*; CRC Press: New York, 1995; pp 317-375.
22. Kant, J.; O'Keeffe, W. S.; Chen, S.-H.; Farina, V.; Fairchild, C.; Johnston, K.; Kadow, J. F.; Long, B. H.; Vyas, D. *Tetrahedron Lett.* **1994**, *35*, 5543-5546.
23. Chen, S.-H.; Fairchild, C.; Mamber, S. W.; Farina, V. *J. Org. Chem.* **1993**, *58*, 2927-2928.
24. Kingston, D. G. I. *In Taxane Anticancer Agents: Basic Science and Current Status; ACS Symp. Ser. 583*; American Chemical Society: Washington, D. C., 1995; pp 203-216.
25. Chaudhary, A. G.; Gharpure, M. M.; Rimoldi, J. M.; Chordia, M. D.; Gunatilaka, A. A. L.; Kingston, D. G. I.; Grover, S.; Lin, C. M.; Hamel, E. *J. Am. Chem. Soc.* **1994**, *116*, 4097-4098.
26. Kingston, D. G. I.; Chaudhary, A. g.; Chordia, M. D.; Gharpure, M.; Gunatilaka, A. A. L.; Higgs, P. I.; Rimoldi, J. M.; Samala, L.; Jagtap, P. G.; Giannakakou, P.; Jiang, Y. Q.; Lin, C. M.; Hamel, E.; Long, B. H.; Fairchild, C. R.; Johnston, K. A. *J. Med. Chem.* **1998**, *41*, 3715-3726.
27. Ojima, I.; Wang, T.; Miller, M. L.; Lin, S.; Borella, C.; Geng, X.; Pera, P.; Bernacki, R. J. *Bioorg. Med. Chem. Lett.* **1999**, *9*, 3423-3428.
28. Appendino, G.; Gariboldi, P.; Gabetta, B.; Pace, R.; Bombardelli, E.; Viterbo, D. *J. Chem. Soc., Perkins Trans 1* **1992**, 2925-2929.
29. Ojima, I.; Fenoglio, I.; Park, Y. H.; Pera, P.; Bernacki, R. J. *Bioorg. Med. Chem. Lett.* **1994**, *4*, 1571-1576.
30. Ojima, I.; Park, Y. H.; Sun, C.-M.; Fenoglio, I.; Appendino, G.; Pera, P.; Bernacki, R. J. *J. Med. Chem.* **1994**, *37*, 1408-1410.
31. Ojima, I.; Fenoglio, I.; Park, Y. H.; Sun, C.-M.; Appendino, G.; Pera, P.; Bernacki, R. J. *J. Org. Chem.* **1994**, *59*, 515-517.
32. Ojima, I.; Lin, S.; Chakravarty, S.; Fenoglio, I.; Park, Y. H.; Sun, C.-M.; Appendino, G.; Pera, P.; Veith, J. M.; Bernacki, R. J. *J. Org. Chem.* **1998**, *63*, 1637-1645.
33. Ojima, I.; Slater, J. S.; Kuduk, S. D.; Takeuchi, C. S.; Gimi, R. H.; Sun, C.-M.; Park, Y. H.; Pera, P.; Veith, J. M.; Bernacki, R. J. *J. Med. Chem.* **1997**, *40*, 267-278.
34. Ojima, I.; Slater, J. C. *Chirality* **1997**, *9*, 487-494.
35. Ojima, I.; Inoue, T.; Slater, J. C.; Lin, S.; Kuduk, S. C.; Chakravarty, S.; Walsh, J. J.; Gilchrist, L.; McDermott, A. E.; Cresteil, T.; Monsarrat, B.; Pera, P.; Bernacki, R. J. *In "Asymmetric Fluoroorganic Chemistry: Synthesis, Application, and Future Directions"; ACS Symp.*

Ser. 746; American Chemical Society: Washington, D. C., 1999; pp Chapter 12, pp 158-181.
36. Ojima, I.; Lin, S.; Slater, J. C.; Wang, T.; Pera, P.; Bernacki, R. J.; Ferlini, C.; G., S. *Bioorg. Med. Chem.* **2000**, *8*, 1576-1585.
37. (a) Distefano, M.; Scambia, G.; Ferlini, C.; Gaggini, C.; Vincenzo, R. D.; Riva, A.; Bombardelli, E.; Ojima, I.; Fattorossi, A.; Panici, P. B.; Mancuso, S. *Int. J. Cancer* **1997**, *72*, 844-850. (b) Ferlini, C.; Distefano, M.; Pierelli, L.; Bonanno, G.; Riva, A.; Bombardelli, E.; Ojima, I.; Mancuso, S.; Scambia, G. *Oncol. Res.* **1999**, *11*, 471-478.
38. Polizzi, D.; Pratesi, G.; Tortoreto, M.; Supino, M.; Riva, A.; Bombardelli, E.; Zunino, F. *Cancer Res.* **1999**, *59*, 1036-1040.
39. Nicoletti, M. I.; Colombo, T.; Rossi, C.; Monardo, C.; Stura, S.; Zucchetti, M.; Riva, A.; Morazzoni, P.; Donati, M. B.; Bombardelli, E.; D'Incalici, M.; Giavazzi, R. *Cancer Res.* **2000**, *60*, 842-846.
40. Polizzi, D.; Pratesi, G.; Monestiroli, S.; Tortoreto, M.; Zunino, F.; Bombardelli, E.; Riva, A.; Morazzoni, P.; Colombo, T.; D'Incalici, M.; Zucchetti, M. *Clin. Cancer Res.* **2000**, *6*, 2070-2074.
41. Vredenburg, M. R.; Ojima, I.; Veith, J.; Pera, P.; Kee, K.; Cabral, F.; Sharma, A.; Kanter, P.; Bernacki, R. J. *J. Nat'l. Cancer Inst.* **2001**, in press.
42. Kirschner, L. S.; Greenberger, L. M.; Hsu, S. I.-H.; Yang, C.-P. H.; Cohen, D.; Piekarz, R. L.; Castillo, G.; Han, E. K.-H. H.; Yu, L.; Horwitz, S. B. *Biochem. Pharm.* **1992**, *43*, 77-78.
43. Gottesman, M. M.; Pastan, I. *Annu. Biochem.* **1993**, *62*, 385-427.
44. Sikic, B. I. *J. Clin. Oncol.* **1993**, *11*, 1629-1635.
45. Ferlini, C.; Distefano, M.; Pignatelli, F.; Lin, S.; Riva, A.; Bombardelli, E.; Mancuso, S.; Ojima, I.; Scambia, G. *Brit. J. Cancer* **2001**, in press.
46. Ojima, I.; Duclos, D.; Dormán, G.; Simonot, B.; Prestwich, G. D.; Rao, S.; Lerro, K. A.; Horwitz, S. B. *J. Med. Chem.* **1995**, *38*, 3891-3894.
47. Ojima, I.; Bounaud, P.-Y.; Ahern, D. G. *Bioorg. Med. Chem. Lett.* **1998**, *9*, 1189-1194.
48. Ojima, I.; Bounaud, P.-Y.; Takeuchi, C. S.; Pera, P.; Bernacki, R. J. *Bioorg. Med. Chem. Lett.* **1998**, *8*, 189-194.
49. Boge, T. C.; Himes, R. H.; Vander Velde, D. G.; Georg, G. I. *J. Med. Chem.* **1994**, *37*, 3337-3343.
50. Ojima, I.; Bounaud, P.-Y.; Bernacki, R. J. *CHEMTECH* **1998**, *28*, 31-36.
51. Bollag, D. M.; McQueney, P. A.; Zhu, J.; Hensens, O.; Koupal, L.; Liesch, J.; Goetz, M.; Lazarides, E.; Woods, C. M. *Cancer Res.* **1995**, *55*, 2325-2333.
52. Kowalski, R. J.; Giannakakou, P.; Hamel, E. *J. Biol. Chem.* **1997**, *272*, 2534-2541.
53. Lindel, T.; Jensen, P. R.; Fenical, W.; Long, B. H.; Casazza, A. M.; Carboni, J.; Fairchild, C. R. *J. Am. Chem. Soc.* **1997**, *119*, 8744-8745.

54. ter Haar, E.; Kowalski, R. J.; Hamel, E.; Lin, C. M.; Longley, R. E.; Gunasekera, S. P.; Rosenkranz, H. S.; Day, B. W. *Biochemistry* **1996**, *35*, 243-250.
55. Kowalski, R. J.; Giannakakou, P.; Gunasekera, S. P.; Longley, R. E.; Day, B. W.; Hamel, E. *Mol. Pharmacol.* **1997**, *52*, 613-622.
56. Giannakakou, P.; Sackett, D. L.; Kang, Y.-K.; Zhan, Z.; Buters, J. T. M.; Fojo, T.; Poruchynsky, M. S. *J. Biol. Chem.* **1997**, *272*, 17118-17125.
57. Kowalski, R. J.; ter Haar, E.; Longley, R. E.; Gunasekera, S. P.; Lin, C. M.; Day, B. W.; Hamel, E. *Mol Biol. Cell* **1995**, *6*, 368a.
58. Hung, D. T.; Chen, J.; Schreiber, S. L. *Chem. Biol.* **1996**, *3*, 287-293.
59. Ojima, I.; Kuduk, S. D.; Pera, P.; Veith, J. M.; Bernacki, R. J. *J. Med. Chem.* **1997**, *40*, 279-285.
60. Ojima, I.; Chakravarty, S.; Inoue, T.; Lin, S.; He, L.; Horwitz, S. B.; Kuduk, S. D.; Danishefsky, S. J. *Proc. Natl. Acad. Sci. USA* **1999**, *96*, 4256-4261.
61. Su, D.-S.; Meng, D.; Bertinato, P.; Balog, A.; Sorensen, E. J.; Danishefsky, S. J. *Angew. Chem. Int. Ed. Engl.* **1997**, *36*, 757-833.
62. Su, D.-S.; Balog, A.; Meng, D.; Bertinato, P.; Danishefsky, S. J.; Zheng, Y.-H.; Chou, T.-C.; He, L.; Horwitz, S. B. *Angew. Chem. Int. Ed. Engl.* **1997**, *36*, 2093-2096.
63. Chou, T.-C.; Zhang, X.-G.; Balog, A.; Su, D.-S.; Meng, D.; Savin, K.; Bertino, J. R.; Danishefsky, S. J. *Proc. Natl. Acad. Sci. USA* **1998**, *95*, 9642.
64. Chou, T. C.; Zhang, X. G.; Harris, C. R.; Kuduk, S. D.; Balog, A.; Savin, K.; Danishefsky, S. J. *Proc. Natl. Acad. Sci* **1998**, *95*, 15798.
65. McDaid, H. M.; Bhattacharya, S. K.; Chen, S.-T.; He, L.; Shen, H.-J.; Gutteridge, C. E.; Horwitz, S. B.; Danishefsky, S. J. *Cancer Chemother. Pharmacol.* **1999**, *44*, 131.
66. Johnson, J.; Kim, S.-H.; Bifano, M.; Dimarco, J.; Fairchild, C.; Gougoutas, J.; Lee, F.; Long, B.; Tokarski, J.; Vite, G. *Org. Lett.* **2000**, *2*, 1537-1540.
67. Nicolaou, K. C.; Kim, S.; Pfefferkorn, J.; Xu, J.; Ohshima, T.; Hosokawa, S.; Vourloumis, D.; Li, T. *Angew. Chem. Int. Ed.* **1998**, *37*, 1418-1421.
68. Ojima, I. Lin. S. Inoue, T.; Miller, M. L.; Borella, C. P.; Geng, X.; Walsh, J.J. *J. Am. Chem. Soc.* **2000**, *122*, 5343-5353.
69. Schuster, M.; Siegfried, B. *Angew. Chem. Int. Ed. Engl.* **1997**, *36*, 2036-2056.
70. Fürstner, A. *Top. Catal.* **1997**, *4*, 285-299.
71. Grubbs, R. H.; Chang, S. *Tetrahedron* **1998**, *54*, 4413-4450.
72. Schwab, P.; France, M. B.; Ziller, J. W.; Grubbs, R. H. *Angew. Chem. Int. Ed. Engl.* **1995**, *34*, 2039-2041.
73. Fürstner, A.; Langemann, K. *Synthesis* **1997**, 792-803.

74. Guéritte-Voegelein, F.; Mangatal, L.; Guénard, D.; Potier, P.; Guilhem, J.; Cesario, M.; Pascard, C. *Acta Crstallogr.* **1990**, *C46*, 781-784.
75. Miller, M. L.; Geng, X.; Lin, S.; Pera, P.; Bernacki, R. J.; Ojima, I. *219th American Chemical Society National Meeting, San Francisco, CA., March 26-30, 2000, Abstracts* **2000**, MEDI 40.
76. Ojima, I.; Slater, J. C.; Pera, P.; Veith, J. M.; Abouabdellah, A.; Bégué, J.-P.; Bernacki, R. J. *Bioorg. Med. Chem. Lett.* **1997**, *7*, 209-214.
77. Ojima, I.; Kuduk, S. D.; Chakravarty, S.; Ourevitch, M.; Bégué, J.-P. *J. Am. Chem. Soc.* **1997**, *119*, 5519-5527.
78. Griffiths, J. M.; Griffin, R. G. *Anl. Chim. Acta* **1993**, *283*, 1081-1101.
79. Beusen, D. D.; McDowell, L. M.; Slomczynska, U.; Schaefer, J. *J. Med. Chem.* **1995**, *38*, 2742-2747.
80. Gilchrist, L.; McDermott, Nakanishi, K.; A.; Kuduk, S. D.; Walsh, J. J.; Chakravarty, S.; Ojima, I. unpublished results.
81. Li, Y.; Poliks, B.; Cegelski, L.; Poliks, M.; Cryczynski, A.; Piszcek, G.; Jagtap, P. G.; Studelska, D. R.; Kingston, D. G. I.; Schaefer, J.; Bane, S. *Biochemistry* **2000**, *39*, 281-291.
82. Rao, S.; Horwitz, S. B.; Ringel, I. *J. Natl. Cancer Inst.* **1992**, *84*, 785-788.
83. Rao, S.; Krauss, N. E.; Heerding, J. M.; Swindell, C. S.; Ringel, I.; Orr, G. A.; Horwitz, S. B. *J. Biol. Chem.* **1994**, *269*, 3132-3134.
84. Rao, S.; Orr, G. A.; Chaudhary, A. G.; Kingston, D. G. I.; Horwitz, S. B. *J. Biol. Chem.* **1995**, *270*, 20235-20238.
85. Rao, S.; He, L.; Chakravarty, S.; Ojima, I.; Orr, G. A.; Horwitz, S. B. *J. Biol. Chem.* **1999**, *274*, 37990-37994.
86. Nogales, E.; Wolf, S. G.; Downing, K. H. *Nature* **1998**, *391*, 199-203.
87. Giannakakou, P.; Gussio, R.; Nogales, E.; Downing, K. H.; Zaharevitz, D.; Bollbuck, B.; Poy, G.; Sackett, D.; Nicolaou, K. C.; Fojo, T. *Proc. Natl. Acad. Sci. USA* **2000**, 2904-2909.
88. He, L.; Jagtap, P. G.; Kingston, D. G. I.; Shen, H.-J.; Orr, G.; Horwitz, S. B. *Biochemistry* **2000**, *39*, 3972-3978.
89. Dubois, J.; Thoret, S.; Guéritte, F.; Guénard, D. *Tetrahedron Lett.* **2000**, *41*, 3331-3334.
90. Wu, Q.; Bounaud, P.-Y.; Kuduk, S. D.; Yang, C.-P.; Ojima, I.; Horwitz, S. B.; Orr, G. A. *Biochemistry* **1998**, *37*, 11272-11279.

Chapter 5

Discodermolide and Taxol: A Synergistic Drug Combination in Human Carcinoma Cell Lines

Susan Band Horwitz[1], Laura A. Martello[1], Chia-Ping H. Yang[1], Amos B. Smith, III[2], and Hayley M. McDaid[1]

[1]Department of Molecular Pharmacology, Albert Einstein College of Medicine, 1300 Morris Park Avenue, Bronx, NY 10461
[2]Department of Chemistry, University of Pennsylvania, Philadelphia, PA 19104

New natural products with Taxol-like activities have been identified during a search for compounds with the same mechanism of action as Taxol, but with better therapeutic properties. The epothilones, eleutherobin and discodermolide, like Taxol, all enhance the polymerization of stable microtubules. Careful analyses of these compounds have indicated that Taxol and discodermolide have differential effects in cells. The presence of low concentrations of Taxol significantly increased the cytotoxicity of discodermolide. Median effect analysis, using the combination index method, revealed a schedule-independent synergistic interaction between Taxol and discodermolide in human carcinoma cell lines, suggesting that these two drugs could represent an important drug combination in the treatment of cancer.

Introduction

Taxol is an effective antitumor drug that has been approved by the FDA for the treatment of ovarian, breast and lung carcinomas (1) and is under evaluation for the treatment of Kaposi's sarcoma (2). It is being used both as a single agent and in combination chemotherapy. The microtubule polymer is the cellular target for Taxol, specifically the β-tubulin subunit with which Taxol interacts (3). The drug binds specifically, but not covalently, to microtubules with a stoichiometry approaching one mole of Taxol per mole of tubulin heterodimer (4,5). Two hallmarks of this antimitotic agent are (i) the enhancement of assembly of microtubules, even in the absence of GTP that is normally required for *in vitro* tubulin assembly (6), and (ii) the formation of stable bundles or parallel arrays of microtubules that results from a reorganization of the microtubule cytoskeleton (7). The normal assembly/disassembly dynamics of microtubules are disrupted by Taxol and cells are arrested at metaphase. This perturbation of normal tubulin kinetics results in cell death. High Taxol concentrations result in sustained mitotic arrest, while low Taxol concentrations which inhibit microtubule dynamics also cause cell death in the absence of an apparent mitotic block. Low concentrations of Taxol have a major effect on microtubule dynamics, inhibiting this process even at nM concentrations of drug (8-10).

Extensive photoaffinity labeling studies have been done with Taxol analogs bearing photoreactive groups at the C-2, C-3' or C-7 positions to identify the sites of interaction between Taxol and β-tubulin (11-13). In an early study (3), we demonstrated by direct photolabeling, using radiolabeled Taxol, that the drug interacted specifically with β-tubulin. Further studies with [^{3}H]3'-(p-azidobenzamido)Taxol (where the arylazide was incorporated into the C-13 side chain) resulted in the isolation of a photolabeled peptide containing amino acid residues 1-31 of β-tubulin (11). Studies with [^{3}H]2-(m-azidobenzoyl)Taxol (where the photoreactive group is attached to the B ring of the taxoid nucleus) have shown that a peptide containing amino acid residues 217-233 of β-tubulin also is involved in Taxol binding (12). More recently we have utilized a photoreactive Taxol analog containing a benzoyldihydrocinnamoyl (BzDC) substituent at the C-7 position of the drug (13) that allowed the determination of the single amino acid in tubulin which interacts with Taxol, a limitation with arylazide containing Taxol analogs. Studies done with the BzDC analog indicated that it photo-crosslinked to Arg^{282} of β-tubulin. The information gained from our photoaffinity labeling studies and from electron crystallographic studies done by Nogales and colleagues (14) has allowed the visualization of the binding pocket for Taxol in β-tubulin.

The success of Taxol in the clinic plus the desire to improve on the properties of the drug, have prompted the search for new natural products with Taxol-like activity. It would be advantageous to have Taxol-like compounds with a broad range of tumor efficacy, with improved aqueous solubility, and most importantly, the ability to retain potency in Taxol-resistant tumors. In the past five years, at least three new natural products, the epothilones, (15,16) eleutherobin (17,18), and discodermolide (19-21) have been isolated, each from distinct natural sources, a bacterium, marine coral and sponge, respectively. These drugs share the same mechanism of action as Taxol, although the structures of each of the four compounds is unique (Fig. 1). One focus of our laboratory is to delineate the mechanistic similarities and differences between these molecules, with particular emphasis on Taxol and discodermolide.

In vitro Polymerization of Microtubules and Bundle Formation

An *in vitro* assay has been used extensively to monitor the polymerization of soluble tubulin dimers to form the microtubule polymer. This assay requires the presence of purified bovine brain tubulin, GTP and MES buffer and is carried out at 37°C. The turbidity or change in absorbance (light scattering) at 350 nm correlates with the initiation and elongation of microtubules (22) and is routinely confirmed by electron microscopy. Normal microtubules assembled in the presence of GTP depolymerize in the presence of mM concentrations of Ca^{++}. Since all of the small molecules used in this assay are hydrophobic and are solubilized in DMSO, the control also contains DMSO. The four drugs that were compared all share the same ability as Taxol to stabilize microtubules. Each was assayed at 10 μM and shown to assemble tubulin, in the absence of GTP, into stable microtubules that are not depolymerized by Ca^{++} or cold temperatures (Fig. 2). The inset depicts the first two minutes of polymerization, from which it is clear that epothilone B, and particularly discodermolide, have an initial slope that is extremely steep, exceeding that of Taxol. This indicates that these two drugs have a major effect on the initiation of tubulin polymerization. The microtubules formed in the presence of discodermolide are distinctly shorter than normal microtubules or even Taxol-induced microtubules (23), emphasizing the effect that this drug has on initiation of microtubule polymerization.

Although both Taxol and discodermolide enhance the assembly of stable microtubules, there are distinct differences in the way the two drugs reorganize the microtubule cytoskeleton. As depicted in Figure 3, both drugs enhance the formation of microtubule bundles, but with unique morphologic characteristics. Taxol-treated cells have microtubule bundles, often centered around the nucleus but also localized throughout the cells, whereas discodermolide treatment induces short bundles, predominantly at the periphery of the cells. The latter are

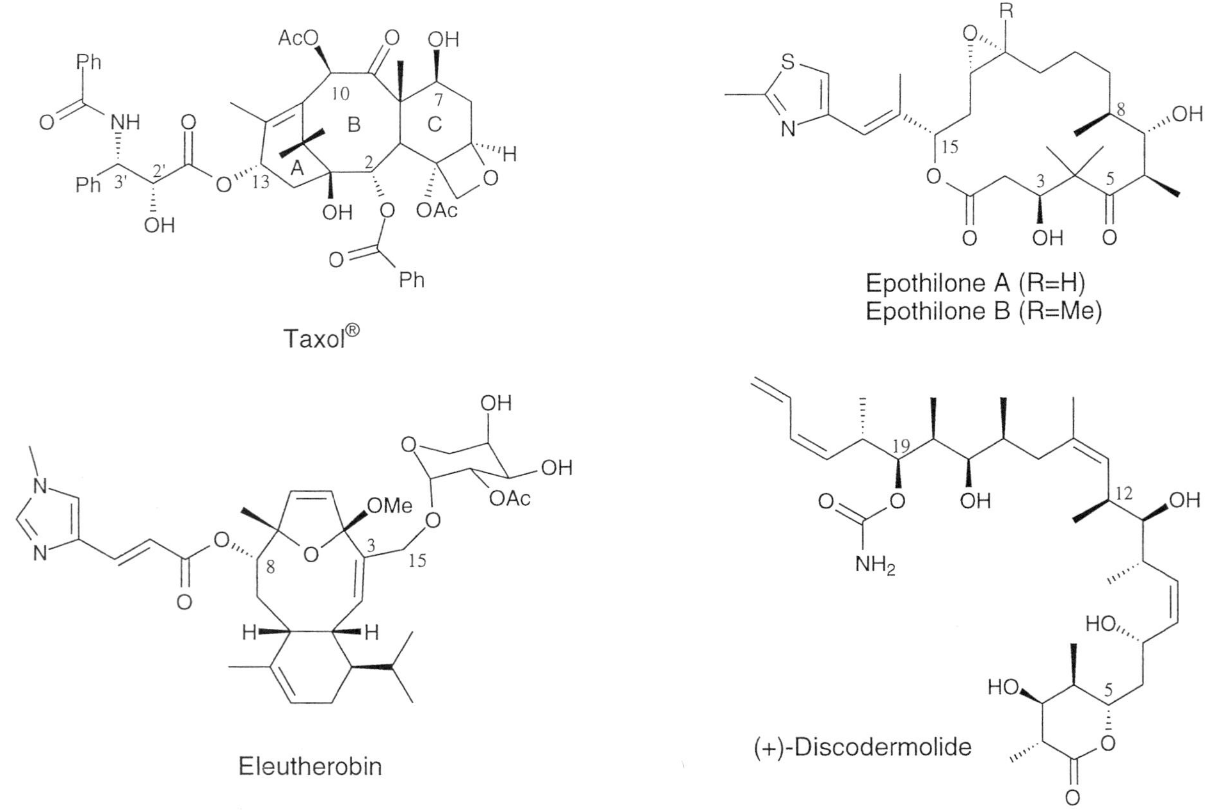

Figure 1. Chemical structures of Taxol, epothilone A and B, eleutherobin, and discodermolide. (Reproduced with permission from reference 21).

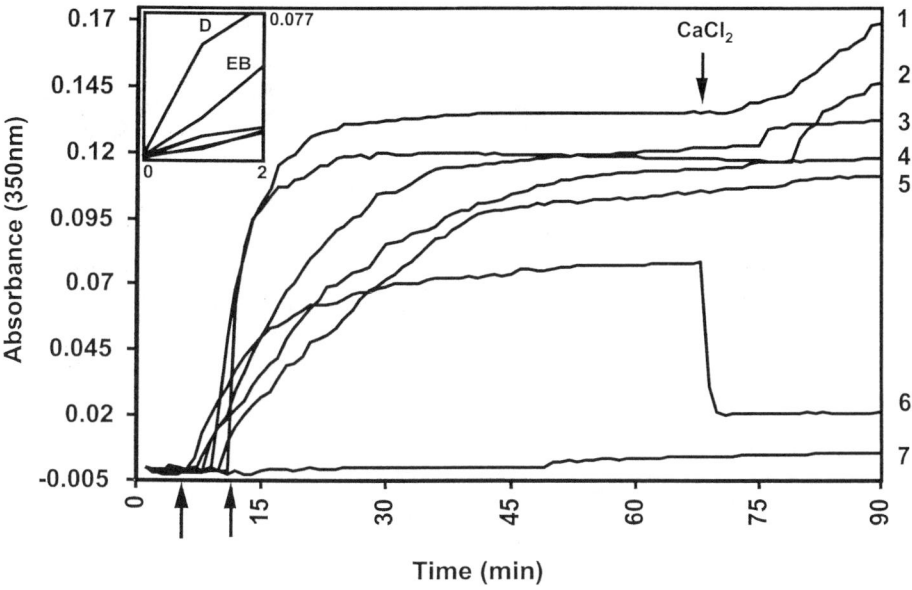

Figure 2. *In vitro* activity of microtubule agents in a tubulin polymerization assay. Microtubule protein (MTP) was diluted to 1 mg/ml in MES buffer containing 3 M glycerol. The compounds (10 µM) to be evaluated were then added in 1 min intervals (bottom arrows) to the MTP at 37° C. Changes in absorbance at 350 nm, a characteristic of tubulin assembly, were monitored for 70 min and then 8 mM $CaCl_2$ (top arrow) was added. 1, epothilone B; 2, discodermolide; 3, epothilone A; 4, Taxol; 5, eleutherobin; 6, GTP (1 mM); 7, DMSO (control). Inset, initial slope (0-2 min) representing the initial activity of each compound (D, discodermolide; EB, epothilone B).

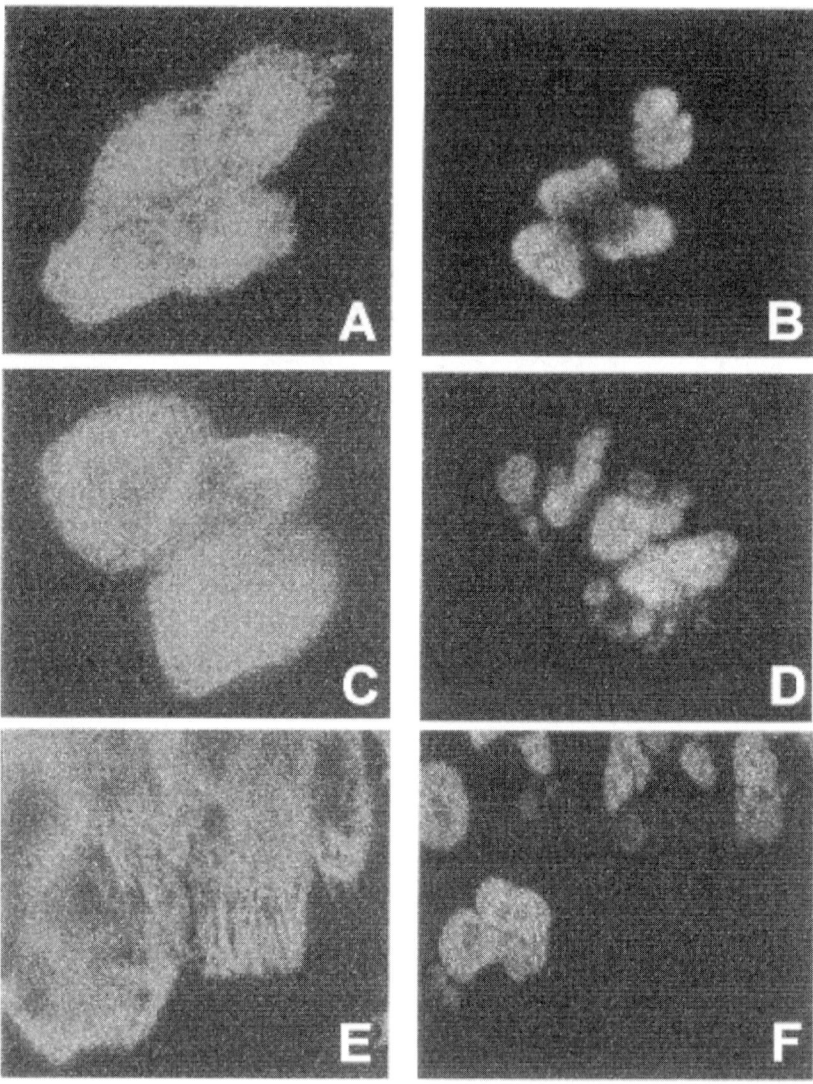

Figure 3. After incubation with Taxol or discodermolide, differences in bundle formation were observed. A549 cells were plated on glass coverslips, allowed to attach for 24 h, and then treated with the indicated drug concentrations for 24 h. The samples were prepared for immunofluorescence microscopy and the slides viewed on a Zeiss Axiophot microscope (rhodamine and DAPI filters) at X100 magnification. A, B, controls (+DMSO) ; C, D, 12 nM Taxol; E, F, 24 nM Disco. A, C, E, stained with α-tubulin antibody. B, D, F, stained for DNA.

reminiscent of the short microtubules that are formed during *in vitro* polymerization.

Cross Resistance and Cytotoxicity of Discodermolide in Taxol-resistant Cell lines

The cross-resistance of discodermolide has been examined in two different types of Taxol-resistant cell lines. In a highly Taxol-resistant SKOV3 ovarian carcinoma cell line that was selected with vinblastine and overproduces P-glycoprotein, there was low cross-resistance to epothilones A and B, and discodermolide, but definite cross-resistance to eleutherobin (Table I). These results suggest that eleutherobin, but not epothilone A and B or discodermolide, is a substrate for P-glycoprotein. Other published data corroborates these findings in drug resistant cell lines (18,24,25). Further evidence indicated that the epothilones and discodermolide did not reverse the inhibition of Taxol accumulation in a P-glycoprotein expressing cell line. Steady state accumulation of [^{3}H]Taxol was reduced approximately 13-fold in the multidrug resistant cell line J7-T3-1.6 (T3) compared to the drug sensitive parental cell line, J774.2 (J7) (Fig. 4). This inhibition could be reversed by verapamil and vinblastine as has been previously demonstrated (26). Neither epothilone A, epothilone B, or discodermolide were able to reverse the very low level of Taxol in the T3 drug-resistant cells. The addition of eleutherobin caused a 2-fold increase in the steady state accumulation of Taxol in the J7 resistant cell line.
The fact that the epothilones and discodermolide do not appear to be substrates for P-glycoprotein and therefore have the potential to be active in Taxol-resistant cell lines that overexpress P-glycoprotein, makes them extremely attractive for development as antitumor drugs.

In A549-T12 cells, a human lung carcinoma cell line that is approximately 9-fold resistant to Taxol and does not express P-glycoprotein, there was a lack of cross-resistance to discodermolide (Table II). The mechanism of Taxol resistance in A549-T12 cells is currently under investigation. Although cross-resistance was observed with epothilones A and B, and eleutherobin, there was no cross-resistance observed with discodermolide, indicating that the latter has properties that are distinct from the other microtubule stabilizing drugs. It is interesting to note that there is no cross-resistance to either vinblastine or colchicine, two antimitotic agents that interact with the tubulin dimer to inhibit microtubule polymerization.

The A549-T12 cells are a particularly interesting cell line because although they were selected for resistance to Taxol, they have an unusual requirement for a low level of Taxol (2-6 nM) for normal cell growth (21). The ability of the

Table I: Cytotoxicity of antimitotic agents in a drug-resistant cell line that overexpresses P-glycoprotein

Cell lines	IC$_{50}$ (nM) [a]					
	Taxol	VBL[b]	EpoA	EpoB	Disco	Eleu
SKOV3	2.0 ± 0.1 [c]	1.1 ± 0.1	10.3 ± 4.5	0.5 ± 0.09	9.7 ± 0.3	8.2 ± 3.8
SKVLB	13333 ± 1154	1880.3 ± 347.5	190 ± 14	9.5 ± 2.1	575 ± 233	>5000
Fold resistant [d]	6666.5	1709.4	18.5	19.0	59.3	>609.8

[a] IC$_{50}$, drug concentration that inhibits cell division by 50% after 6 days.
[b] VBL, vinblastine; EpoA, epothilone A; EpoB, epothilone B; Disco, discodermolide; Eleu, eleutherobin.
[c] Mean ± SE.
[d] Ratio of IC$_{50}$ for SKVLB resistant cell line to that for SKOV3 sensitive cell line.
SOURCE: Reproduced with permission from reference 21. Copyright 2000 American Association for Cancer Research.

Table II: Cytotoxicity of antimitotic agents in a Taxol-resistant cell line that does not express P-glycoprotein

Cell lines	IC$_{50}$ (nM) [a]						
	Taxol	EpoA [b]	EpoB	Eleu	Disco	VBL	CLC
A549	2 ± 0.14 [c]	6.4 ± 2.0	0.7 ± 0.1	14 ± 2.8	8.1 ± 0.14	1.8 ± 0.21	33 ± 10.6
A549-T12 [d]	18.7 ± 0.9	34 ± 3.1	3.3 ± 0.72	69 ± 11.3	6.5 ± 2.4	1.8 ± 0.18	33 ± 12.7
Fold resistant [e]	9.4	5.3	4.7	4.9	0.8	1.0	1.0

[a] IC$_{50}$, drug concentration that inhibits cell division by 50% after 72 h.
[b] EpoA, epothilone A; EpoB, epothilone B; Eleu, eleutherobin; Disco, discodermolide; VBL, vinblastine; CLC, colchicine.
[c] Mean ± SE.
[d] Cells were maintained in 2 nM Taxol during cross resistance experiments.
[e] Ratio of IC$_{50}$ for A549-T12 resistant cell line to that for A549 sensitive cell line.
SOURCE: Reproduced with permission from reference 21. Copyright 2000 American Association for Cancer Research.

epothilones, eleutherobin and discodermolide to substitute for Taxol in sustaining the growth of A549-T12 cells was evaluated and it was determined that only discodermolide could not substitute (21). This result further substantiated our other findings, suggesting a distinct difference between Taxol and discodermolide.

Due to their requirement for Taxol, the experiments to determine cytotoxicity and cross resistant profiles in A549-T12 cells were carried out in the presence of 2 nM Taxol. In the absence of 2nM Taxol, it was found that the A549-T12 cells were approximately 20-fold less sensitive to discodermolide. Over a range of discodermolide concentrations, maximum cytotoxicity was observed in the presence of 2 nM Taxol (Fig. 4). In contrast, the presence of Taxol did not significantly potentiate the cytotoxicities of the epothilones or eleutherobin (21). These observations led us to the idea that there may be an interaction between Taxol and discodermolide.

Taxol and Discodermolide are a Synergistic Drug Combination

Median effect analysis using the combination index (CI) method of Chou and Talalay (27) was used to analyze the nature of the interaction, if any, between Taxol and discodermolide. This method resolves the degree of synergy, additivity or antagonism at various levels of cytotoxicity. The type of interaction is expressed as a combination index where 1=additivity, >1=antagonism and <1=synergism. Since it could not be determined whether the interactions between Taxol and other microtubule (MT) binding agents were mutually exclusive or non-exclusive, CI values were calculated using both methods. The data presented in Fig. 5 illustrates the results from two human carcinoma cell lines, A549 (lung) and MCF-7 (breast), based on the more conservative assumption of mutual nonexclusion. Similar results were also observed in the ovarian cell line SKOV3 and the breast cell line MDA-MB-231 (21) (data not shown). The data are represented as fractional cell growth inhibition (FA) as a function of the combination index (CI). The CI values for the combination of concurrent Taxol with discodermolide were significantly less than 1 in both A549 and MCF-7 cells, indicating a synergistic drug interaction. This interaction was effective over a 3-4 fold log concentration of either drug and was not influenced by drug schedule. A control reaction utilizing current exposure to Taxol and epothilone B, demonstrated a CI value of 1, indicating additivity between these two drugs.

Subsequent molecular analysis indicated that the observed synergism between Taxol and discodermolide augmented apoptosis. The cleavage of the nuclear repair enzyme PARP to a 24kDa product is an accurate marker for apoptosis that

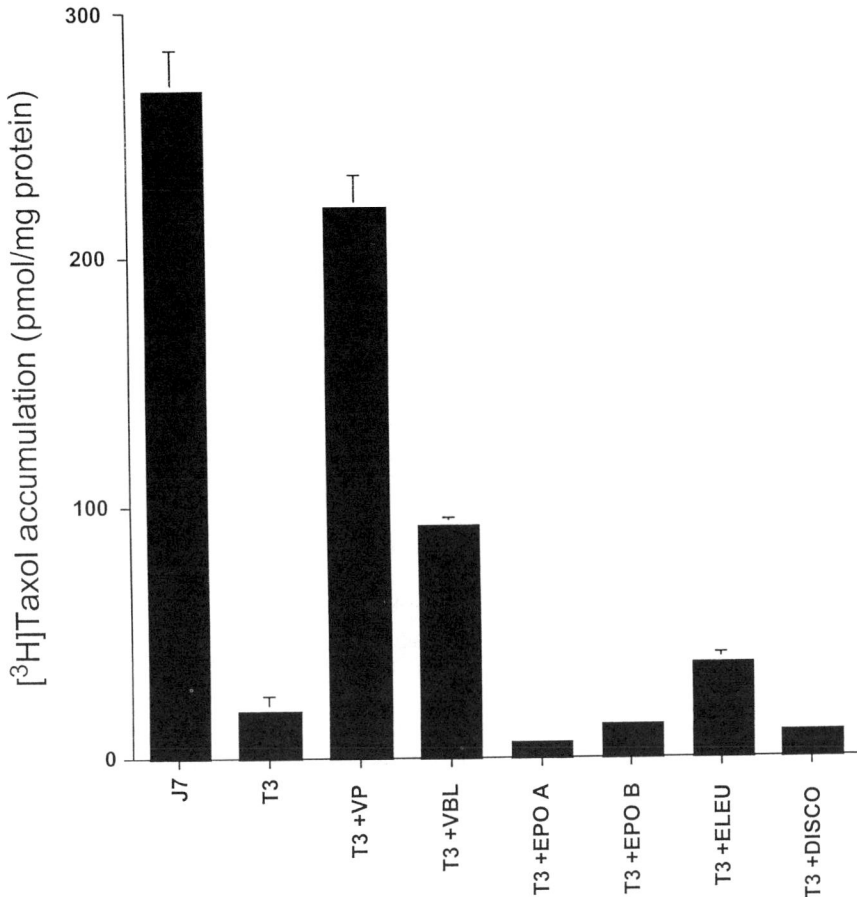

Figure 4. Epothilones and discodermolide do not reverse the inhibition of Taxol accumulation in a P-glycoprotein expressing cell line. The steady state accumulation of [^{3}H]Taxol was measured by plating 6×10^5 cells in 12-well plates. After a 3 h incubation in drug-free medium, fresh medium containing [^{3}H]Taxol (2.5 µM, Specific Activity=130 mCi/mmol) and the test drugs (50 µM) were added. Following incubation at 37°C for 1 h, the cells were washed twice with 3 ml of cold PBS and lysed in 1N NaOH at room temperature for 16 h. Total protein was determined and for each drug, incorporation of [^{3}H]Taxol, as measured by liquid scintillation counting, was normalized for protein concentration. VP, verapamil; VBL, vinblastine; EPO A, epothilone A; EPO B, epothilone B; ELEU, eleutherobin; DISCO, discodermolide.

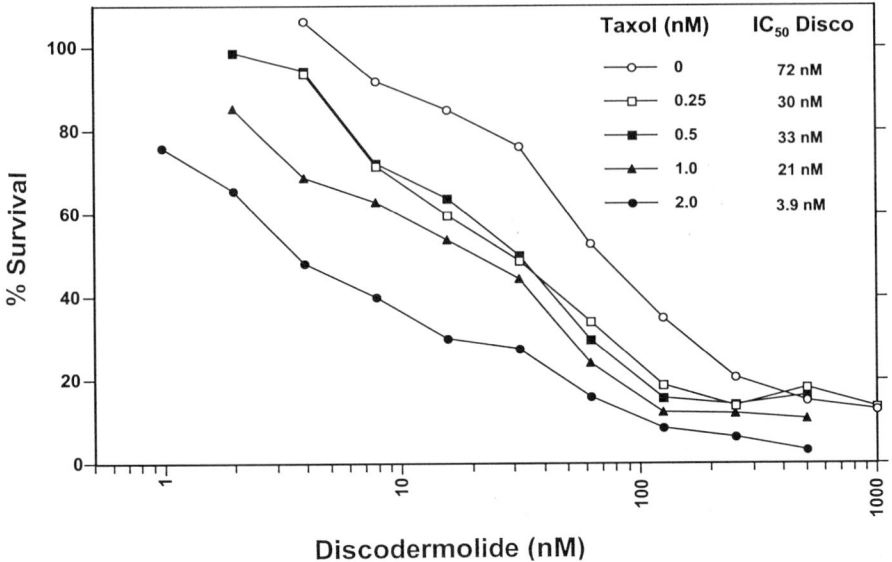

Figure 5. Taxol enhanced the cytotoxicity of discodermolide in A549-T12 cells. For the drug combination assay, the cells were analyzed after a 72 h incubation in the absence or presence of various concentrations of Taxol plus a range of discodermolide concentrations using a cell proliferation assay. (Reproduced with permission from reference 21).

precedes DNA strand break formation. Concurrent exposure of A549 cells to Taxol and discodermolide resulted in increased cleavage of PARP compared to either drug alone (Fig. 6). These results demonstrated that the synergism observed between Taxol and discodermolide enhanced apoptosis. The hyperphosphorylation of Bcl-X_L, an apoptotic regulator, is associated with the accumulation of cells in mitosis and is also a molecular marker of cytotoxicity for microtubule binding drugs. Concurrent exposure of A549 cells to both Taxol and discodermolide resulted in earlier hyperphosphorylation of Bcl-X_L that was not evident when the single drugs were used. Additionally, as the concentration of drug increased, the expression of Bcl-X_L diminished.

Conclusion

Although at least three new compounds with Taxol-like activity have been described in the literature and shown to have many similarities to Taxol, careful analyses have indicated that at least one of these, discodermolide, has distinct differences from Taxol. Taxol and discodermolide are synergistic in their ability to kill cells and it is suggested that they may constitute a promising chemotherapeutic combination.

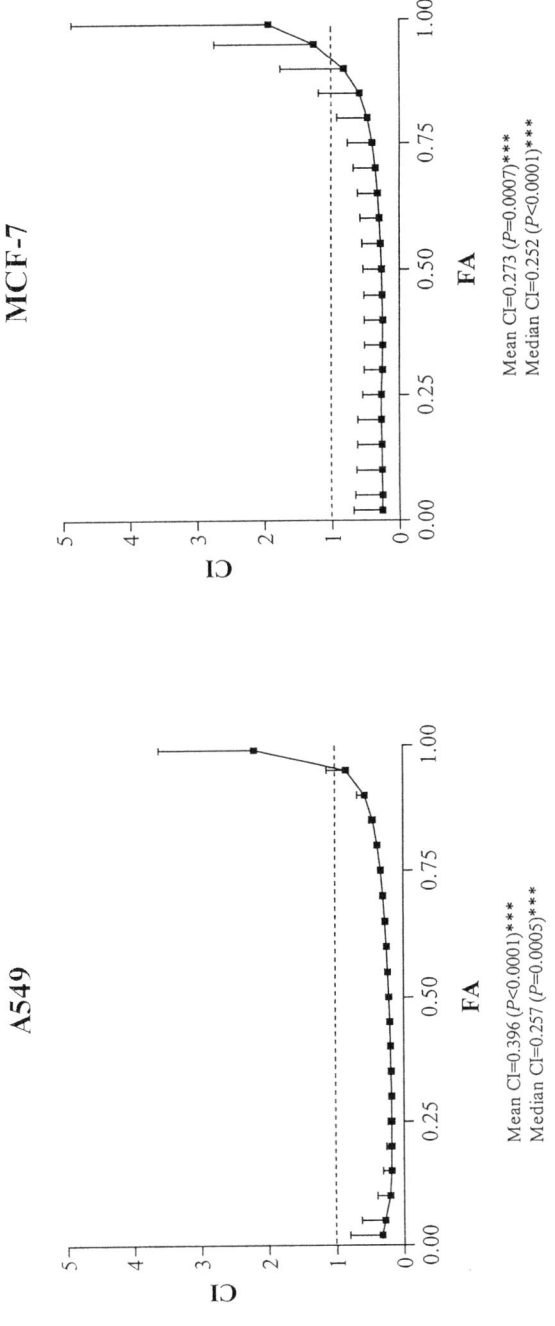

Figure 6. Taxol and discodermolide are a synergistic drug combination in human carcinoma cell lines. Cells were incubated with different drug concentrations at their equipotent ratios for 72 h after which cell counts were determined. Data points represent the mean CI values, based on the mutually nonexclusive assumption, ± SE from at least three independent experiments. Probabilities (P) indicate the level of significance of the mean and median CI values compared to a CI=1. (Reproduced with permission from reference 21).

Continued on next page.

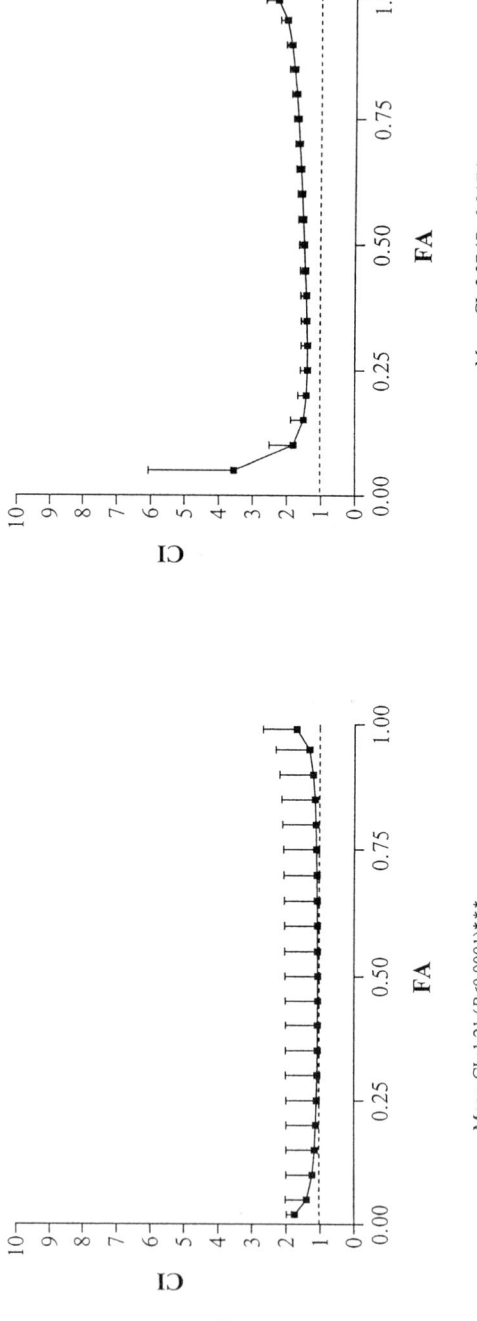

Figure 6. Continued.

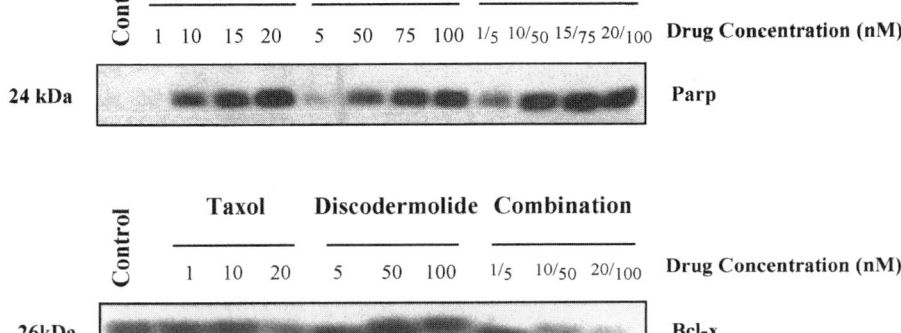

Figure 7. Potentiation of apoptosis by Taxol and discodermolide. A549 cells were incubated with Taxol, discodermolide, or both drugs at their equipotent ratios for 18 h and cell lysates prepared. Proteins were electrophoresed, blotted onto nitrocellulose, and subjected to immunodetection by specific antibodies.

Classically one would not expect a synergistic interaction between two drugs with a similar mechanism of action. Most cancer chemotherapeutic regimens include drugs with distinct mechanisms of action and toxicities. The results reported in this paper indicate that one must be open to new strategies when developing combination therapies. This potentially interesting drug combination must now be explored in animals.

References

1. Rowinsky, E.K *Annu Rev Med.* **1997**, *48*, 353-74.
2. Gill, P.S.; Tulpule, A.; Espina, B.M.; Cabriales, S.; Bresnahan, J.; Ilaw, M.; Louie, S.; Gustafson, N.F.; Brown, M.A.; Orcutt, C.; Winograd, B.; Scadden, D.T. *J Clin Oncol.* **1999**, *17*, 1876-83.
3. Rao, S.; Horwitz, S.B.; Ringel, I. *J. Natl. Cancer Inst.* **1992**, *84*, 785-788.
4. Parness, J.; Horwitz, S.B. *J. Cell Biol.* **1981**, *91*, 479-487.
5. Diaz, J.F.; Andreu, J.M. *Biochemistry* **1993**, *32*, 2747-2755.
6. Schiff, P.B.; Fant, J.; Horwitz, S.B. *Nature(Lond.)*, **1979**, *277*, 665-667.
7. Schiff, P.B.; Horwitz, S.B. *Proc. Natl. Acad. Sci. USA*, **1980**, *77*, 1561-1565.
8. Jordan, M.A.; Toso, R.J.; Thrower, D; Wilson, L. *Proc. Natl. Acad. Sci. USA*, **1993**, *90*, 9552-9556.
9. Jordan, M.A.; Wendell, K.; Gardiner, S.; Derry, W.B.; Copp, H.; Wilson, L. *Cancer Res.*, **1996**, *56*, 816-825.
10. Torres, K; Horwitz, S.B. *Cancer Res.*, **1998**, *58*, 3620-2626.

11. Rao, S.; Krauss, N.E.; Heerding, J.M.; Swindell, C.S.; Ringel, I.; Orr, G.A.; Horwitz, S.B. *J. Biol. Chem.,* **1994,** *69,* 3132-3134.
12. Rao, S.; Orr, G.A.; Chaudhary, A.G.; Kingston, D.G.I.; Horwitz, S.B. *J. Biol. Chem.,* **1995,** *270,* 20235-20238.
13. Rao, S.; He, L.; Chakravarty, S.; Ojima, I.; Orr, G.A.; Horwitz, S.B. *J. Biol. Chem.,* **1999,** *274,* 37990-37994.
14. Nogales, E.; Wolf, S.G.; Downing, K.H. *Nature,* **1998,** *391,* 199-203.
15. Bollag, D.M.; McQueney, P.A.; Zhu, J.; Hensens, O.; Koupal, L.; Liesch, J.; Goetz, M.; Lazarides, E.; Woods, C.M. *Cancer Res.,* **1995,** *55,* 2325-2333.
16. Su, D-S.; Balog, A.; Meng, D.; Bertinato, P.; Danishefsky, S.J.; Zheng, Y-H.; Chou, T-C.; He, L.; Horwitz, S.B. *Chem. Int. Ed. Engl.* **1997,** *36,* 2093-2096,
17. Lindel, T.; Jensen, P.R.; Fenical, W.; Long, B.H.; Casazza, A.M.; Carboni, J.; Fairchild, C.R. *J. Am. Chem. Soc.,* **1997,** *119,* 8744-8745.
18. McDaid, H.M.; Bhattacharya, S.K.; Chen, X-T.; He, L.; Shen, H-J.; Gutteridge, C.E.; Horwitz, S.B.; Danishefsky, S.J. *Cancer Chemother. Pharmacol.,* **1999,** *44,* 131-137.
19. Ter Haar, E.; Kowalski, R.J.; Hamel, E.; Lin, C.M.; Longley, R.E.; Gunasekera, S.P.; Rosenkranz, H.S.; Day, B.W. *Biochemistry,* **1996,** *35,* 243-250.
20. Hung, D.T.; Chen, J.; Schreiber, S.L. *Chem. Biol.,* **1996,** *3,* 287-293.
21. Martello, L.A.; McDaid, H.M.; Regl, D.L.; Yang, C-P.H.; Meng, D.; Pettus, T.R.R.; Kaufman, M.D.; Arimoto, H.; Danishefsky, S.J.; Smith, A.B.; Horwitz, S.B. *Clin. Cancer Res.* **2000,** *6,* 1978-1987.
22. Gaskin, F.; Cantor, C.R.; Shelanski, M.L. *J. Mol. Biol.,* **1974,** *89,* 737-758.
23. Kowalsiki, R.J.; Giannakakou, P.; Gunasekera, S.P.; Longley, R.E.; Day, B.W.; Hamel, E. *Mol. Pharm.,* **1997,** *52,* 612-622.
24. Lee, F.Y.F.; Vite, G.D.; Borzilleri, R.M.; Arico, M.A.; Clark, J.L.;, Fager, K. L.; Kan, D.; Kennedy, K.A.; Kim, A.S-H.; Smykla, R.A.; Wen, M.-L., Kramer, R.A. *Proceedings of the 11th NCI-EORTC-AACR* **2000***, Symposium on New Drugs in Cancer Therapy section 573,* 2000.
25. Giannakakou, P.; Gussio, R.; Nogales, E.; Downing, K.H.; Zaharevitz, D.; Bollbuck, B.; Poy, G.; Sackett, D.; Nicolaou, K.C.; & Fojo, T. *Proc. Natl. Acad. Sci. USA* **2000,** *97,* 2904 - 2909.
26. *Reversal of multidrug-resistance in tumor cells.* In: *Chemotherapeutic Synergism and Antagonism.* Yang, C.-P,H; Greenberger, L.M; and Horwitz, S.B. D. Rideout and T.-C. Chou, Eds.; Academic Press, N.Y. 1991; p. 311-338.
27. Chou, T.-C.; Talalay, P.; *Adv. Enzyme Regul.* **1984,** *22,* 27-55.

Chapter 6

Highly Efficient Semisynthesis of Biologically Active Epothilone Derivatives

Gregory D. Vite, Robert M. Borzilleri, Soong-Hoon Kim,
Alicia Regueiro-Ren, W. Griffith Humphreys, and Francis Y. F. Lee

Pharmaceutical Research Institute, Bristol-Myers Squibb Company, Route 206 and Province Line Road, Princeton, NJ 08543–4000

Novel epothilone derivatives were prepared by both total synthesis and semisynthesis. Comparison of the two strategies suggests that a semisynthesis approach has several practical advantages including ease of preparation, stereochemical control, and potential for scale-up. Synthetic chemistry for efficient deoxygenation of epothilones, preparation of epoxide bioisosteres, and an efficient lactone-to-lactam conversion are presented. *In vitro* biological data for the new epothilone analogues are provided, along with preliminary *in vivo* data for clinical candidate BMS-247550.

Introduction

The epothilone class of natural products, exemplified by epothilones A and B (**1** and **2**, Figure 1), has generated considerable enthusiasm among researchers in the anticancer field primarily due to the mechanism of action behind the potent tumor cell-killing effects of these agents.[2,3] Stabilization of microtubules leading to mitotic arrest is a clinically validated biological target emanating from the groundbreaking therapeutic and commercial success of paclitaxel (**3**, Taxol). While the epothilones (**1-2**) and taxanes (**3-4**) share this common mechanism of action, it is the unattenuated cytotoxicity of the epothilones against taxane-resistant tumor cell lines that is driving the vigorous research efforts in this area. A goal of these efforts is to extrapolate this observed advantage in cellular activity to human conditions where tumors are either innately refractory or developmentally resistant to taxanes, thereby expanding upon the therapeutic utility of microtubule stabilization agents.

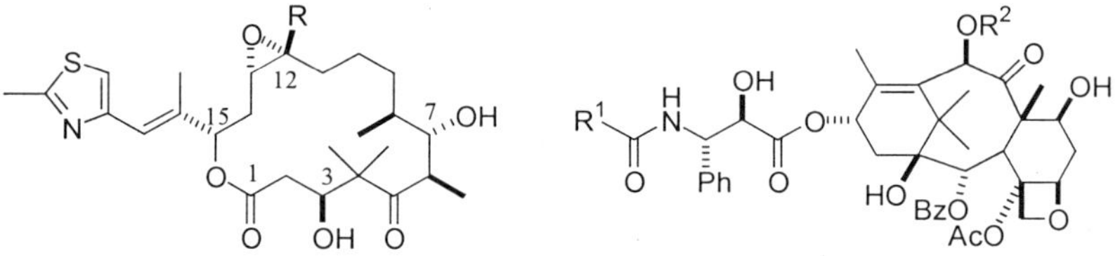

1 R=H Epothilone A
2 R=Me Epothilone B

3 R^1=Ph, R^2=Ac Paclitaxel
4 R^1=O-tBu, R^2=H Docetaxel

Figure 1. Structures of representative epothilones and taxanes.

Structurally, the epothilones are sixteen-membered ring macrolides obtained from the fermentation of a myxobacterium (*Sorangium cellulosum*), as described by Reichenbach and Höfle.[2] The polyketide-derived framework is composed of multiple propionate and acetate fragments ending in a heterocyclic side-chain derived from cysteine.[4] The epothilones were independently discovered by Bollag and co-workers who elucidated their mechanism of action.[3] Unequivocal structural assignment by single crystal X-ray diffraction[5] preceded the first total syntheses of epothilone A by the laboratories of Danishefsky,[6] Nicolaou,[7] and Schinzer.[8] These worthwhile studies highlighted the power of total synthesis for the preparation of the structurally complex epothilones and related analogues.

From a pharmaceutical perspective, the epothilones appear to have advantages over prototypical taxanes. For example, improved solubility of the epothilones versus taxanes may allow for drug formulations with improved patient tolerability. However, preclinical experiments in these laboratories using epothilone B suggested that the natural product has two distinct flaws that could hamper its clinical development. First, epothilone B was unstable in murine plasma ($T_{1/2}$ ~ 40 min) and liver microsomal preparations. This instability was attributed to cleavage of the lactone linkage by esterases since addition of a general esterase inhibitor, BNPP (bis-(*p*-nitrophenyl)phosphate), stabilized the molecule in these biological milieus. Importantly, demonstrating antitumor effects of epothilone B in murine models of cancer, a prerequisite for clinical evaluation of a drug candidate, proved to be difficult. Consequently, stabilization of the lactone linkage of the epothilones through structural modification was identified as an important objective. Secondly, early drug safety assessment demonstrated that epothilone B was highly toxic when administered to rodents and lower primates. Therefore, a second and equally important objective was to identify epothilone analogues that have an improved safety profile relative to that of the natural product.

Epothilone Derivatives via Total Synthesis

From the start, total synthesis of epothilone analogs was viewed as a valuable complement to semisynthesis of epothilones starting from the natural product. Total synthesis could provide access to a greater number of structurally diverse analogues, and potentially lead to novel, biologically active compounds. Conversely, semisynthesis, while not as flexible, could allow for shorter synthetic routes and provide "unlimited" scale-up potential. The practicality of total synthesis was evaluated following Nicolaou's approach.[7] Large scale preparation of advanced intermediate 5 and subsequent ring-closing metathesis (RCM) of 7 established that the poor stereoselectivities of both the aldol condensation and the RCM reaction presented significant but perhaps manageable challenges (Scheme 1).

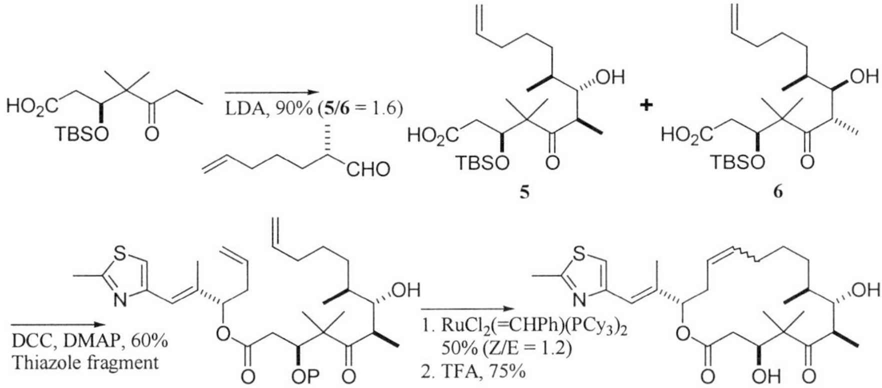

Scheme 1. Evaluation of Nicolaou's total synthesis route to epothilones.

In the context of the pharmaceutical objectives outlined above, an esterase-resistant lactam derivative **11** was prepared by total synthesis (Scheme 2). Thus, thiazole fragment **9**, derived from commercially available L-allylglycine was coupled to the polypropionate acid **5** using EDCI-HOBt [EDCI = 1-(3-

dimethylaminopropyl)-3-ethylcarbodiimide, HOBt = 1-hydroxybenzotriazole] to afford olefin metathesis precursor **10** in good yield (Scheme 1).[9] RCM reaction of the bis-olefin **10** with the Grubbs ruthenium-based catalyst (~0.7 mol% $RuBnCl_2(PCy_3)_2$) afforded the protected macrocycle with the less desired 12E geometry, along with a minor amount of the desired 12Z-olefin. Interestingly, Schinzer and coworkers have shown that this selectivity was reversed upon protection of the 7-hydroxyl group as a silyl ether.[10] Subsequent removal of the t-butyldimethylsilyl ether of the major isomer with trifluoroacetic acid (TFA) afforded lactam **11(E)**. In general, these preliminary results indicated that it was unlikely that the RCM approach would provide sufficient quantities of material needed to sustain a significant medicinal chemistry effort, even if improved conditions could be found for the aldol condensation or the RCM reaction.

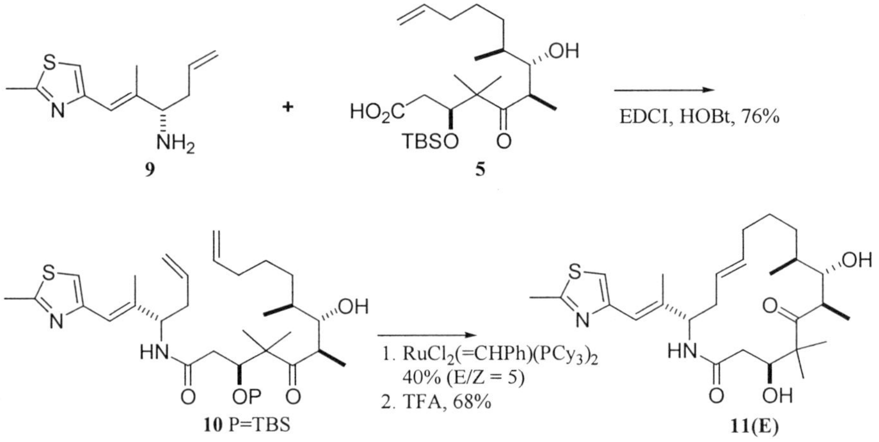

Scheme 2. Total synthesis of epothilone lactams.

Epothilone Derivatives via Degradation of Natural Product

In view of these results, semisynthesis approaches to epothilone analogues were investigated. Progress in this area was greatly facilitated by the collaborative efforts between Bristol-Myers Squibb and Gesellschaft für Biotechnologische Forschung (GBF). The GBF team provided fermentation technology, gram-quantities of natural epothilones, and scientific expertise in the epothilone field. With regard to semisynthesis, two tactics were followed: 1) direct derivatization of the natural product, and 2) degradation of the natural product to a versatile intermediate that could be utilized in a relay synthesis of epothilone analogues. It was postulated that a successful degradation strategy would hinge on the ability to deoxygenate the 12,13-oxirane to provide 12,13-olefinic epothilones (**12**, **13**) which are only minor components of the fermentation process. Such a conversion when coupled with oxidative cleavage of the olefin and hydrolysis of the lactone would provide a C1-C12 polypropionate fragment for further manipulation (Figure 2).

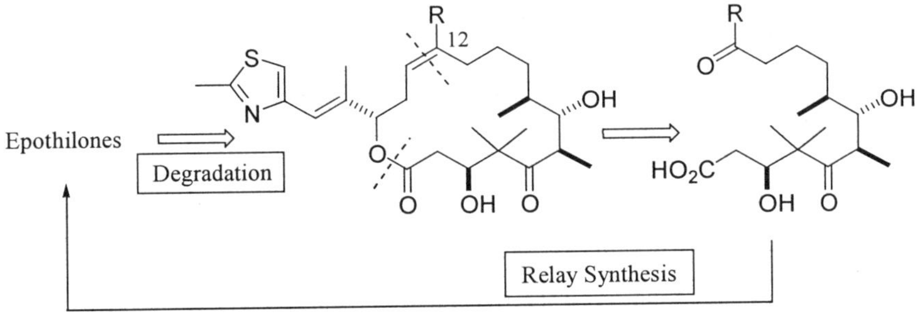

Figure 2. Degradation/relay synthesis strategy.

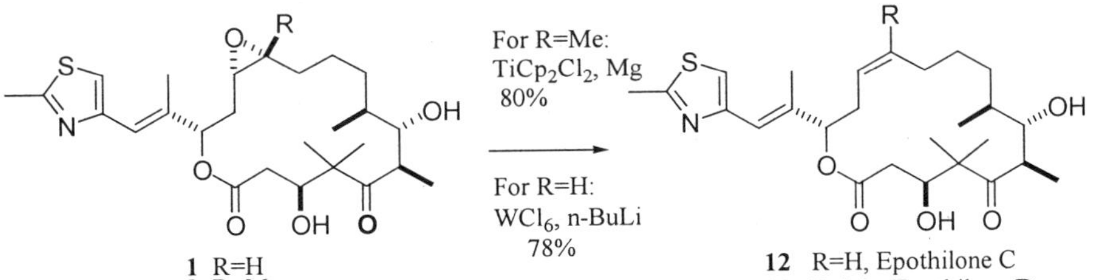

1 R=H
2 R=Me

12 R=H, Epothilone C
13 R=Me, Epothilone D

Scheme 3. Efficient deoxygenation of oxiranyl epothilones.

Thus, the first goal of this semisynthetic approach was the stereoselective conversion of the epoxide to the olefin (Scheme 3).[11] Remarkably, the deoxygenation of *unprotected* **1** with Cp_2TiCl_2/Mg[12] occurred in 80% yield to give **12** as a single olefin isomer. The analogous reaction using unprotected **2**, gave only trace amounts of **13**. However, the reduction of **2** using the procedure of Sharpless (WCl_6/n-BuLi)[13] cleanly afforded **13** stereoselectively in 78% yield. The success of these critical transformations provided the basis for continuation of the degradation/relay synthesis strategy outlined above.

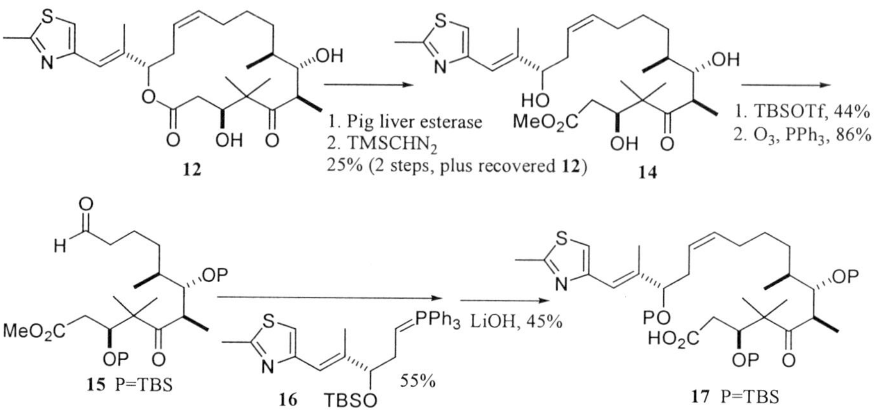

Scheme 4. *Degradation of epothilone C (12) to C1-C12 fragment (15) and subsequent conversion to known epothilone precursor 17.*

An unoptimized degradation of **12** is shown in Scheme 4. Unprotected **12** was converted to ring-opened product **14** by treatment with pig liver esterase[14] followed by esterification with trimethylsilyl diazomethane. Silylation of the hydroxyl groups of **14** with *t*-butyldimethylsilyl triflate occurred in 44% yield, and subsequent ozonolysis of the 12,13-olefin provided the key degradation intermediate **15** in 86% yield. The feasibility of relay synthesis was then demonstrated by conversion of **15** to known epothilone A precursor **17**[15] via Wittig olefination (55% yield) with readily obtained ylide **16**, followed by saponification of the methyl ester with lithium hydroxide (45% yield).

The utility of this approach was tested in a relay synthesis of analogues where the 12,13-oxirane ring was replaced with a C-N bond (Scheme 5). Racemic amine **20**, readily prepared in two steps from known aldehyde **19**,[16] was combined with **15** in a reductive alkylation (NaBH(OAc)$_3$, 74% yield) to give the corresponding tertiary amine **21**. Alternatively, **20** was coupled with acid **18** using EDCI-HOBt to provide the corresponding tertiary amide **22** in 79% yield. In both cases, **21** and **22** were saponified (LiOH, 86% yield), cyclized with Yamaguchi's reagent (2,4,6-trichlorobenzoyl chloride, 37% yield), and deprotected (TFA, 66% yield) to afford epothilone analogues **23** and **24** as mixtures of diastereomers. Secondary amine **25** and secondary amide **26** were also prepared using an analogous sequence, except that in the preparation of **25** N-protection of the secondary amine was required prior to macrolactonization. Unfortunately, **23-26** were poorly active in a tubulin polymerization assay and efforts in this area were discontinued.

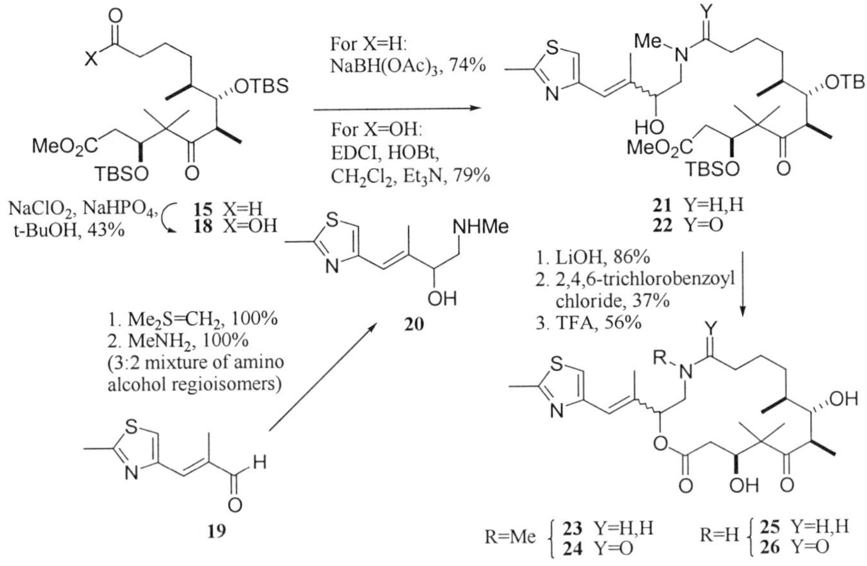

Scheme 5. Replacement of 12,13-oxirane with C-N bond.

Epothilone Derivatives via Direct Semisynthesis

A second semisynthesis tactic involving direct derivatization of the epothilone ring system was investigated. Ready access to **12** and **13** provided an opportunity for direct aziridination or cyclopropanation of the 12,13-olefin. Several attempts at the direct aziridination of **12** or **13** using published procedures for the generation of reactive nitrenes[17] were either unsuccessful or provided only trace amounts of desired aziridine adduct. In addition, a survey of numerous reaction conditions for the stereoselective epoxidation of **12** to provide 12,13-epi epothilone A as a precursor to aziridinyl analogs provided the undesired "natural" 12R,13S-epoxide as the major product.

Therefore, a less direct synthesis of aziridinyl analogues requiring double inversion of stereochemistry at both C12 and C13 was investigated (Scheme 6).[18] Thus, **1** was converted to the corresponding bis-triethylsilyl (bis-TES) ether by treatment with triethylsilyl chloride in the presence of diisopropylethylamine, and epoxide opening using MgBr$_2$/OEt$_2$ at low temperature (-20° to -5° C) afforded a regioisomeric mixture of bromohydrins (**27a/27b** =20) in 45% yield (67% yield based on recovered **1**, bis-TES). Nucleophilic substitution of the 13-bromide with sodium azide in DMF completed the double inversion at C13, affording azido alcohol **28** in 60% yield. Inversion of the stereochemistry at C12 was carried out using *p*-nitrobenzoic acid under modified Mitsunobu conditions (diethylazodicarboxylate, triphenylphosphine).[19] Selective cleavage of the *p*-nitrobenzoate with methanolic ammonia provided **29** in quantitative yield. Azido alcohol **29** was then converted to the corresponding mesylate **30**. Azide reduction with triphenylphosphine and spontaneous ring-closure to the 12R,13S-aziridine ring proceeded in 96% yield. Removal of the TES groups using TFA in CH_2Cl_2 furnished the parent aziridinyl analogue **31** in 90% yield. Despite the 8 step sequence, the overall yield for the process was excellent (27%), and gram-quantities of **31** were obtained.

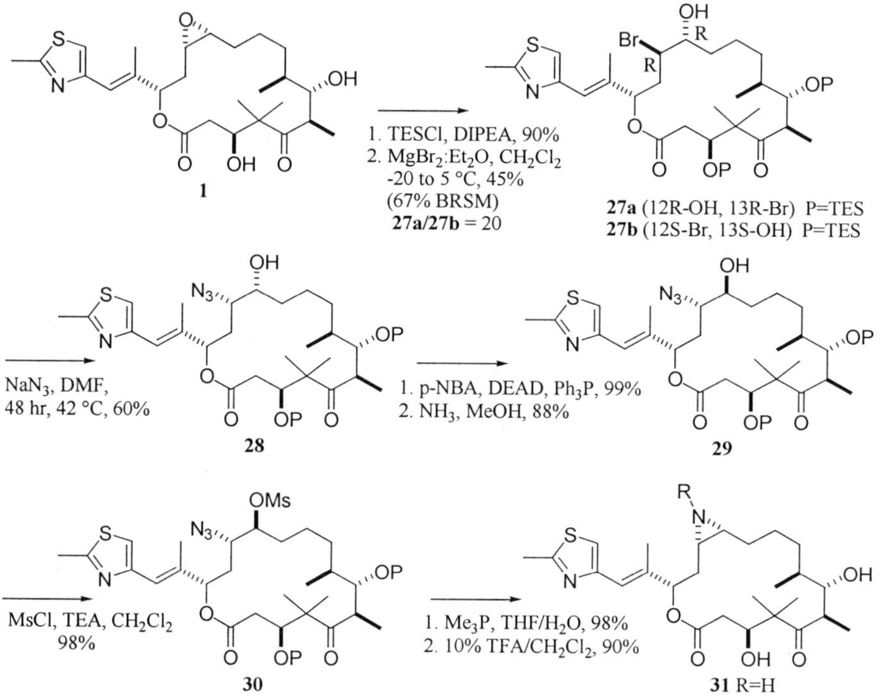

Scheme 6. Conversion of epothilone A to 12,13-aziridinyl epothilones.

Starting from either **31** or its 3,7-bis-triethylsilyl ether, several N-substituted aziridinyl analogues were prepared by standard alkylation or acylation protocols. Table 1 shows the *in vitro* biological data for representative aziridinyl epothilones **31-35**. While the parent epothilone A aziridine (**31**) is a less effective tubulin polymerization agent when compared to epothilone A (**1**), **31** is equally effective compared to **1** in a cytotoxcity assay using a human colon cancer cell line (HCT-116). N-Alkylation (**32**) and N-acylation (**33**) appear to improve both *in vitro* biological activities when compared to the parent aziridine **31**. Interestingly, while the ethyl carbamate functionality (**34**) is well tolerated, the closely related ethyl urea functionality (**35**) is inactive.

Table 1. *In vitro* data for aziridinyl epothilones 31-35

#	Compound	Tubulin $EC_{0.01}$ $(\mu M)^a$	HCT-116 IC_{50} $(nM)^b$
1	Epothilone A	2.3	3.2
3	Paclitaxel	5.0	2.3
31	Aziridine R=H	14	2.7
32	Aziridine R=Me	2.6	0.13
33	Aziridine R=Ac	1.2	0.9
34	Aziridine R=CO$_2$Et	1.2	7.7
35	Aziridine R=C(O)NHEt	>1000	NT

NOTE: [a] Tubulin polymerization assay performed using the method described in ref. 9, and references therein. [b] HCT-116, a human colon carcinoma cell line cytotoxicity assay performed using the method described in ref. 9, and references therein. NT = not tested.

While direct aziridination of olefins **12** and **13** was problematic, the analogous direct cyclopropanation was less troublesome. As noted by Nicolaou and co-workers for a related substrate,[20] cyclopropanation of a protected epothilone C (**36**) using Simmons-Smith or related conditions gave only trace amounts of the desired adduct.[11] However, treatment of **36** with NaOH/CHBr$_3$ under phase transfer conditions (triethybenzylammonium chloride) afforded the dibromocyclopropane derivative **37** in 12% yield as a single diastereomer (Scheme 7). The yield of this carbene cycloaddition was improved (30% yield) in the case of a more electron-rich 12-methyl olefin. The 12,13-configuration of **37** was confirmed by single crystal X-ray structure determination. The desired cyclopropane **39** was obtained in 68% yield after reduction of **38** with tri-*n*-butyltin hydride and deprotection with trifluoroacetic acid. The unequivocal structural assignment and the potent *in vitro* biological data for **39** (vide infra) support a correction to previously published results for 12,13-cyclopropyl epothilones.[20] Other cyclopropanes, including the cyclopropyl analogue of epothilone B (**40**), were obtained following the same route (Table 2).

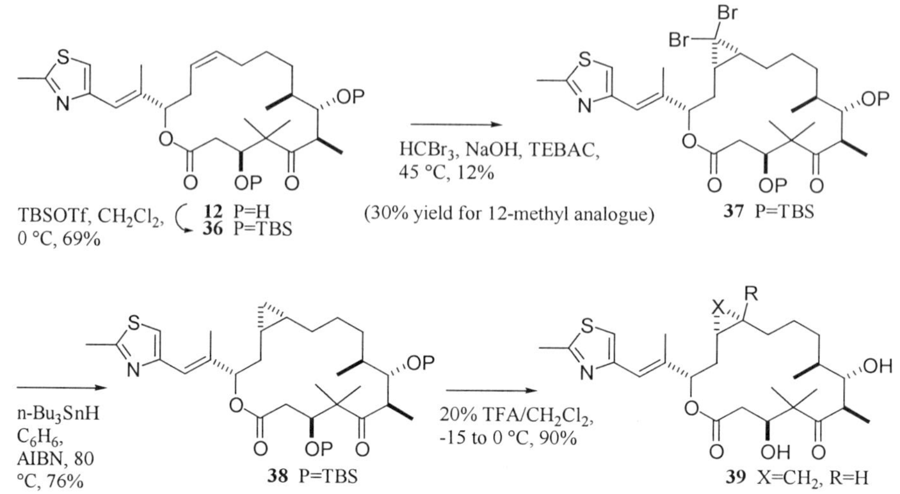

Scheme 7. Stereoselective synthesis of 12,13-cyclopropyl epothilones.

As shown in Table 2, all of the cyclopropyl analogues, including the bulky dihalocyclopropanes **41** and **42**, are active in both the tubulin polymerization and cytotoxicity assays. The comparable activities of the cyclopropyl analogues indicate that the role of the epoxide in the context of tubulin binding is largely conformational. Clearly, non-bonded interactions between the ether oxygen of epothilone and the tubulin protein, if present, do not appear to be critical for binding. Furthermore, as demonstrated above for the aziridinyl analogs, the tubulin-binding site accommodates comparatively larger substituents at C12-C13.

Our successes in utilizing the direct semisynthesis approach led to a re-investigation into the preparation of lactam analogues. Careful analysis of the epothilone structure, led to the critical observation that the lactone is actually an allylic ester and may be susceptible to palladium-catalyzed ring opening to afford an intermediate π-allylpalladium complex (Figure 3). This latent species could then be trapped in situ with a "soft" external nucleophile, such as azide, to form an allylic azide. If successful, this sequence of events would provide the requisite nitrogen atom at C15 and a readily accessible coupling partner (carboxylic acid) at C1. Subsequent reduction of the azide followed by macrocyclization would provide the epothilone lactam. However, one concern about this key reaction was the regio- and stereochemical outcome of the Pd(0)-catalyzed nucleophilic substitution reaction involving a highly functionalized macrolactone substrate.

Table 2. *In vitro* data for cyclopropyl epothilones 39-42

#	Compound	Tubulin $EC_{0.01}$ $(\mu M)^a$	HCT-116 IC_{50} $(nM)^b$
1	Epothilone A	2.3	3.2
3	Paclitaxel	5.0	2.3
39	Epo A cyclopropane ($X=CH_2$, R=H)	1.4	1.4
40	Epo B cyclopropane ($X=CH_2$, R=Me)	2.1	0.7
41	Epo B cyclopropane ($X=CCl_2$, R=Me)	1.7	1.9
42	Epo B cyclopropane ($X=CBr_2$, R=Me)	1.6	3.8

NOTE: See note in Table 1.

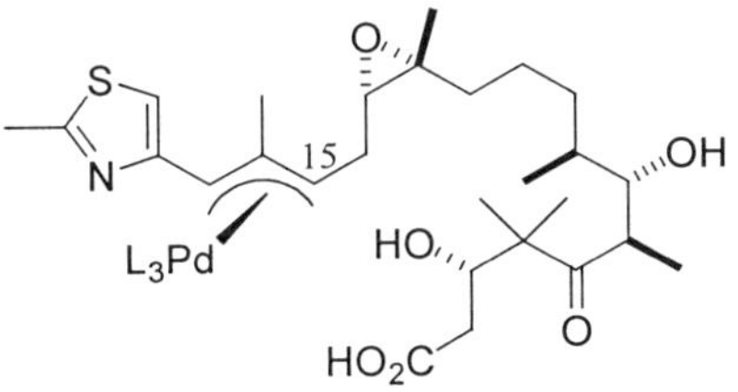

Figure 3. Epothilone B-derived π-allylpalladium complex.

In the event, treatment of *unprotected* epothilone B (**2**) with a catalytic amount of tetrakis(triphenylphosphine)palladium(0) and sodium azide afforded the azido acid **43** as a single diastereomer in good yield (Scheme 8).[9] Mechanistically, this reaction can be considered to proceed via anti-attack of palladium(0) on the allylic lactone to form an initial π-allylpalladium complex (see Figure 3). Subsequent attack by azide at C15, from the opposite face of the palladium/ligands, provides **43** with the desired regio- and stereochemistry. In effect, net retention of configuration at C15 was observed. Surprisingly, all of the potentially labile functionality found in the epothilone core, including the epoxide, remained intact. Synthesis of lactam **45** was completed by reduction of the azide under hydrogenation conditions with stoichiometric amounts of Adams' catalyst (PtO_2) affording amino acid **44** in 61% yield, followed by macrolactamization of **44** with diphenylphosphoryl azide (DPPA, 40% yield).

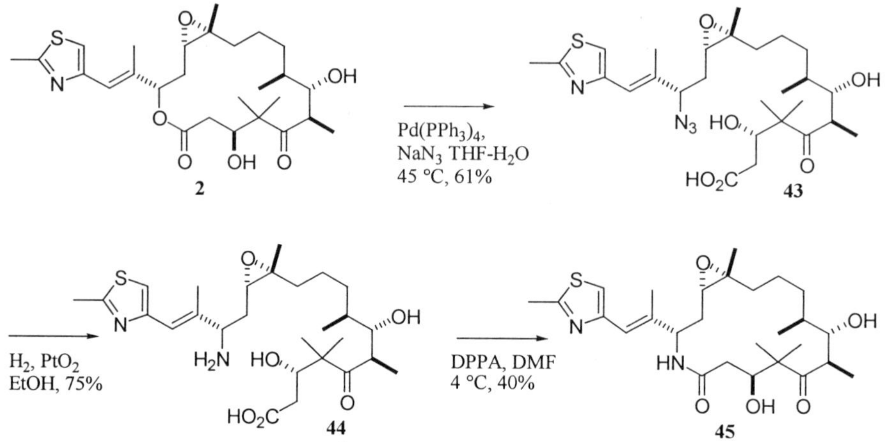

Scheme 8. Three-step semisynthesis of epothilone B lactam.

A more efficient semisynthesis of **45** resulted from efforts to overcome the difficult isolation and purification of polar intermediates **43** and **44**. Changing the azide reducing agent to trimethylphosphine and changing the macrolactamization agent to EDCI-HOBt allowed the entire process to be carried out in a single reaction vessel without the need for isolation of intermediates (Scheme 9). This "one pot" process required only one day and gave a 20-25% yield of lactam **45** from epothilone B (**2**). The epothilone A lactam **46** was prepared in analogous fashion. Deoxygenation of **45** and **46** using the procedures described (vide supra) provided the corresponding epothilone C and epothilone D lactams **11(Z)** and **47**, respectively.

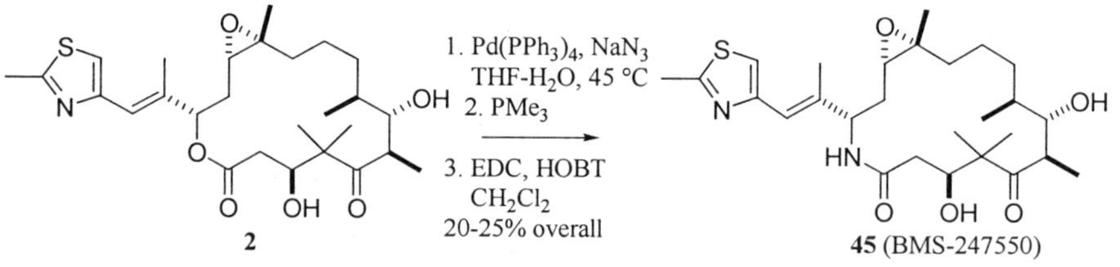

Scheme 9. Efficient "one pot" synthesis of clinical candidate BMS-247550.

The *in vitro* biological activities of the lactam analogues **11(Z)** and **45-47** are shown in Table 3. The epothilone B lactam **45** is ~10-fold more potent than the corresponding epothilone A derivative **46** in both a tubulin polymerization assay and a cytotoxicity assay. While the epothilone D lactam **47** is well tolerated in the tubulin polymerization assay, this analogue is considerably less active than **45** in the cell assay. Based on these and other *in vitro* studies, lactam **45** was further evaluated in murine tumor models. In a representative study, **45** was dosed (iv, q2dx5) in immunocompromised mice bearing a clinically derived tumor sample, characterized as paclitaxel-resistant (model Pat-7). Figure 4 shows that **45**, at its maximally tolerated dose (MTD, 6.3 mpk), caused significant tumor growth delay in this model when compared to optimal dosing of paclitaxel. In related studies, *in vivo* efficacy was demonstrated at doses below the MTD suggesting that, unlike epothilone B (**2**), the lactam analogue has an improved safety profile. Further details of the *in vivo* biological evaluation of **45** is described elsewhere.[21] In comparison to epothilone B, lactam **45** (BMS-247550) met preclinical objectives of improved metabolic stability, reduced toxicity in rodents, and a broad spectrum of antitumor activity in animal models of cancer. Accordingly, BMS-247550 was chosen for drug development and clinical evaluation. BMS-247550 is currently in phase I clinical trials sponsored by Bristol-Myers Squibb and the National Cancer Institute.

Table 3. *In vitro* data for lactam analogues

#	Compound	Tubulin $EC_{0.01}$ $(\mu M)^a$	HCT-116 IC_{50} $(nM)^b$
2	Epo B	1.4	0.42
3	Paclitaxel	5.0	2.3
45	Epo B-lactam	3.8	3.6
46	Epo A-lactam	12	130
11(Z)	Z-Epo C-lactam	110	NT
47	Z-Epo D-lactam	5.5	65

NOTE: See note for Table 1.

Conclusion

In summary, the work described here demonstrates that, in the case of epothilones, semisynthesis has distinct advantages over total synthesis. Arguably, the short synthetic routes and the scale-up potential offered by semisynthesis offset the better access to the diverse structural modifications afforded by total synthesis. That fermentation provides multigram-quantities of epothilones for use as starting materials was a key factor in this successful semisynthetic approach. In this account, dexterous application of organic chemistry combined with clear objectives for target molecule design provided a vehicle for maximizing the drug discovery opportunities afforded by the epothilones.

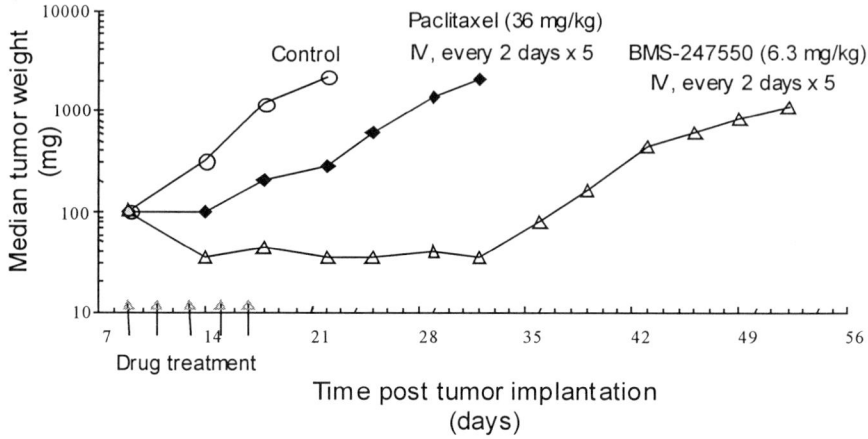

Figure 4. Tumor growth delay for BMS-247550 versus paclitaxel in clinically derived tumor xenograft model (model Pat-7).

References

1. For a preliminary account of the work described herein, see: Vite, G.; Borzilleri, R.; Fooks, C.; Johnson, J.; Kim, S.-H.; Leavitt, K.; Mitt, T.; Regueiro-Ren, A.; Schmidt, R.; Zheng, P.; Lee, F. "Epothilones A and B: Springboards for the synthesis of promising antimitotic agents," Book of Abstracts, 219th National Meeting of the American Chemical Society, San Francisco, CA, March 26-30, 2000; #286.
2. Gerth, K.; Bedorf, N.; Höfle, G.; Irschik, H.; Reichenbach, H. *Journal of Antibiotics* **1996**, *49*, 560-563.
3. Bollag, D.M.; McQueney, P.A.; Zhu, J.; Hensens, O.; Koupla, L.; Liesch, J.; Goetz, M.; Lazarides, E.; Woods, C.M; *Cancer Research* **1995**, *55*, 2325-2333.
4. Molnar, I.; Schupp, T.; Ono, M.; Zirkle, R.; Milnamow, M.; Nowak-Thompson, B.; Engel, N.; Toupet, C.; Stratmann, A.; Cyr, D.D.; Gorlach, J.; Mayo, J.M.; Hu, A.; Goff, S.; Schmid, J.; Ligon, J.M. *Chemistry and Biology* **2000**, *7*, 97-109.
5. Höfle, G.; Bedorf; N.; Steinmetz, H.; Schomburg, D.; Gerth, K.; Reichenbach, H. *Angew. Chem., Int. Ed.* **1996**, *35*, 1567-1569.

6. Balog, A.; Meng, D.; Kamenecka, T.; Dertinato, P.; Su, D.-S.; Sorensen, E.J.; Danishefsky, S.J.; *Angew. Chem., Int. Ed.* **1996**, *35*, 2801-2803.
7. Yang, Z.; He, Y.; Vourloumis, D.; Vallberg, H.; Nicolaou, K.C.; *Angew. Chem., Int. Ed.* **1997**, *26*, 166-168.
8. Schinzer, D.; Limberg, A.; Bauer, A.; Boehm, O.M.; Cordes, M. *Angew. Chem., Int. Ed.* **1997**, *36*, 523-524.
9. Borzilleri, R.; Zheng, X.; Schmidt, R.; Johnson, J.; Kim, S.-H.; DiMarco, J.; Fairchild, C.; Gougoutas, J.; Lee, F.; Long, B.; Vite, G. *J.Am. Chem. Soc.* **2000**, *122*, 8890-8897.
10. Schinzer, D.; Altmann, K.-H.; Stuhlmann, F.; Bauer, A.; Wartmann, M. *ChemBioChem* **2000**, *1*, 67-70.
11. Johnson, J.; Kim, S.-H.; Bifano, M.; DiMarco, J.; Fairchild, C.; Gougoutas, J.; Lee, F.; Long, B.; Tokarski, J.; Vite, G. *Org. Lett.* **2000**, *2*, 1537-1540.
12. Schobert, R.; Hoehlein, U. *Synlett* **1990**, *8*, 465.
13. Sharpless, K.; Umbreit, M.; Nieh, M.; Flood, T. *J.Am. Chem Soc.* **1972**, *94*, 6538.
14. Höfle, G.; Kiffe, M. PCT Int. Appl. (1997) WO9719086.
15. Nicolaou, K.; Sarabia, F.; Ninkovic, S.; Vourlomis, D.; Yang, Z. *Angew. Chem., Int. Ed.* **1997**, *36*, 525-527.
16. Meng, D.; Sorensen, E.J.; Bertinato, P.; Danishefsky, S.J. *J. Org. Chem.* **1996**, *61*, 7998-7999.
17. Evans, D.A.; Bilodeau, M.T.; Faul, M.M. *J. Am. Chem. Soc.* **1994**, *116*, 2742; and references therein.
18. Regueiro-Ren, A.; et al. (Bristol-Myers Squibb) *in preparation.*
19. Martin, S.F.; Dodge, J.A. *Tetrahedron Lett.* **1991**, *32*, 3017-3020.
20. Nicolaou, K.; Finlay, M.; Ninkovic, S.; King, N.; He, Y.; Li, T.; Sarabia, F.; Vourlomis, D. *Chemistry and Biology*, **1998**, *5(7)*, 365-372.
21. Lee, F.Y.F.; et al. (Bristol-Myers Squibb) *in preparation.*

Chapter 7

Synthetic and Semisynthetic Analogs of Epothilones: Chemistry and Biological Activity

Karl-Heinz Altmann[1], Marcel J. J. Blommers[2], Giorgio Caravatti[1], Andreas Flörsheimer[1], Kyriacos C. Nicolaou[3], Terrence O'Reilly[1], Alfred Schmidt[2], Dieter Schinzer[4], and Markus Wartmann[1]

[1]Novartis Pharma AG, TA Oncology Research, WKL–136.4.21, CH–4002 Basel, Switzerland
[2]Novartis Pharma AG, Core Technology Area, CH–4002 Basel, Switzerland
[3]Department of Chemistry, The Scripps Research Institute, 10550 North Torrey Pines Road, La Jolla, CA 92037
[4]Chemisches Institut, Otto-von-Gueriche-Universität, Magdeburg, Universitätsplatz 2, D–39106, Magdeburg, Germany

Abstract

Epothilones A and B are naturally occurring microtubule depolymerization inhibitors, which exhibit potent *in vitro* antiproliferative activity. Epothilone B is a 3-30-fold more potent inhibitor of human cancer cell growth than paclitaxel in paclitaxel-sensitive cancer cell lines and in paclitaxel-resistant lines exceeds paclitaxel activity by 10^2 - 10^3-fold. In addition, epothilone B exhibits potent *in vivo* antitumor activity even in multidrug-resistant tumor models. In order to gain a better understanding of the structural requirements for epothilone-mediated cytotoxicity and antitumor activity and to discover analogs with similar potency but perhaps better tolerability *in vivo*, we have investigated a series of structural modifications involving the epoxide site (C12/C13) and the heterocyclic side-chain of epothilones. In this paper we present the synthesis of these analogs and we discuss the impact of such modifications on tubulin polymerization activity as well as cytotoxicity *in vitro*.

Introduction

Epothilones are 16-membered cytotoxic macrolides produced by the myxobacterium *Sorangium cellulosum* Sc 90, which were first reported by Reichenbach and Höfle in 1993 (1) and then re-discovered in 1995 by a group at Merck & Co in a screen for tubulin-polymerizing compounds (2). There are two major variants of this class of natural products, termed epothilone A and epothilone B, which are distinguished by the absence or presence of a methyl group at C12 of the 16-membered macrocycle (Figure 1).

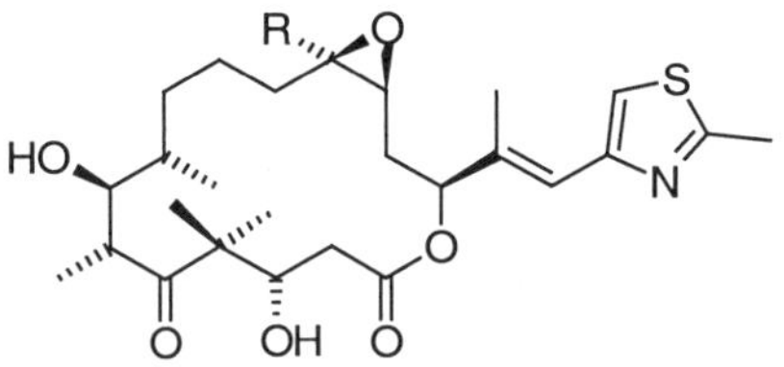

Figure 1. Structures of Epothilones A (R = H) and B (R = CH$_3$).

Epothilones are potent cytotoxic agents which inhibit the growth of a variety of human cancer cell lines *in vitro* with nM or even sub-nM IC50s (2 - 4). (Table I). Mechanistically, cytotoxicity is believed to be related to the ability of these compounds to interfere with cellular microtubule dynamics, resulting in abnormal spindle formation during mitosis and concomitant G2/M arrest followed by apoptosis (2 - 4). In a cell-free *in vitro* system, epothilones induce polymerization of tubulin heterodimers into polymers and they can stabilize preformed microtubules against cold- or Ca^{2+}-induced depolymerization (2, 4). As first recognized by Bollag *et al.* (2) their mechanism of action is thus similar to that of the prominent anticancer drug paclitaxel (Taxol®). Based on displacement studies with radioactively labeled paclitaxel it appears that microtubule binding of epothilones and taxanes occurs at the same, or at least largely overlapping, site(s) on β-tubulin (2). In contrast to paclitaxel, however, epothilones also inhibit the growth of P-gp overexpressing multidrug-resistant cancer cell lines (Table I; *cf.* also refs. 2 - 4), indicating that they are poor substrates for the P-gp efflux pump. Epothilone B, even on drug-sensitive lines, is generally a more potent inhibitor of cancer cell growth than paclitaxel, while epothilone A and paclitaxel exhibit similar potencies (Table I). Recent data obtained in our laboratories also demonstrate that epothilones, like paclitaxel (5), accumulate intracellularly several hundred-fold over external medium concentrations, such that nM medium concentrations translate into μM concentration levels inside cells.

Table I: IC50-values [nM] for net growth inhibition of human carcinoma cell lines by epothilone A and B in comparison to paclitaxel[a,b]

Cell Line	Epothilone A	Epothilone B	Paclitaxel
NCI-H460 (lung)	3.2	0.3	5.7
HCT-116 (colon)	2.5	0.3	2.8
Du145 (prostate)	4.9	0.3	4.1
MCF-7 (breast)	1.5	0.2	1.8
MCF-7/ADR[c]	27.5	2.9	9105
KB-31 (epidermoid)	2.1	0.2	2.3
KB-8511[d]	1.9	0.2	533

[a]Cells were exposed to drugs for 3-5 days, allowing for at least two population doublings. Cell numbers were estimated by quantification of protein content of fixed cells by methylene blue staining (6). Data are presented as the mean of at least 3 experiments.
[b]Multidrug-resistant cell lines are underlined.
[c]Multiple resistance mechanisms/MDR.
[d]P-gp overexpression/MDR (7).

These findings resolve the apparent discrepancy between the fact that tubulin polymerization and microtubule stability *in vitro* are only affected at µM epothilone concentrations, while cancer cell growth inhibition is associated with low nM IC50s.

Based on its very attractive *in vitro* profile we have extensively studied the antitumor activity of epothilone B in human tumor models. In general, the compound was found to exhibit potent antitumor activity (including regressions) at tolerated dose levels, but it is also characterized by a narrow therapeutic window. As an example Figure 2 illustrates the effect of epothilone B on the growth of Taxol®-resistant human epidermoid carcinoma KB-8511 together with its effects on mouse body weight (as a gross indicator for overall toxicity). Weekly treatment with 4 mg/kg of epothilone B in this model resulted in profound tumor regressions, such that 5 out of 8 mice were cured at the end of this experiment (i. e. they were tumor-free up to 60 days subsequent to the treatment period, which ended on d27).

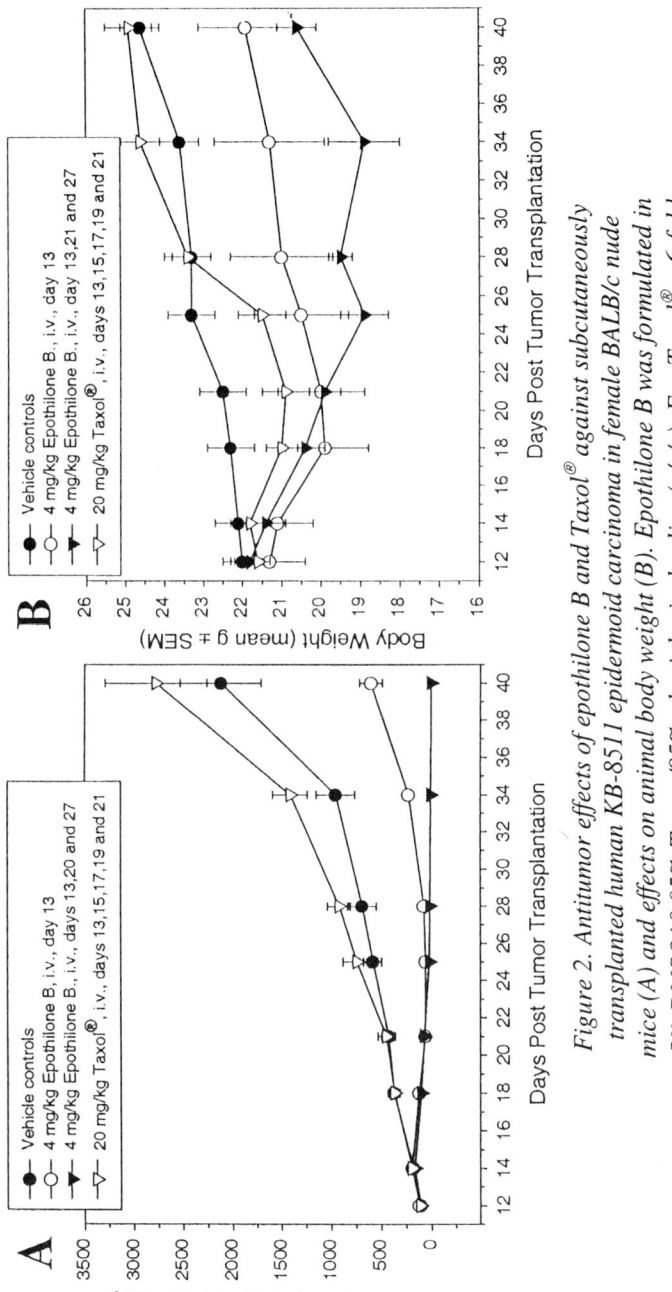

Figure 2. Antitumor effects of epothilone B and Taxol® against subcutaneously transplanted human KB-8511 epidermoid carcinoma in female BALB/c nude mice (A) and effects on animal body weight (B). Epothilone B was formulated in 5% DMSO/ 0.05% Tween/95% physiological saline (v/v/v). For Taxol® a 6-fold dilution of the clinical formulation with 5% w/v glucose in water was used. (Reprinted from Biochimica et Biophysica Acta 1470, Epothilones and related analogs – a new class of microtubule inhibitors with potent in vivo anti-tumor activity, Altmann, K.-H.; Wartmann, M.; O'Reilly, T., p. M79-M91, **2000** (ref. 3) with permission from Elsevier Science)

Synthesis and Biological Activity of Epothilone Analogs

Synthesis of Epothilone Fragments

The potent biological activity of epothilones combined with their *relatively* simple structures (at least in comparison to paclitaxel) has led to widespread interest in these natural products throughout the scientific community. Since the first disclosure of their absolute stereochemistry in 1996 (8), several total syntheses of epothilones have been published in the literature (9 - 11) and the methodology developed in the course of those studies has also been exploited for the synthesis of a host of synthetic analogs (9, 10). Our own work on synthetic epothilone analogs has primarily relied on the strategy developed by Schinzer *et al.* for the synthesis of epothilone B (12), which is outlined in Scheme 1.

While this approach does not represent the shortest known route to the natural products themselves, it has provided a very efficient framework for the synthesis of a variety of synthetic analogs for biological studies. Most importantly, we have been able to develop very efficient syntheses for the requisite building blocks utilized in this approach (or for related structures), which have been carried out on a multiple 100 g scale (Schemes 2 - 4).

As illustrated in Scheme 2 the key step in the synthesis of *C1-C6 fragments* entails aldol reaction between Oppolzer sultam **4** and β-keto aldehyde **3**, which provided **5** as a single diastereoisomer in 46% yield after simple recrystallization (13). TBS-protection of the 3-OH group, removal of the chiral auxiliary by basic hydrolysis, reduction of acid **7** with $BH_3 \times Me_2S$, and finally acetalization with acetone/CF_3COOH led to Schinzer ketone **9** (12, 14) in 36% overall yield (from **5**).

The synthesis of *C7-C11 fragments* is summarized in Scheme 3 and was again based on a chiral auxiliary approach, which in the stereogenic step involved methylation of chiral acyl-oxazolidinone **12** with methyl iodide (*cf.* also ref. 12). The reaction proceeded with *ca.* 13/1 stereoselectivity and diastereomerically pure **13** was obtained from the resulting mixture of isomers through recrystallization in 46% yield. Reduction of **13** with $LiAlH_4$ gave monoprotected diol **14**, which was then further elaborated into alkyl iodide **16**.

In contrast to the two cases discussed above, our strategy for the synthesis of *thiazole fragments C12-C15* utilized a chiral pool compound, (S)-malic acid, as the unequivocal source of chirality (Scheme 4) (15). Thus, malic acid was transformed into known α-hydroxy lactone **18**, which was subsequently elaborated into ketone **21** in 86% yield (from **18**). Horner-Emmons reaction of **21** with phosphonate **22** gave *trans* olefin **23**, which was selectively deprotected to **24** in 81% overall yield. This compound was then further converted to aldehyde **25** and acid **26**, which served as starting materials for the synthesis of *trans*-epothilones A and amide-modified deoxyepothilone B analogs, respectively (*vide infra*).

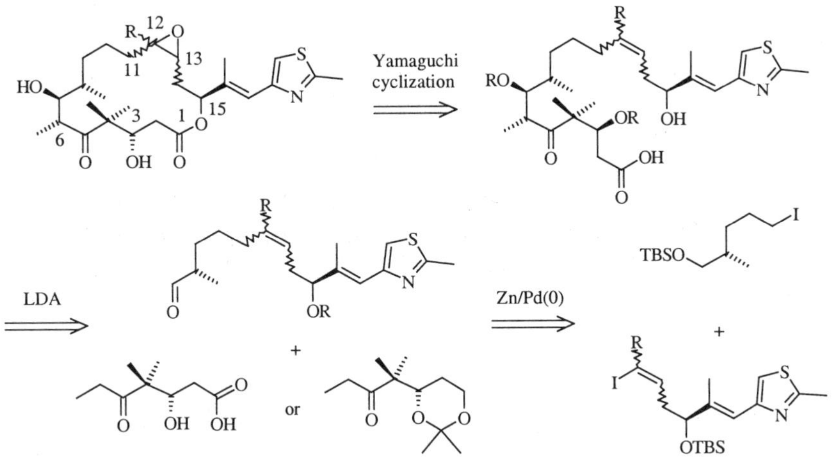

Scheme 1. Synthesis of epothilones and analogs - Generalized retrosynthesis

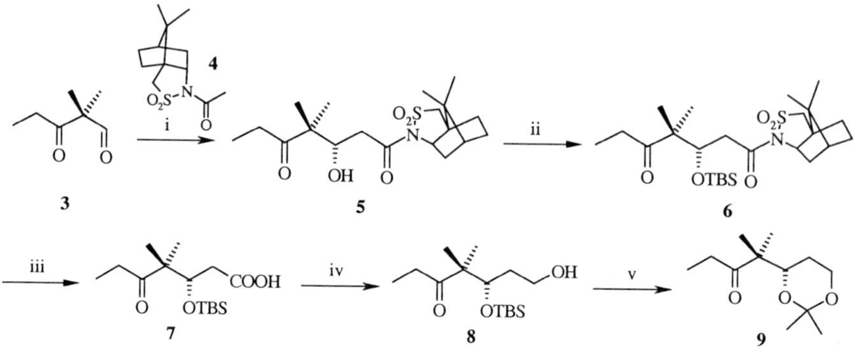

Scheme 2. i. a. Et$_2$B-OTf, (i-C$_3$H$_7$)$_2$NEt, 0°, 40 min. b. + 3, -78°, 2h, 46%. ii. TBS-OTf, -78°, 1 h, -78°→RT, 16h, 91%. iii. LiOH, H$_2$O$_2$, THF/H$_2$O 4/1, RT, 7h, 76%. iv. BH$_3$xMe$_2$S, B(OMe)$_3$, 0°, 2h, 15°, 6h, 56%. v. 5% CF$_3$COOH, acetone, 35°, 18h, 93%.

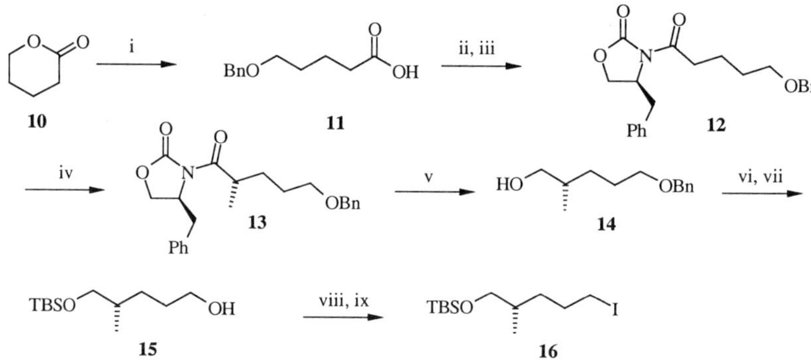

Scheme 3. i. KOH, Bn-Cl, toluene, refl., 16h, 75%. ii. (COCl)$_2$, DMFcat, toluene, RT, 2h, 50-60°, 1.75h, 95%. iii. (S)-4-Benzyl-oxazolidin-2-one, n-BuLi, THF, -78°→ 0°, 30 min, 81%. iv. a. NaHMDS, THF, -75°- -70°, 2.5h. b. CH$_3$I, THF, -78°, 2h, 46%. v. LAH, 5°, 20 min, RT, 1h, 71% vi. TBS-Cl, imidazole, DMF, RT, 89%. vii. H$_2$, Pd-C, MeOH/cyclohexane, 97%. viii. Mes-Cl, Et$_3$N, CH$_2$Cl$_2$, 0°, 15 min. ix. NaI, acetone, refl., 93% (2 steps).

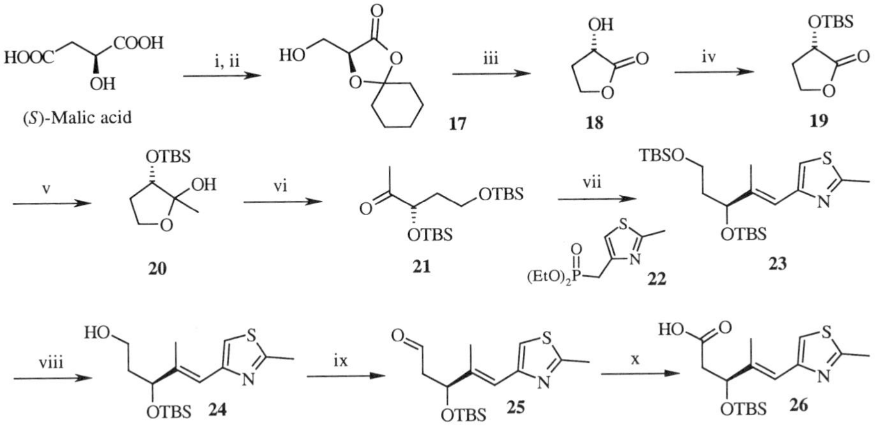

Scheme 4. i. Cyclohexanone, BF_3xEt_2O, Et_2O, RT, 95%. ii. BH_3xMe_2S, THF, RT, 24h. iii. p-TosOH, CH_2Cl_2, RT, 2.5h, 72% (2 steps). iv. TBS-Cl, DBU, CH_2Cl_2, 0°-3°, 1h, 92%. v. CH_3Li, Et_2O, -78°, 3h, 99%. vi. TBS-Cl, CH_2Cl_2, 0°-4°, 4h, 95%. vii. n-BuLi, THF, -70°→RT. viii. HF (40%), CH_3CN/Et_2O 1/1, glas powder, 5h, 81% (2 steps). ix. $(COCl)_2$, DMSO, Et_3N, CH_2Cl_2, -78°, 2h, 93%. x. $NaClO_2$, iso-butene, NaH_2PO_4, THF/ tert.-$BuOH/H_2O$, RT, 4h, 78%.

C12/C13 Amide- and Imidazole-Containing Epothilone Analogs

One of the most significant early observations on the epothilone SAR was the fact that *deoxyepothilones*, which incorporate a C12/C13 *cis* double bond in place of an epoxide moiety, are virtually equipotent inducers of tubulin polymerization *in vitro* as the corresponding parent epoxides and exhibit only moderately (*ca*. 10-fold) reduced antiproliferative activity (16, 17; M. Wartmann *et al.*, unpublished data). Based on these findings we hypothesized that the replacement of the epoxide moiety by structural units that would resemble the geometry of a *cis* C=C double bond should result in epothilone analogs with similar potencies as those of deoxyepothilones. The structural moieties that we felt most appropriate to test this hypothesis were N-alkyl amides (assuming a preference of the C-N partial double bond for a *cis* conformation) and 1,2-disubstituted heterocycles, such as imidazole (Figure 3). These polar double bond substitutes were also expected to result in improved aqueous solubility over the very lipophilic deoxyepothilone B.

As an example, Scheme 5 summarizes the synthesis of N-methyl amide analog **35**, which is largely based on the general principles outlined in Scheme 1. Thus, one of the key steps consists in the highly diastereoselective aldol reaction between aldehyde **27** and ketone **9** (*cf*. Scheme 2), which provided the desired

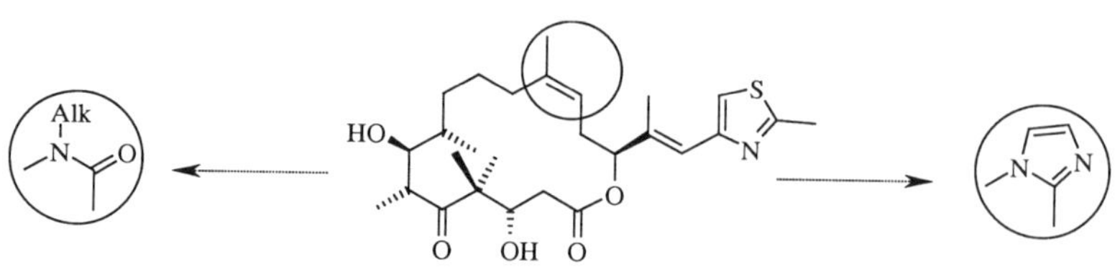

Figure 3. Epothilone analogs incorporating 12/13 N-alkyl amide and imidazole groups as potential deoxyepothilone B mimetics.

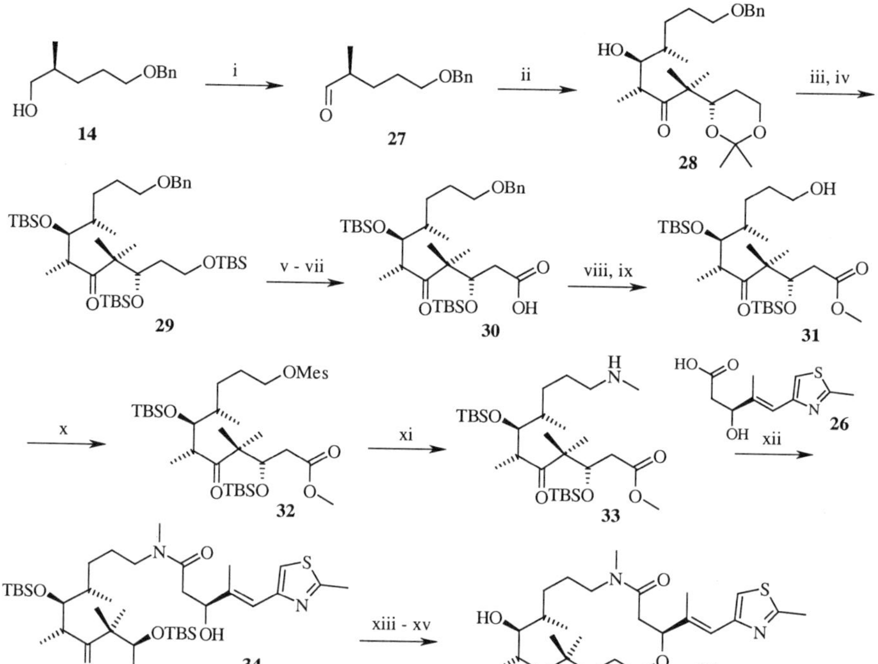

Scheme 5. i. (COCl)$_2$, DMSO, Et$_3$N, CH$_2$Cl$_2$, -78°, 67%. ii. a. **9** (2 equiv.), LDA, THF, -78°, 80 min; b. + aldehyde **27** (1 equiv.), -78°, 75 min, 82%. iii. PPTS, MeOH, RT, 22h, 83%. iv. TBS-OTf, lutidine, CH$_2$Cl$_2$, RT, 84%. v. CSA, MeOH/CH$_2$Cl$_2$, 0°, 1h, 80%. vi. (COCl)$_2$, DMSO, Et$_3$N, CH$_2$Cl$_2$, -78°, 93%. vii. NaClO$_2$, iso-butene, NaH$_2$PO$_4$, THF/ tert.-BuOH/H$_2$O, RT, 4h, 93%. viii. DCC, DMAP, MeOH, CH$_2$Cl$_2$, -20°→ RT, 4h, 71%. ix. H$_2$, Pd-C, MeOH, RT, atm. pressure, 1h, 80%. x. Mes-Cl, Et$_3$N, CH$_2$Cl$_2$, 0°, 1h, 91%. xi. CH$_3$NH$_2$, MeOH, 50°, 4h. xii. DCC, HOBt, (i-C$_3$H$_7$)$_2$NEt, DMF, 28% (2 steps). xiii. NaOH, MeOH/H$_2$O, 58%. xiv. 2.4.6-Cl$_3$C$_6$H$_2$C(O)Cl, Et$_3$N, DMAP, THF/toluene, 15 min, 60%. xv. HFxpyridine, THF, RT, 26 h, 90%.

diastereoisomer **28** in 82% yield after FC. Elaboration of **28** into N-methyl amine **33** and DCC/HOBt-mediated coupling with acid **26** (*cf.* Scheme 4) gave methyl ester **34** in 7% overall yield for the 9-step sequence from **28**. The synthesis was completed by ester saponification, Yamaguchi macrolactonization (18) of the resulting seco acid and removal of the TBS protecting groups with HF x pyridine to provide **35** in 31% yield (based on **34**).

Compound **35** as well as other C12/C13 amide- and imidazole-based analogs were evaluated for their ability to induce tubulin polymerization *in vitro* at 5 μM compound concentration relative to the effect of 25 μM of epothilone B. In addition, antiproliferative activity was assessed against human epidermoid cancer cell lines KB-31 and KB-8511, respectively, with KB-8511 being a P-gp overexpressing, multidrug-resistant sub-line of the KB-31 parental line (7). The results of these experiments are summarized in Table II. None of the analogs shown in Table II (including an N-unsubstituted secondary amide, which would mimic a *trans* olefin geometry) shows any appreciable tubulin polymerization or antiproliferative activity. This is in spite of the fact that preliminary NMR studies with compound **35** in DMSO/water indicate that the preferred conformation about the 12/13 N-methyl amide bond is indeed *cis*, i. e. with the methyl group and the carbonyl oxygen being located on the same side of the partial C-N double bond (*cis/trans* ratio ~ 4/1). It thus appears that the lack of activity of these analogs is related to the increase in steric bulk in the 12/13 region as compared with deoxyepothilone B (in particular at position 13) and/or the increase in polarity associated with either amide- or imidazole-based modifications. In this context it should be noted that Danishefsky's laboratory has recently reported a deoxyepothilone B analog incorporating a 12/13 *ortho*-phenylene moiety in place of the olefinic double bond, which was found to be several hundred-fold less active than deoxyepothilone B (20). This finding indicates that the steric component plays a significant, if not dominant, role for the reduced activity of our N-alkyl amide- and imidazole-based analogs (*cf.*, however, data in Table IV and the related discussion).

Trans-Epothilones A

Another intriguing feature of the epothilone SAR that was revealed during early SAR studies was the fact that even C12/C13 *trans*-analogs exhibit very potent tubulin polymerizing as well as antiproliferative activity (16, 17, 21). According to literature data it appears that *trans*-deoxyepothilone A is only slightly less active than deoxyepothilone A, whereas in the B series the activity difference is more pronounced. At the epoxide level contradictory data exist with regard to the relative activity of *trans*-epothilone B *vs.* epothilone B (17, 21); *trans*-epothilone A was reported to be virtually equipotent to epothilone A on an

Table II: Induction of tubulin polymerization and growth inhibition of human carcinoma cell lines by C12/C13 amide- and imidazole-modified epothilone analogs

A–B group	%Tubulin Polymerization[a]	IC50 KB-31 [nM][b]	IC50 KB-8511 [nM][b]
isopropenyl (C=CH−CH₃)	85	2.70	1.44
N-methyl amide	< 10	3670[c]	> 10000[c]
N-ethyl amide	< 10	5160[c]	> 10000[c]
imidazole	< 10	4652[c]	> 10000[c]
NH amide	< 10	> 10000[c]	> 10000[c]

[a]Induction of polymerization of porcine brain microtubule protein by 5 μM of test compound relative to the effect of 25 μM of epothilone B, which gave maximal polymerization (85% of protein input). Tubulin polymerization was determined using a centrifugation assay essentially as described in ref. 19. [b]IC50 values for growth inhibition of human epidermoid carcinoma cell lines KB-31 and KB-8511. For experimental conditions and methods cf. Table I. [c]Single determinations.

ovarian (1A9) and a breast cancer (MCF-7) cell line. However, in neither case was the absolute stereochemistry of the (active) epoxide isomer reported and the *trans*-isomers were obtained as minor components during the synthesis of the natural *cis* isomers rather than being the result of a directed synthetic effort. In view of the interesting biological features of *trans*-epothilones we embarked on a project directed at the stereoselective synthesis of *trans*-epothilones A, the determination of the absolute stereochemistry of the bioactive isomer, and a more exhaustive biological characterization of this compound.

The stereoselective synthesis of *trans*-deoxyepothilone A **39** is summarized in Scheme 6. While synthesis of this olefin proved to be rather straightforward, the critical step in the preparation of *trans*-epothilones was the stereo- and regioselective epoxidation of the C12/C13 *trans* double bond. This problem could be solved by the use of a fructose-based epoxidation catalyst developed by

Shi (22). Using the two enantiomeric forms of this catalyst, either isomer of *trans*-epothilone A was obtained with *ca.* 8/1 - 10/1 selectivity (Scheme 7).

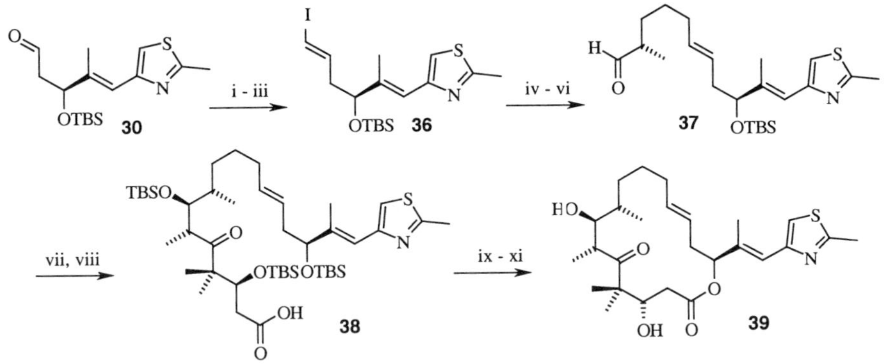

*Scheme 6. i. CBr$_4$, Ph$_3$P, CH$_2$Cl$_2$, 10°, 53%. ii. n-BuLi, THF, -78°, 63%. iii. a. Cp$_2$Zr(H)Cl, THF; b. I$_2$, 74%. iv. **16**, Zn-Cu, Pd(Ph$_3$P)$_4$, benzene, 60 - 80°, 64%. v. CSA/MeOH, 67%. vi. (COCl)$_2$/DMSO, CH$_2$Cl$_2$, - 78°, 92%. vii. **7**, LDA, -78°. viii. a. TBS-OTf, lutidine; b. K$_2$CO$_3$, MeOH, 30% (2 steps). ix. TBAF, THF, 42%. x. 2,4,6-Cl$_3$C$_6$H$_2$C(O)Cl, Et$_3$N, DMAP, THF/toluene, 61%. xi. CF$_3$COOH/ CH$_2$Cl$_2$, 91%.*

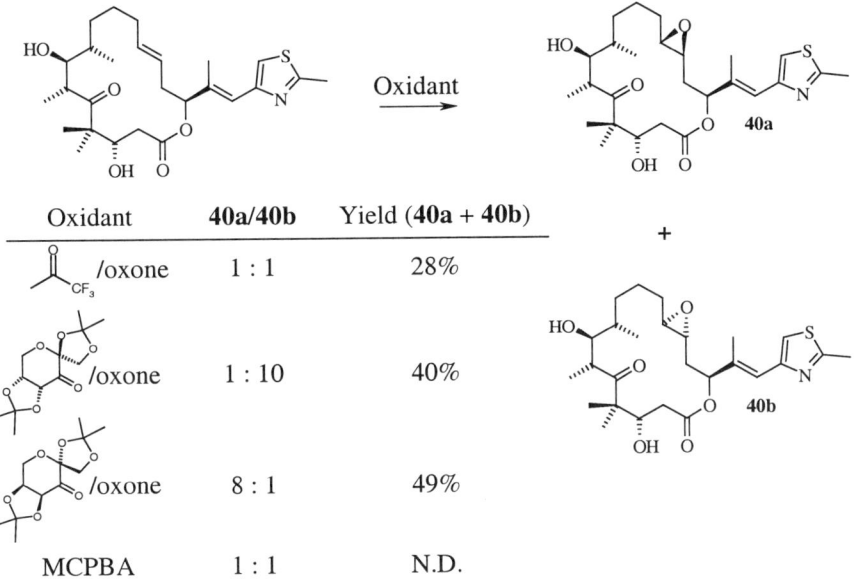

Oxidant	40a/40b	Yield (40a + 40b)
/oxone (CF$_3$)	1 : 1	28%
/oxone	1 : 10	40%
/oxone	8 : 1	49%
MCPBA	1 : 1	N.D.

Scheme 7: Synthesis of trans-epothilones A - Epoxidation selectivity.

The absolute stereochemistry of **40a** and **40b** was originally inferred from the known stereochemical preferences of the epoxidation catalysts (22) and in the case of **40a** was subsequently confirmed by X-ray crystallography (23).

Table III summarizes the biological activity of compounds **40a/b** in comparison to epothilones A and B. It is immediately clear that **40a**, which retains the natural stereochemistry at C13, is a strong inducer of tubulin polymerization *in vitro* and exhibits potent antiproliferative activity, whereas **40b** is at least 500-fold less active than **40a**. In fact, **40a** has shown slightly higher growth inhibitory activity than epothilone A across a wide range of human cancer cell lines, which makes this compound an interesting candidate for *in vivo* profiling.

Table III: Induction of tubulin polymerization and growth inhibition of human carcinoma cell lines by C12/C13 *trans*-epothilones A.[a]

A-B	%Tubulin Polymerization[b]	Growth inhibition (IC50 [nM])			
		KB-31	KB-8511	PC-3M[c]	A549[d]
[e]	85	0.19	0.19	0.53	0.09
[f]	63	2.00	1.79	4.30	1.07
[g]	85	1.00	0.87	2.15	0.39
[h]	< 10	523[j]	305[j]	N.D.[i]	N.D.[i]

[a]For experimental conditions and methods *cf.* Tables I and II. [b]Maximal polymerization (85% of protein input) was achieved with 25 μM epothilone B. [c]Human prostate carcinoma cell line. [d]Human lung carcinoma cell line. [e]Epothilone B. [f]Epothilone A. [g]**40a**. [h]**40b**. [i]Not determined. [j]Single determinations.

Analogs Derived from Epoxide Ring Opening

Scheme 8 summarizes some of the fundamental reactions involving the epoxide moiety of epothilone A, including acid-catalyzed hydrolysis, deoxygenation with 3-methyl-2-selenoxo-benzothiazole (24), *cis*-dihydroxylation of the deoxy derivative, and acetalization of the various diols. Analogous chemistry has been independently reported by Höfle *et al.* (25).

The various 12/13 *trans* and *cis* diols are devoid of any appreciable biological activity (IC50s > 1000 nM, data not shown), which may be a consequence of increased conformational flexibility and/or the presence of polar substituents at the 12/13 site. On the other hand, the antiproliferative activity of acetonides **42a** and **44a** is within one order of magnitude of that of epothilone A (Table IV), thus indicating that the ring fused to the C12-C13 bond may be significantly enlarged without dramatic loss in biological potency (26). This finding could represent an interesting starting point for the design of new epothilone analogs.

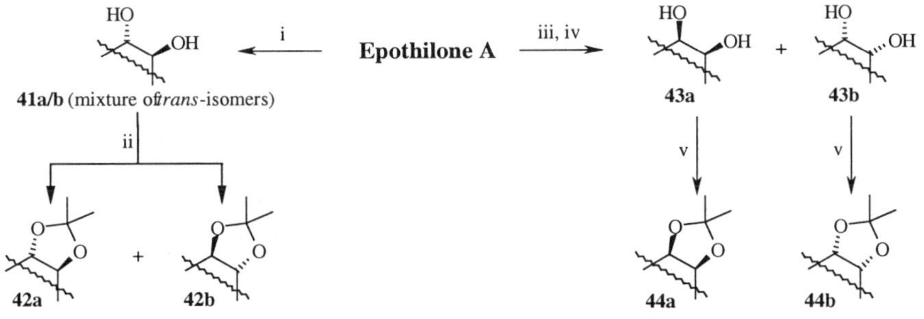

*Scheme 8. i. CF$_3$COOH/H$_2$O, 64% (mixture of isomers). ii. p-TosOH, acetone, **42a**: 28%; **42b**: 15%. iii. 3-Methyl-2-selenoxo-benzothiazole, CH$_2$Cl$_2$/CF$_3$COOH, 40%. iv. OsO$_4$, NMO, **43a**: 54%; **43b**: 8%. v. p-TosOH, acetone, **44a**: 43%; **44b**: 41%.*

Heterocycle Modifications

Modification of the heterocyclic side-chain of epothilones represents a potentially promising approach to modulate the physico-chemical, and perhaps also the pharmacokinetic properties of these natural poducts. One of the aspects that we were particularly focused on in this context was to what extent the replacement of the 5-membered thiazole moiety by different types of 6-membered aromatic rings would be tolerated and how the presence of methyl substituents on these rings would affect biological activity. We have thus prepared a series of variously substituted pyridine-based analogs of epothilone B,

Table IV: IC50-values [nM] for growth inhibition of human carcinoma cell lines by C12/C13-modified epothilone A analogs.[a]

Compound	IC50 KB-31 [nM]	IC50 KB-8511 [nM]
42a	23	10
42b	> 1000	> 1000
44a	30	17
44b	> 1000	> 1000

[a]For experimental conditions cf. legend of Table I.

whose synthesis was based on the Pd(0)-catalyzed coupling of the corresponding 2-pyridyl-stannanes with vinyl iodide 45 (Scheme 9) (27). The biological properties of pyridine-based analogs 46 are summarized in Table V and the data illustrate that the presence of a 5-membered heterocycle attached to C18 is not a prerequisite for epothilone B-like activity. Thus, compared to epothilone B, unsubstituted pyridine analog 46a displays less than two-fold decreased potency against KB-31 and KB-8511 cells. Most significantly, methylated analogs 46c and 46d are even more potent antiproliferative agents than 46a, and we have observed this rank order of activity across a broad range of human cancer cell lines. In contrast, methylation at position 3 (46b) significantly reduces the ability of this analog to induce tubulin polymerization and inhibit tumor cell growth. This finding is in line with recent solution NMR studies on the tubulin-bound conformation of epothilone A, which indicate that the olefinic double bond between C16 and C18 and the aromatic C=N double bond within the thiazole ring assume a transoid arrangement. In contrast, for epothilone A free in solution the transoid as well as the cisoid conformation are equally populated (28). Presumably the methyl group at the 3-position of the pyridine ring in 46b due to unfavorable steric interactions with C17 prevents the required transoid arrangement of the C16-C18 and the aromatic C=N double bond and thus leads to reduced biological activity.

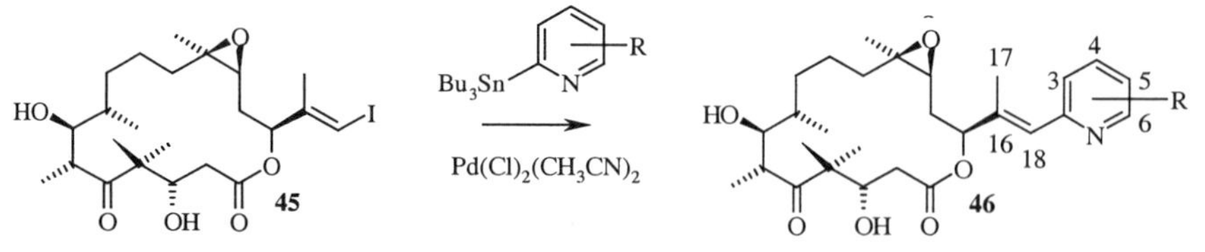

Scheme 9. Synthesis of side-chain modified epothilone B analogs.

Table V: Pyridine-based analogs of epothilone B: Biological activity[a]

R^b	Cpd.	%Tubulin Polymerization	IC50 KB-31 [nM]	IC50 KB-8511 [nM]
H	**46a**	79.5	0.30	0.30
3-Methyl	**46b**	12.0	39.4	50.5
4-Methyl	**46c**	90.3	0.16	0.16
5-Methyl	**46d**	89.3	0.11	0.10
6-Methyl	**46e**	52.8	3.7[c]	2.3[c]

[a]For experimental details cf. Tables I and II. [b]For structures cf. Scheme 9. [c]Single determination. Cf. also ref. 29.

In a second step we have investigated the importance of nitrogen positioning within the pyridine ring and we have also assessed the activity of pyrimidine- rather than pyridine based analogs. The results of these studies are summarized in Table VI.

It is clear from these data that for pyridine-based analogs the location of the ring N-atom *ortho* to the attachment point of the linker between the heterocycle

Table VI: Pyridine and Pyrimidine-based Analogs of Epothilone B: Importance of Nitrogen Positioning [a]

W, X, Y, Z	%Tubulin Polymerization	IC50 KB-31 [nM]	IC50 KB-8511 [nM]
CH, CH, N, CH	41.7	4.32	16.5
CH, CH, CH, N	34.5	11.8	34.7
CH, CH, CH, CH	74.4	2.88	1.27
N, CH, N, CH	47.3	8.78	17.6
CH, N, CH, N	28.0	14.9	49.5

[a]For experimental details cf. Table I.

and the macrocyclic skeleton is critical for epothilone B-like antiproliferative activity (i.e. sub-nM IC50s for growth inhibition). However, even with a N-atom in this obligatory position, incorporation of a second nitrogen either at the 3- or the 4- position of the 6-membered ring results in a profound decrease in antiproliferative activity. In fact, the corresponding analogs are even less potent than a compound whose side-chain incorporates a plain phenyl group rather than a heterocyclic ring (Table VI, W = X = Y = Z = CH). The origin of these effects is unclear and their true understanding will require detailed stuctural information on complexes between β-tubulin and the above epothilone analogs.

Conclusions

Epothilones A and B are potent inhibitors of human cancer cell growth *in vitro* with equal activity against drug-sensitive and multidrug-resistant cell lines. Antiproliferative activity is based on interference with microtubule functionality, which results in mitotic arrest and subsequent cell demise through apoptosis. Epothilone B in our hands has proven to possess potent *in vivo* antitumor activity at tolerated dose levels in several human xenograft models. Based on these data Novartis has initiated Phase I clinical trials with epothilone B which are currently ongoing. In the course of SAR studies a number of analogs have been identified which exhibit antiproliferative activity comparable to that of epothilones A/B, e. g. *trans*-epothilone A **40a** or pyridine-based analogs **46a**, **46c**, and **46d**. In addition, it is important to note that several of the compounds described in this report, although significantly less active than the corresponding parent epothilones, are still highly potent antiproliferative agents (e. g. **42a**, **44a**, **46e**). Whether any of these analogs may be superior drug candidates over epothilone B will primarily depend on their therapeutic index, i.e. on the *combination* of potency and tolerability, rather than plain potency alone. Studies to address this question are currently in progress in our laboratories.

References and Notes

1. Höfle, G.; Bedorf, N.; Gerth, K.; Reichenbach, H. German patent disclosure DE 4138042, May 5, 1993 (Priority Nov. 19, 1991). b) Gerth, K.; Bedorf, N.; Höfle, G.; Irschik, H.; Reichenbach, H. *J. Antibiotics* **1996**, *49*, 560-564.
2. Bollag, D. M.; McQueney, P. A.; Zhu, J.; Hensens, O.; Koupal, L.; Liesch, J.; Goetz, M.; Lazarides, E.; Woods, C. A. *Cancer Res.* **1995**, *55*, 2325-2333.

3. Altmann, K.-H.; Wartmann, M.; O'Reilly, T. *Biochimica et Biophysica Acta* **2000**, *1470*, M79-M91.
4. Kowalski, R. J.; Giannakakou, P.; Hamel, E. *J. Biol. Chem.* **1997**, *272*, 2534-2541.
5. Jordan, M. A.; Wendell, K.; Gardiner, S.; Derry, W. B.; Copp, H.; Wilson, L. *Cancer Res.* **1996**, *56*, 816-825.
6. Meyer, T.; Regenass, U.; Fabbro, D.; Alteri, E.; Rösel, J.; Müller, M.; Caravatti, G.; Matter, A. *Int. J. Cancer* **1989**, *43*, 851-856.
7. Akiyama, S.; Fojo, A.; Hanover, J. A.; Pastan, I.; Gottesmann, M. M. *Somat. Cell. Mol. Genet.* **1985**, *11*, 117-126.
8. Höfle, G.; Bedorf, N.; Steinmetz, H.; Schomberg, D.; Gerth, K.; Reichenbach, H. *Angew. Chem.* **1996**, *108*, 1671-1673; *Angew. Chem. Int. Ed. Engl.* **1996**, *35*, 1567-1569.
9. Nicolaou, K. C.; Roschangar, F.; Vourloumis, D. *Angew. Chem.* **1998**, *110*, 2120-2153; *Angew. Chem. Int. Ed. Engl.* **1998**, *37*, 2014-2045.
10. Harris, C. R.; Danishefsky, S. J. *J. Org. Chem.* **1999**, *64*, 8434-8456.
11. Mulzer, J.; Martin, H. J.; Berger, M. *J. Heterocycl. Chem.* **1999**, *36*, 1421-1436.
12. Schinzer, D.; Bauer, A.; Schieber, J. *Chem.-Eur. J.* **1999**, *5*, 2492-2500.
13. This strategy was originally suggested by de Brabander *at al.*: De Brabander, J.; Rosset, S.; Bernardinelli, G. *Synlett*, **1997**, 824-826. However, it should be noted that contrary to what is reported in this paper, the preparation of the desired (*3S*)-enantiomer 7 requires the use of the (*2R*)-bornane-10,2-sultam as chiral auxiliary.
14. Schinzer, D.; Limberg, A.; Böhm, O. M. *Chem. Eur. J.* **1996**, *2*, 1477-1482.
15. A related approach has been reported by Mulzer: Mulzer, J.; Mantoulidis, A.; Öhler, E. *Tetrahedron Lett.* **1998**, *98*, 8633-8636.
16. D.-S. Su, A. Balog, D. Meng, P. Bertinato, S. J. Danishefsky, Y.-H. Zheng, T.-C. Chou, L. He, S. B. Horwitz, *Angew. Chem* **1997**, *109*, 2178-2181; *Angew. Chem. Int. Ed. Engl.* **1997**, *36*, 2093-2096.
17. K. C. Nicolaou, D. Vourloumis, T. Li, J. Pastor, N. Winssinger, Y. He, S. Ninkovic, F. Sarabia, H. Vallberg, F. Roschangar, N. P. King, M. R. V. Finlay, P. Giannakakou, P. Verdier-Pinard, E. Hamel, *Angew. Chem* **1997**, *109*, 2181-2187; *Angew. Chem, Int. Ed.* **1997**, *36*, 2097-2103.
18. Inanaga, J.; Hirata, K.; Saeki, H.; Katsuki, T.; Yamaguchi, M. *Bull. Chem. Soc. Jap.* **1979**, *52*, 1989-1993.
19. Lin, C. M.; Yian, Y. Q.; Chadhary, A. G.; Rimoldi, J. M.; Kingston, D. G. I.; Hamel, E. *Cancer Chem. Pharm.* **1996**, *38*, 136-140.
20. Glunz P. W.; He L.; Horwitz S. B.; Chakravarty, S.; Ojima I.; Chou T.-C.; Danishefsky, S. J. *Tetrahedron Lett.* **1999**, *40*, 6895-6898.

21. Nicolaou, K. C.; Winssinger, N.; Pastor, J.; Ninkovic, S.; Sarabia, F.; He, Y.; Vourloumis, D.; Yang, Z.; Li, T.; Giannakakou, P.; Hamel, E. *Nature (London)* **1997**, *387*, 268-272.
22. Y. Tu, Z.-X. Wang, Y. Shi, Yian, *J. Am. Chem. Soc.* **1996**, *118*, 9806-9807.
23. Caravatti, G.; Rihs, G. unpublished data.
24. Calo, V.; Lopez, L.; Mincuzzi, A.; Pesce, G. *Synthesis* **1976**, 200-201.
25. Sefkow, M.; Kiffe, M.; Höfle, G. *Bioorg. Med. Chem. Lett.* **1998**, *8*, 3031-3036.
26. The absolute stereochemistry of compounds **42a/b**, and **44a/b** has *not* been explicitly determined. The stereochemical assignments shown in Scheme 8 are simply inferred from a comparison of the biological data obtained for **44a/b** and **42a/b** with the relative activities of the epothilone A/*epi*-epothilone A (the inactive (*12S, 13R*) isomer of epothilone A) pair and the *trans*-epothilone A pair **40a/40b** (Table II), respectively. It is assumed that for each relative stereochemsitry (*cis* or *trans*) the active isomers possess the same absolute stereochemistry at C12 and C13.
27. Nicolaou, K. C.; Scarpelli, R.; Bollbuck, B.; Werschkun, B.; Manuela, M.; Pereira, A.; Wartmann, M.; Altmann, K.-H.; Zaharevitz, D.; Gussio, R.; Giannakakou, P. *Chemistry & Biology* **2000**, in press.
28. Blommers, M. J. J. *et al.*, unpublished data.
29. We have previously reported 3-4-fold higher IC50-values for **46e** in the KB-31 and KB-8511 cell lines (27). The reason for this internal discrepancy is not entirely clear, but we consider the data provided in Table VI to be more reliable. The basic conclusions reported in ref. 27, however, remain unchanged.

Chapter 8

Synthesis and Biological Activity of Epothilones

Ulrich Klar, Werner Skuballa, Bernd Buchmann,
Wolfgang Schwede, Thomas Bunte, Jens Hoffmann, and
Rosemarie B. Lichtner

Research Laboratories of Schering AG, Müllerstrasse 170, D–13342 Berlin, Germany

The total synthesis and biological activity of epothilone analogs are described. Selected SAR data indicate the possibility to improve activity and selectivity by structural modifications. The new compounds may help to elucidate the therapeutic potential of this class of anticancer drugs.

Introduction

The discovery that the novel natural product class of epothilones parallels the biological activity of paclitaxel (PT) regarding its action on the tubulin system has stimulated intensive research activities in chemistry, pharmacology, and medicine. Epothilones, which have been isolated from myxobacterial strain *sorangium cellulosum* and characterized by the groups of Reichenbach and Höfle (1),stabilize microtubules by inhibiting their depolymerization, analogous to PT (2). As a consequence, cell cycle is blocked in G2/M phase driving the cell into apoptosis.

Epothilone A (R=H)
Epothilone B (R=CH₃)

Epothilone C (R=H)
Epothilone D (R=CH₃)

In contrast to PT, epothilones display their antiproliferative effects in so called multi-drug-resistant (MDR) cells *in vitro* and *in vivo* (3). This has been demonstrated for epothilone D (epo D) very impressively by the group of Danishefsky with a variety of xenograft models (4, 5, 6, 7).

Epothilone B (epo B), which is more potent than PT *in vitro* and *in vivo* displays significant toxicity at therapeutically relevant doses (7). Thus, an epothilone analog possessing an improved therapeutic window would be favorable.

Syntheses of Epothilone B and D Analogs

Due to the reduced structural complexity of epothilones compared to the taxoids, total syntheses offer the opportunity for extensive structural modifications most of which cannot be performed using a partial synthetic approach. Like other groups we made use of a highly convergent strategy using three modular building blocks designated as A, B and C which represent ring carbons 1 to 6, 7 to 12, and 13 to 15 respectively (8, 9).

As a prerequisite, each module should offer the potential for flexible structural modifications as well as a large scale production with high optical purity.

Synthesis of Building Block A (C1 - C6)

(-)-Pantolactone (**1**) was chosen as a readily available optically active starting material which possesses epothilone carbons 2 to 5 including the geminal dimethyl moiety at C-4 (Scheme 1). The hydroxyl group was protected as a tetrahydropyranyl (THP) ether (**2**) and the lactone was reduced to the lactol. After elongation by carbon atom 1 in a Wittig reaction, the remaining primary hydroxyl group was protected as t-butyldiphenylsilyl ether (TBDPS) and the THP ether was cleaved to yield allylic alcohol **4**. Olefin **4** was then subjected to a hydroboration-oxidation sequence giving 1,3-diol **5** along with minor amounts of stereochemically pure 1,2-diol **5a** resulting from a Markovnikov hydration (10). The 1,3-diol **5** was then converted into acetonide **6**. Removal of the silyl ether, Swern-oxidation and subsequent Grignard reaction followed by an oxidation of the resulting alcohol epimers gave ketones **8a** to **8k**. If desired, the acetonide in **8** can be converted into the bis-t-butyldimethylsilyl ether (e.g. **30**). In most cases, a better stereoselectivity in the aldol reaction is observed using the acetonide (Scheme 4).

This chiral pool synthesis of building block A is characterized by the facts that
- no racemization occurs during synthesis;
- all intermediates are chemically stable;
- the sequence can be scaled up;
- there is flexibility for synthetic modifications at C-5.

Synthesis of Building Block B (C7 - C12)

Starting from 1,4-butynediol, triflate **10** was reacted with the Evans oxazolidinone **11** to yield the acyloxazolidinone **12** with high diastereoselectivity (Scheme 2). Transesterification followed by hydrogenation and reduction yielded the primary alcohol **14** in enantiomerically pure form.

To introduce the methyl group at C-12, protecting group manipulation to **15** was followed by oxidation and subsequent reaction with methyllithium to afford key alcohol **16**. By oxidation to ketone **18** a fragment suitable for the connection to building block C was generated. In order to connect to building block A, alcohol **16** was converted to aldehyde **17** by silylation, cleavage of the tetrahydropyranyl ether and subsequent oxidation.

This synthesis of building block B is characterized by the features that
- all intermediates are chemically stable and in particular crystalline;
- the sequence can be scaled up;
- the chiral auxiliary can be recycled;
- there exists high flexibility for synthetic modifications (e.g. at C-8, C-10 to C-12).

Scheme 1

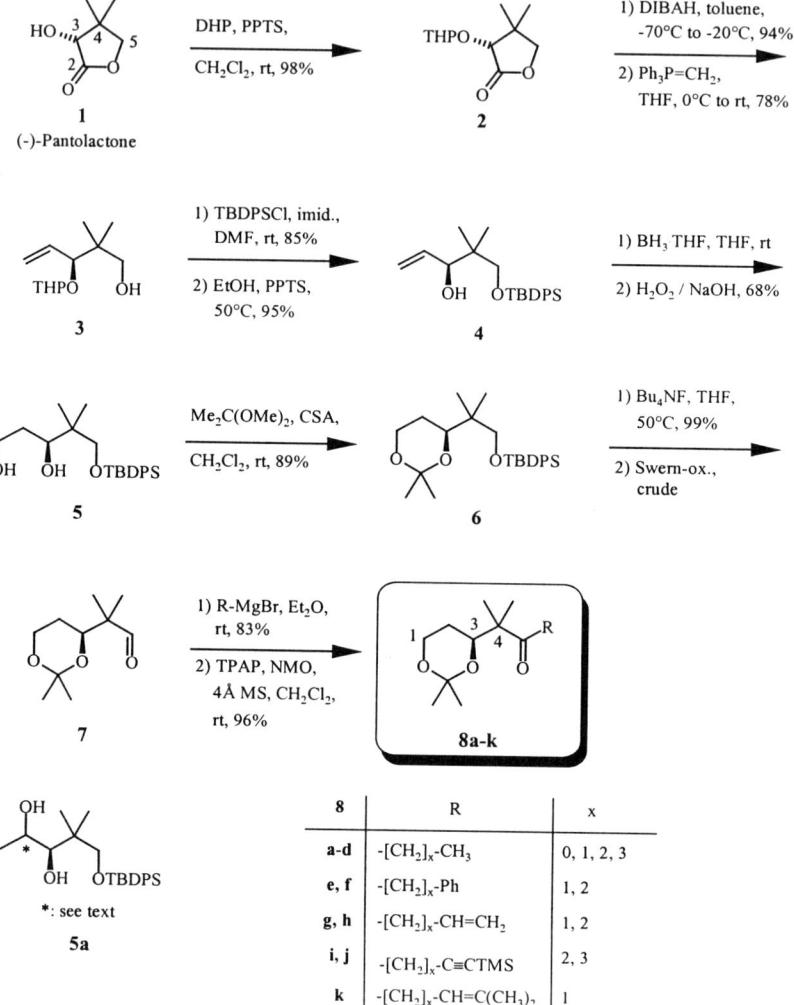

135

Scheme 2

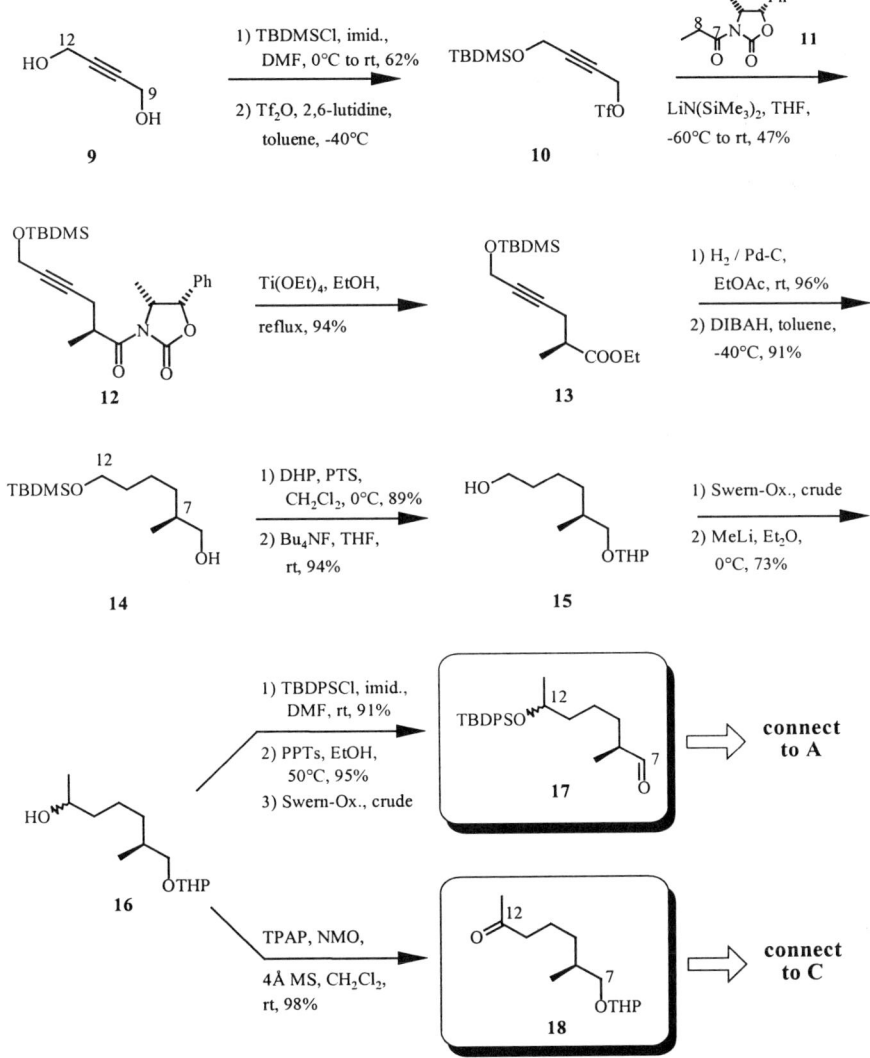

Synthesis of Building Block C (C13 - C17)

As described for building block A, the chiral pool was also used for preparation of building block C (Scheme 3). Starting with **L(-)Malic acid**, (2S)-hydroxybutyrolactone **20** was prepared in 4 steps according to the literature (11). After protection of the hydroxyl group as t-butyldiphenylsilyl ether, addition of methyllithium at −70°C afforded **21** in equilibrium with its open chain isomer. Silylation of the primary hydroxyl function gave methyl ketone **22**. The thiazole moiety was introduced by a Horner-Wittig reaction. Afterwards, the t-butyldimethylsilyl ether was removed under acidic conditions and the primary alcohol **24** converted to phosphonium salt **25** via the corresponding iodide.

Scheme 3

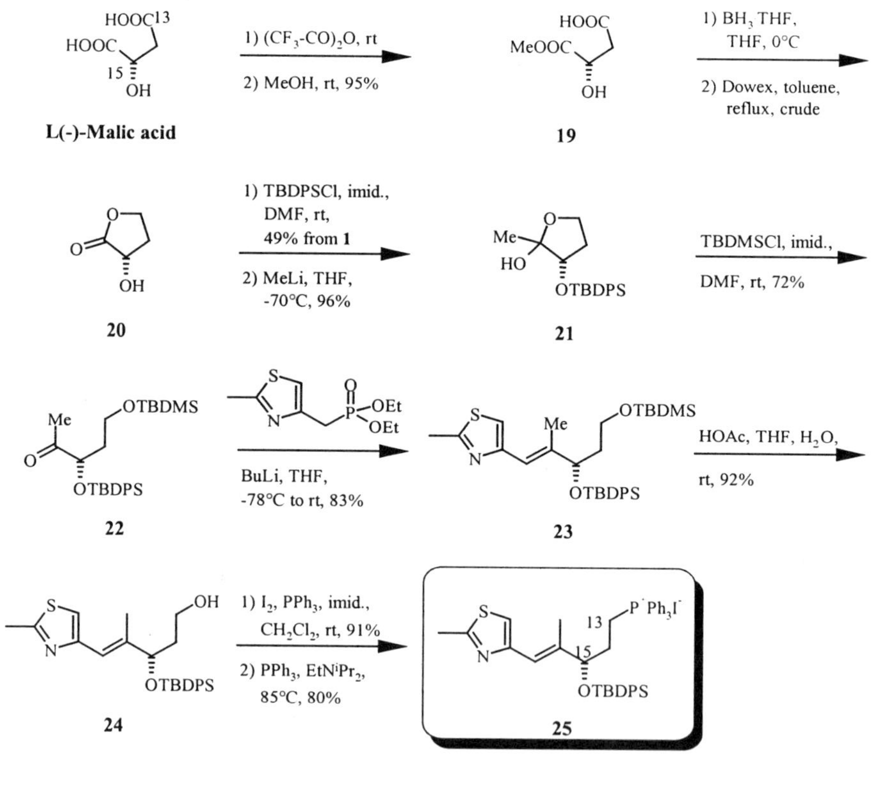

The synthesis of building block C is characterized by the facts that
- no racemization occurs during synthesis;
- all intermediates are chemically stable;
- the sequence can be scaled up;
- there is flexibility for modifications in the side chain (C-16 and aryl).

Construction of the Framework

With all three fragments in hand two routes can be followed to complete the carbon skeleton as depicted in Figure 1. By this strategy the most valuable building block A or C may be introduced last.

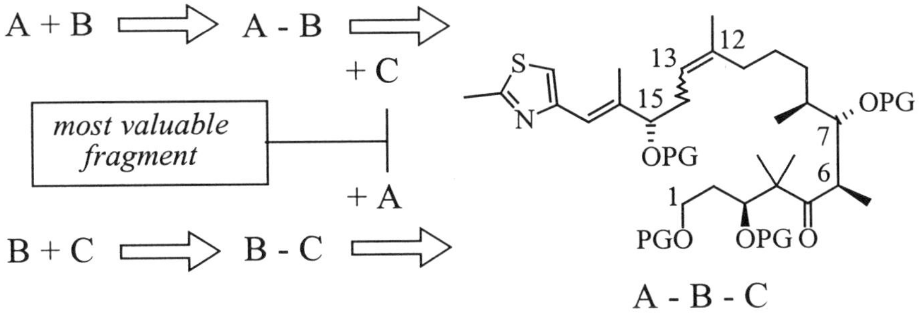

Figure 1

Although this general sequence is well established in several laboratories, the total synthesis of compounds **35** to **37** in which the methylene groups at C-9 to C-11 are replaced by a phenyl ring is shown as an example (Scheme 4). The modified building block B is synthesized in a straight-forward manner similar to the one described above.

Wittig reaction between ketone **27** (12) and the ylide generated from salt **26** afforded the B-C-fragment **28** as a nearly 1:1 mixture of E/Z-isomers. The tetrahydropyranyl ether was cleaved and the alcohol oxidized to aldehyde **29** which was coupled with building block A (**30**) in an aldol reaction to yield **31** with high selectivity along with minor amounts of its diastereoisomer **32** which was removed by chromatography. The stereochemistry was assigned in analogy to the synthesis of the natural product epo B. The newly formed secondary alcohol was protected and the primary silyl ether selectively removed under mild acidic conditions to give **33** which was oxidized in two steps to the corresponding carboxylic acid (**34**). At this stage the Z/E-isomers **34-Z** and **34-E** were separated very easily by chromatography. The double bond configuration can be assigned undoubtedly by NOE experiments. Next, the allylic alcohol was liberated and the crude hydroxy acid was subjected to Yamaguchi cyclization conditions (13). The remaining protecting groups were removed to yield the modified epo D analog **35-Z**. Epoxidation of the double bond afforded the α-epoxide **36-Z** with high stereoselectivity along with minor amounts of stereoisomer **37-Z**. Starting from acid **34-E** the corresponding double bond isomer **35-E** was obtained. In contrast to the Z-series, the epoxidation to **36-E** and **37-E** was less selective (structures not shown). Compared with their natural counterparts, the biological activity of analogs **35** to **37** was reduced.

Scheme 4

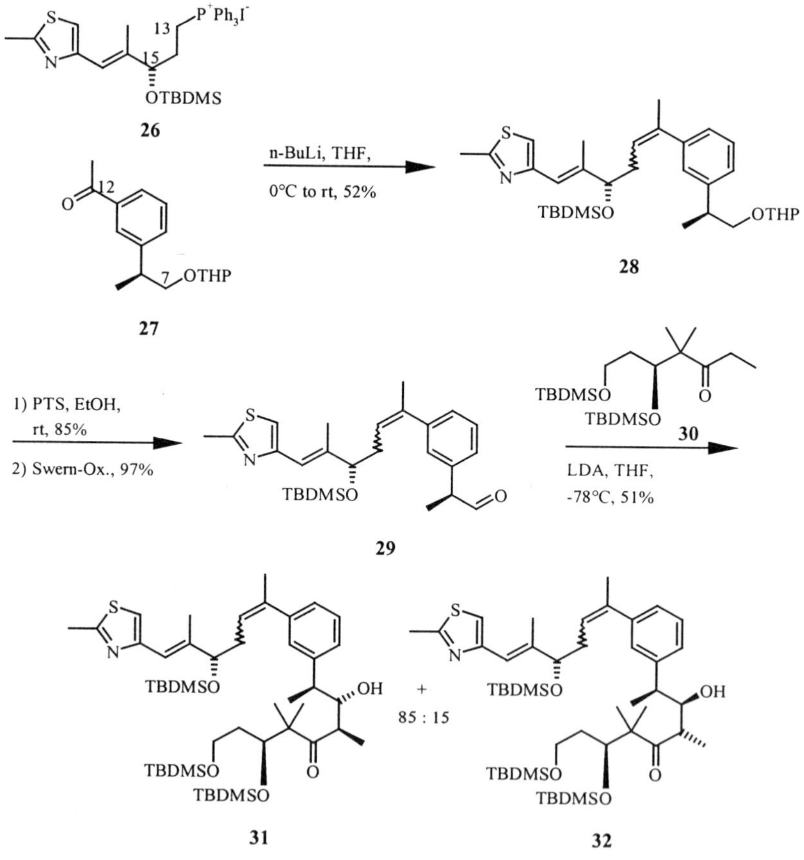

Scheme 4. *Continued.*

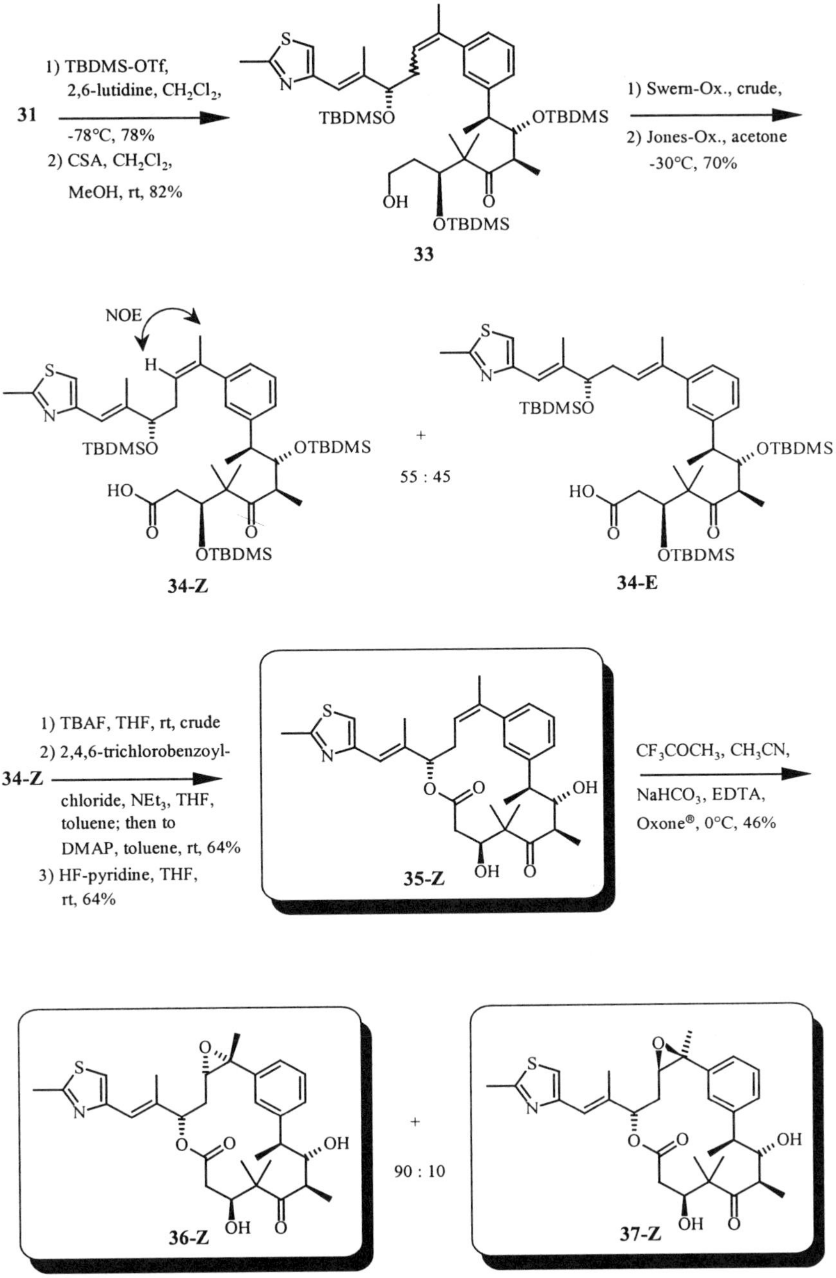

Structure Activity Relationships

Following the above strategy we have synthesized a wide range of analogs of epo B and epo D, most of which cannot be generated by partial synthesis. In this section, some selected aspects regarding structure-activity-relationships will be discussed (14).

Structure and MDR-Sensitivity

One major advantage of epothilones compared with other established anticancer drugs is their ability to overcome MDR. Nevertheless, a closer investigation reveals that this does not apply for all analogs. In the course of our drug-finding program we have synthesized compounds bearing N-aryl-oxides in the side chain.

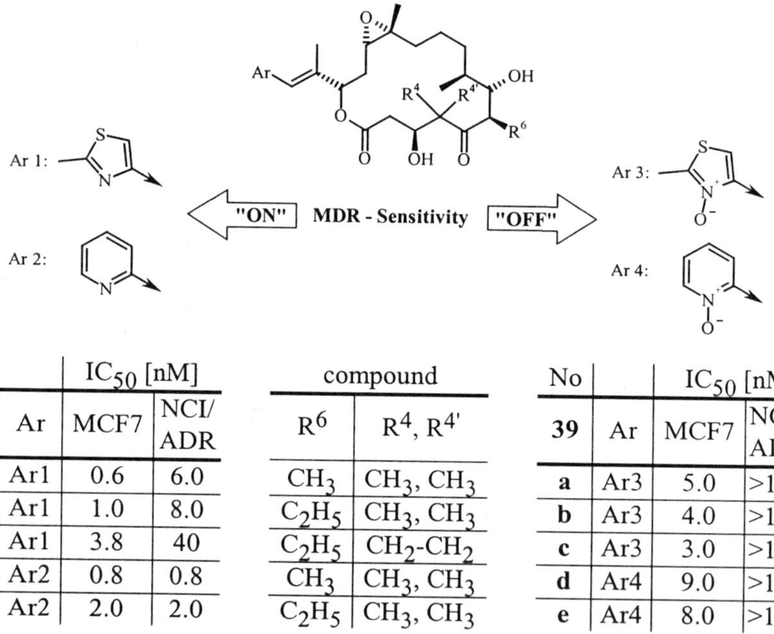

No 38	Ar	IC$_{50}$ [nM] MCF7	NCI/ADR	compound R^6	R^4, R$^{4'}$	No 39	Ar	IC$_{50}$ [nM] MCF7	NCI/ADR
a*	Ar1	0.6	6.0	CH$_3$	CH$_3$, CH$_3$	a	Ar3	5.0	>100
b	Ar1	1.0	8.0	C$_2$H$_5$	CH$_3$, CH$_3$	b	Ar3	4.0	>100
c	Ar1	3.8	40	C$_2$H$_5$	CH$_2$-CH$_2$	c	Ar3	3.0	>100
d	Ar2	0.8	0.8	CH$_3$	CH$_3$, CH$_3$	d	Ar4	9.0	>100
e	Ar2	2.0	2.0	C$_2$H$_5$	CH$_3$, CH$_3$	e	Ar4	8.0	>100

*: **38a** = epo B

Figure 2

As demonstrated in Figure 2, their activities against MCF-7 cells cover a range of 3 to 9 nM compared to 0.6 to 3.8 nM for the corresponding desoxy heterocycles. While the latter ones are also active against the MDR cell line NCI/ADR (0.8 to 8.0 nM) the N-oxides show no inhibition of cell proliferation up to 100 nM. Thus, the introduction of charges leads to a loss in sensitivity against MDR-positive tumor cells.

Interestingly, even an increase in polarity e.g. by replacing the 12-methyl by a hydroxymethylene group or the exchange of the lactone moiety in epo D by a lactam (15) leads to this effect (data not shown) indicating that these analogs represent a better substrate for the efflux pump P-glycoprotein (P-gp).

To address the hypothesis that overcoming P-gp efflux might also be responsible for enhanced toxicity observed for epo B, the following in vivo experiment was performed. Nude mice bearing a human colon carcinoma (LS 174T) were treated with equi-efficient doses of epo B and its corresponding N-oxide. As shown in Figure 3, similar toxicity was observed at therapeutically relevant doses for both compounds indicating that overcoming P-gp efflux may not be responsible for the narrow therapeutic window seen for epo B in this experiment.

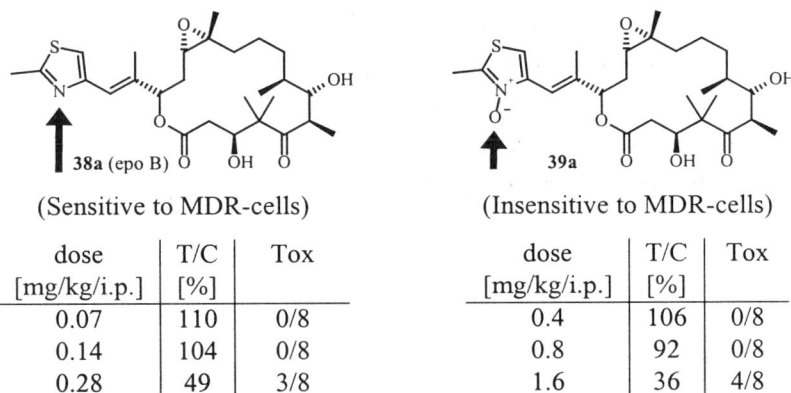

(Sensitive to MDR-cells) (Insensitive to MDR-cells)

dose [mg/kg/i.p.]	T/C [%]	Tox	dose [mg/kg/i.p.]	T/C [%]	Tox
0.07	110	0/8	0.4	106	0/8
0.14	104	0/8	0.8	92	0/8
0.28	49	3/8	1.6	36	4/8

Figure 3

Improvement of Activity

While epo B possesses very impressive antiproliferative effects in vitro the potency can be further improved by simply replacing the methyl at carbon 6 by an ethyl group. Compared to epo B this analog displayed at least a 4-fold enhanced overall activity in a panel of tumor cell lines listed in Figure 4.

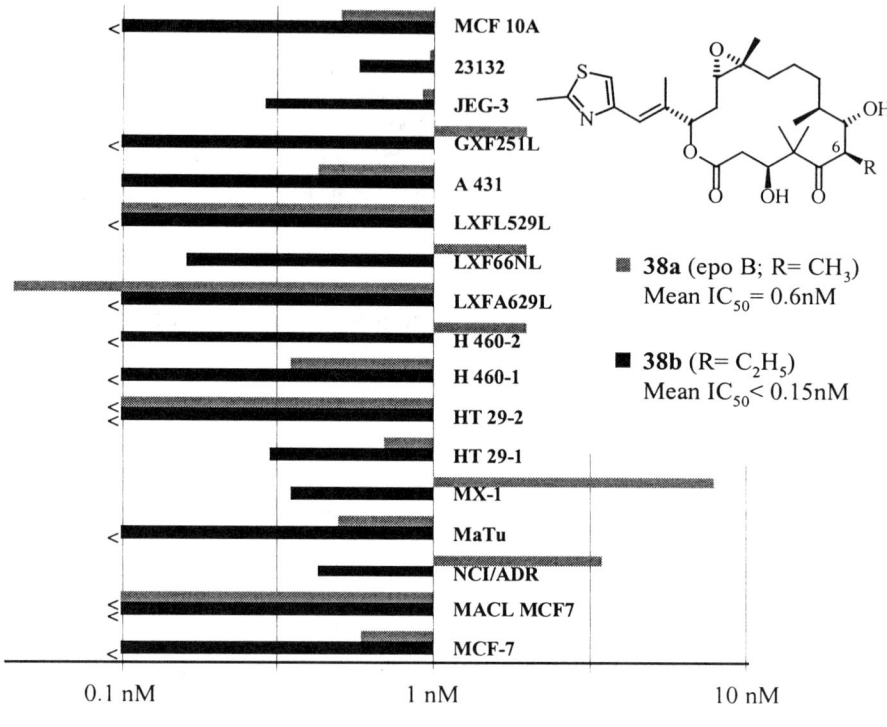

Figure 4

Bioactive Conformation of the Side Chain

For a better understanding of published as well as our own SAR-data we were also interested in getting information about the bioactive conformation of the epothilone side chain. It was already known that removal or shortening of the side chain resulted in a dramatic loss in activity (16).

To understand if the position of nitrogen in the heterocycle is important and if the bioactive conformation is more likely represented by conformer I or II, only three analogs needed to be synthesized (Figure 5):

- In the first analog (**40**) the methylthiazole was replaced by 2-pyridyl. The in vitro activity of this compound is similar or even improved, compared to epo D.
- In the second analog (**41**) a 3-pyridyl moiety is incorporated as heterocycle leading to a significant decrease in activity indicating that the position of nitrogen is crucial.
- In the third analog (**42**) the double bond of the side chain is incorporated into a phenyl ring leading to a fixed conformation I. The high biological activity of this compound which is superior to epo D reveals that the bioactive conformation seems to be well described by conformer I.

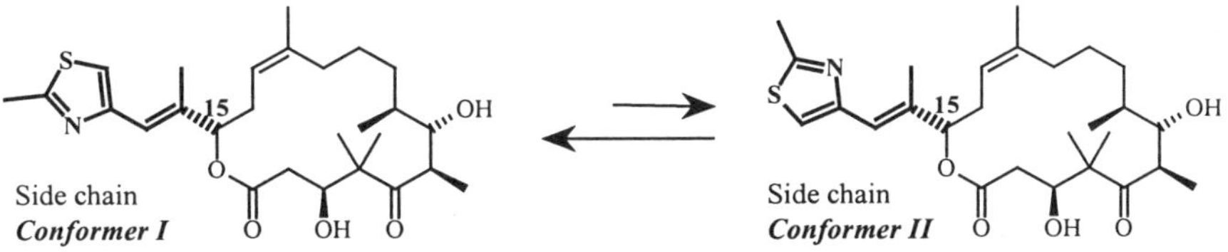

Side chain	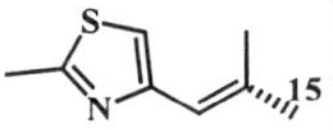 ⟨thiazole⟩	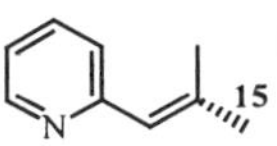 ⟨2-pyridyl⟩	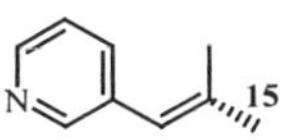 ⟨3-pyridyl⟩	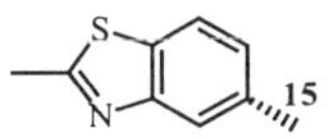 ⟨benzothiazole⟩
Compound	**Epo D**	40	41	42
MCF-7; IC_{50} [nM]	19	6	> 100	3

Figure 5

Improvement in Activity and Profile

In general, the benzothiazole, as an epothilone side chain equivalent, turned out to enhance activity in vitro. Compared to epo D the corresponding analog was 5-, 8- and 14-fold more active on A 431, MCF-7 and NCI/ADR cell lines respectively (Figure 6). Similar results were obtained with the epo B analog **43** showing an 1-, 2- and 9-fold improvement. More interesting, the activity towards the MDR cell line was improved for both analogs. This is demonstrated by the defined selectivity ratio (IC_{50}-MCF7 : IC_{50}-NCI/ADR) which increases for the olefin compound from 0.46 to 0.78 and for the epoxide from 0.17 to 0.75, respectively.

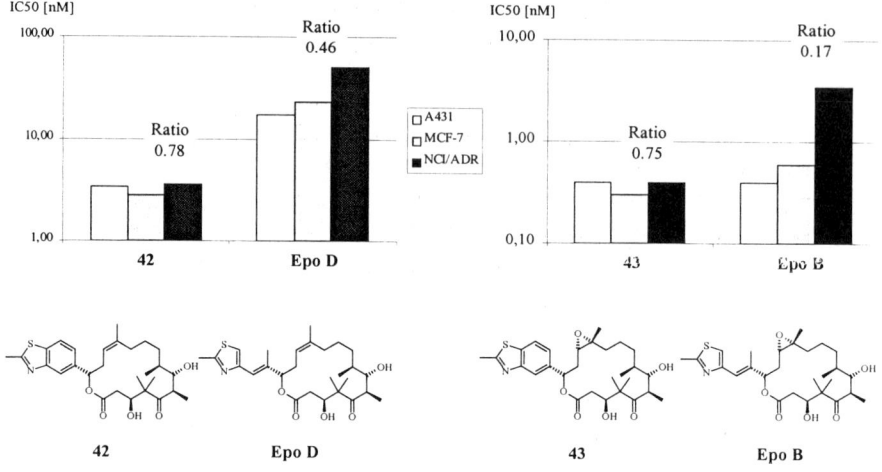

"Ratio" is defined as IC_{50}-MCF7 : IC_{50}-NCI/ADR

Figure 6

Radiolabelled Epothilone B and D Analogs

Data from the literature (7) as well as our own experience suggested that the olefin epo D possesses a broader therapeutic window compared to the epoxide epo B in vivo, the latter being about 10-times more potent in vitro.

To study possible differences (e.g. at the cellular level) between these compounds, we decided to synthesize some closely related analogs of epo D and epo B, which can be labelled at the very last step of the synthesis.

Synthesis and In Vitro Characterization

The synthesis of our target compounds **44** and **46** followed the sequence fragment BC + fragment A as already described before. In these analogs the sterically less hindered mono substituted double bond was hydrogenated or tritiated (17) with high regioselectivity providing **45** and **47** (Scheme 5).

Scheme 5

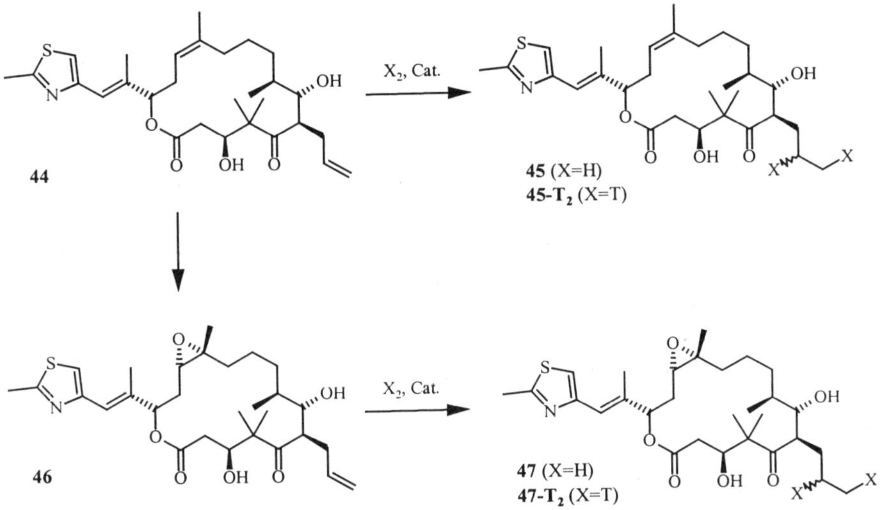

Table 1

Compound	MCF-7 [nM]	NCI/ADR [nM]
44	17	41
45	38	76
epo D	19	50
46	1.2	3.8
47	3.4	4.1
epo B	0.6	3.5

The in vitro characterization revealed that compound **44** was equipotent to epo D while **45** showed a slightly reduced activity in MCF-7 and NCI/ADR cells (Table 1). The epoxides **46** and **47** are equipotent to epo B in the MDR cell line, while their activity was only slightly reduced in MCF-7 cells. Thus, compounds **45** and **47** represent appropriate tools for further studies.

Indications for a Different Mode of Action Between Epothilone B and D

The labelled analogs **45-T$_2$** and **47-T$_2$** were used to study their cellular distribution. As can be seen from Figure 7 (right panel) the total uptake of epoxide **47** is about 4-fold higher compared to olefin **45**. Looking to the relative cellular distribution (left panel) nearly all of olefin **45** is detected in the protein fraction of the cytosol as is to be expected for a compound binding to tubulin. A qualitatively different picture is obtained for epoxide **47** which was also found in the cell nucleus.

Studies are ongoing to figure out whether these results translate to the different activities/selectivities seen between epo D and epo B.

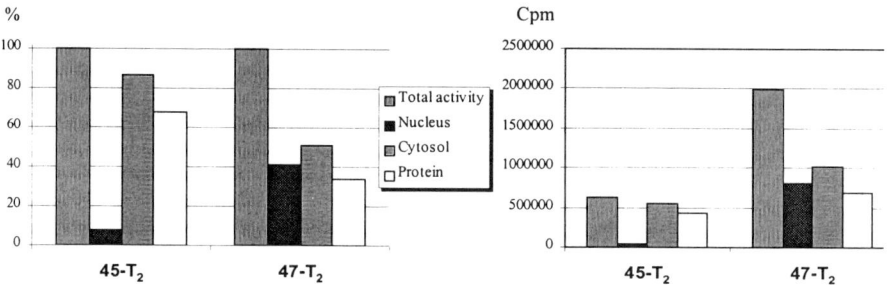

Figure 7

In conclusion, epothilones represent both structurally and mechanistically a truly novel entity among the tubulin active compounds. Analogs obtained by total syntheses may help to elucidate the therapeutic potential of this class of anti cancer drugs.

References

1 Gerth, K.; Bedorf, N.; Höfle, G.; Irschik, H.; Reichenbach, H. *J. Antibiotics* **1996**, *49*, 560-563.
2 Bollag, D. M.; McQueney, P. A.; Zhu, J.; Hensens, O.; Koupal, L.; Liesch, J.; Goetz, M.; Lazarides, E.; Woods, C. M. *Cancer Res.* **1995**, *55*, 2325-2333.
3 Altmann, K.-H.; Wartmann, M.; O'Reilly, T. *Biochim. Biophys. Acta* **2000**, *1470*, M79-M91.
4 Chou, T.-C.; Zhang, X.-G.; Balog, A.; Su, D.-S.; Meng, D.; Savin, K.; Bertino, J. R.; Danishefsky, S. J. *Proc. Natl. Acad. Sci.* **1998**, *95*, 9642-9647.
5 Chou, T.-C.; Zhang, X.-G.; Harris, C. R.; Kuduk, S. D; Balog, A.; Savin, K. A.; Bertino, J. R.; Danishefsky, S. J. *Proc. Natl. Acad. Sci.* **1998**, *95*, 15798-15802.
6 Sepp-Lorenzino, L.; Balog, A.; Su, D.-S.; Meng, D.; Timaul, N.; Scher, H. I.; Danishefsky, S. J.; Rosen, N. *Prostate Cancer Prostatic Dis.* **1999**, *2*, 41-52.
7 Harris, C. R.; Danishefsky, S. J. *J. Org. Chem.* **1999**, *64*, 8434-8456.
8 Mulzer, J. *Monatshefte für Chemie* **2000**, *131*, 205-238.
9 Nicolaou, K. C.; Roschangar, F.; Vourloumis, D. *Angew. Chem. Int. Ed. Engl.* **1998**, *37*, 2014-2045.
10 A similar observation has been recently described for 3,4-dialkoxy-1-alkenes: Jung, M. E.; Karama, U. *Tetrahedron Lett.* **1999**, *40*, 7907-7910.
11 Wünsch, B.; Dieckmann, H.; Höfner, G. *Liebigs Ann. Chem.* **1993**, 1273-1278.
12 Ketone 27 was synthesized using a similar methodology which is described in detail in WO 00/49020.
13 Inanaga, J.; Hirata, K.; Saeki, H.; Katsuki, T.; Yamaguchi, M. *Bull. Chem. Soc. Jpn.* **1979**, *52*, 1989-1993.
14 In vitro cell proliferation assays were performed according to: Kueng, W.; Silber E.; Eppenberger U. *Analyt. Biochem.* **1989**, *182*, 6-19.
15 Stachel, S. J.; Chappell, M. D.; Lee, C. B.; Danishefsky, S. J. *Org. Lett.* **2000**, *2*, 1637-1639.
16 Su, D.-S.; Balog, A.; Meng, D.; Bertinato, P.; Danishefsky, S. J.; Zheng, Y.-H.; Chou, T.-C.; He, L.; Horwitz, S. B. *Angew. Chem. Int. Ed. Engl.* **1997**, *36*, 2093-2096.
17 We thank Dr. Jürgen Gay, Research Laboratories of Schering AG, for the tritiation experiments.

Chapter 9

Epothilones and Sarcodictyins: From Combinatorial Libraries to Designed Analogs

Nicolas Winssinger and K. C. Nicolaou

Department of Chemistry and The Skaggs Institute for Chemical Biology, The Scripps Research Institute, 10550 North Torrey Pines Road, La Jolla, CA 92037

The epothilones and sarcodictyins have shown to be potent cytotoxic compounds with a Taxol™-like mode of action. The clinical importance of Taxol™ has stimulated intense research into finding compounds with prospective superior chemotherapeutic properties to Taxol™. We have developed enabling synthetic technologies for the construction of combinatorial libraries of these two potentially important chemotherapeutics, which led, in both cases to compounds with superior biological activity.

Introduction

Taxol™ (**1**, Figure 1) and its relative Taxotere™ (**2**, Figure 1) have had a striking impact in the treatment of cancer. The discovery by Horwitz and coworkers (1) that the desirable cytotoxic properties of Taxol™ stem from a novel and unique mode of action involving stabilization of microtubules which results in cell cycle arrest and subsequent apoptosis incited intense research efforts in this area (2). Despite their clinical success, cancer treatment with Taxol™ or Taxotere™ has its limitations, however. The most notable short-

This chapter is dedicated to the memory of the late Doretha Hewitt.

comings of these agents are their low aqueous solubility and the development of drug resistance mediated by both the overexpression of P-glycoprotein and mutation of β-tubulin. The search for chemotherapeutics with improved activity remains a high priority and has stimulated an intense search for new compounds with a similar mode of action but better pharmacological profiles (see Figure 1).

Biology of epothilones and sarcodictyins. Epothilones A (**3**, Figure 1) and B (**4**, Figure 1) (3) both have a Taxol™-like mode of action which has been extensively reviewed in the previous chapters and elsewhere (4). Hailed as potentially superior chemotherapeutics to Taxol™, the epothilones have attracted considerable attention in the scientific community (5). Indeed, the epothilones are more water-soluble than Taxol™ and, more importantly, they retain their activity against Taxol™-resistant tumor cells and multi-drug resistant cell lines (MDR). Sarcodictyins A (**6**, Figure 1) and B (**7**, Figure 1) also share these important features. Although the sarcodictyins were isolated more than a decade ago (6), their biological potential was not realized until more recently. In 1997, researchers from Pharmacia-Upjohn (7) drew attention to the sarcodictyins by demonstrating their Taxol™-like mode of action. They showed that these compounds compete with H^3-labeled Taxol™ in a tubulin-binding assay and, more importantly, they found them to have low resistance index to the MDR cell lines suggesting that, as is the case with the epothilones, they are poor substrates for the P-glycoprotein efflux pump. Later studies by Hamel and coworkers confirmed these findings and further demonstrated that the sarcodictyins also retained their activity against Taxol™-resistant cell lines overexpressing β-tubulin (8). In efforts to gain a better understanding of the structural requirements for binding β-tubulin, several groups are exploring the overlap in the pharmacophore between the various members of the tubulin polymerizing family (9).

Epothilones

The attractive biological profile of the epothilones prompted our group to develop a rapid and concise synthesis which would not only provide access to the natural product for initial biological studies but also allow a rapid investigation of their structure-activity relationships. Our first approach is

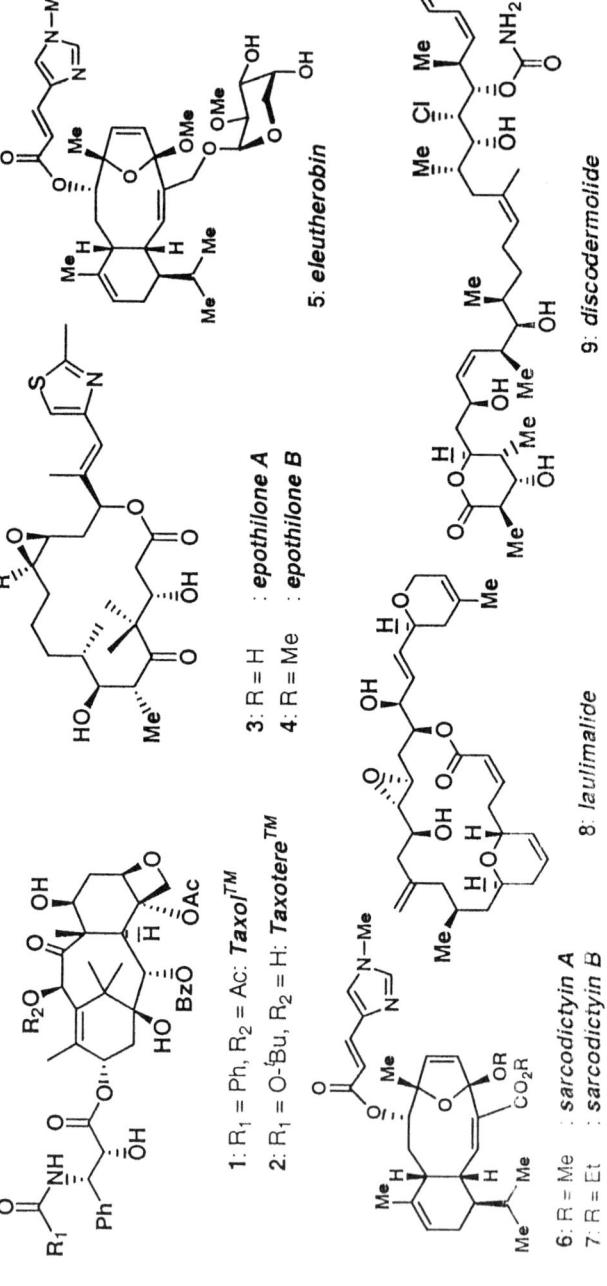

Figure 1. Selected natural products with tubulin polymerization and microtubule stabilization properties: Taxol[TM](1), taxotere[TM] (2), epothilones A (3) and B (4), eleutherobin (5), sarcodictyins A (6) and B (7), laulimal discodermolide (9).

shown in Figure 2 (10). Proceeding in a retrosynthetic direction, removal of the epoxide from **3** afforded the endocyclic olefin **10** which was perceived as an excellent opportunity to evaluate the versatility of the then relatively new reaction, the ring-closing olefin metathesis (11). Thus disconnection of the endocyclic olefin **10** revealed the open chain triene **11**. Further disconnection of the ester and the C7-C8 carbon-carbon bond afforded the three building blocks **12, 13** and **14**. This approach met the set forth criteria and delivered the natural product in a highly convergent and expedient manner (12). Furthermore, it occurred to us that the route shown in Figure 2 was ideally suited for combinatorial chemistry. According to this plan the macrocycle would be assembled in only three synthetic steps from simple building blocks of low molecular complexity and without any protecting group manipulation. The question was how to tether one of the fragments to a solid phase. Additional issues with the synthesis were the lack of stereoselectivity in the aldol condensation leading to an almost equimolar ratio of diastereoisomers and the lack of selectivity in the ring-closing metathesis affording nearly equal amounts of the E and Z isomer. However, for the purpose of molecular diversity, these problems could be considered as advantages since they only expanded the scope of our library, albeit at the cost of some deconvolution. Thus, as shown in Figure 3, we asked whether one of the olefins could be used to link building block **18** to a polymer such that the olefin metathesis could simultaneously

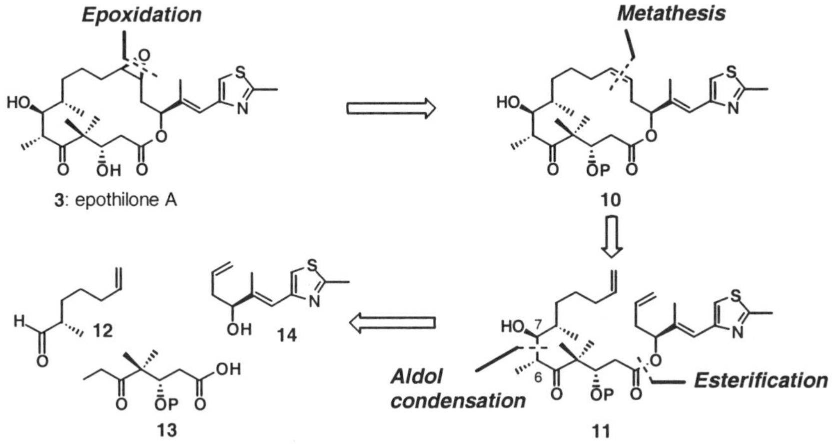

Figure 2. Retrosynthetic analysis of epothilone A (**3**) using a ring-closing olefin metathesis strategy (12).

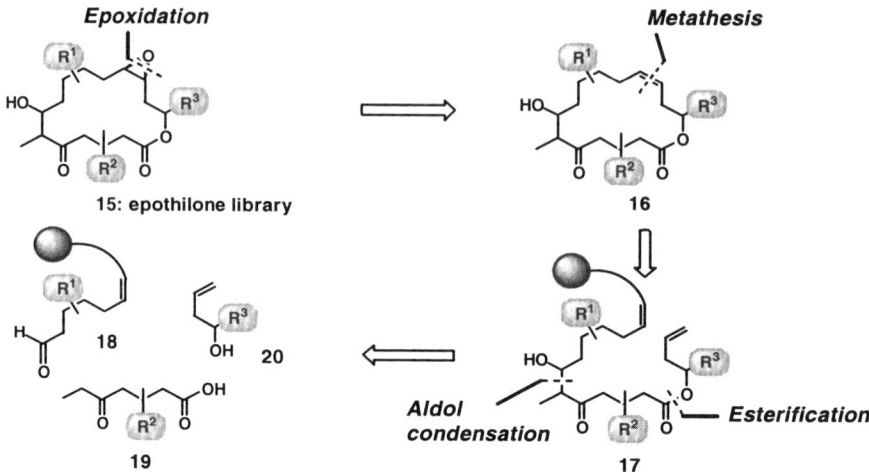

Figure 3. Retrosynthetic analysis of epothilone library 15 by a solid phase olefin metathesis strategy. P = t-BuMe$_2$Si; ● = polystyrene.

affect the desired ring closure and release the substrate from the resin via a cyclorelease reaction (13).

Library synthesis. The chemistry previously developed for our solution phase synthesis was adapted and optimized for solid phase (14). The library was constructed as shown in Scheme 1 (15) using Radiofrequency Encoded Combinatorial (REC™) chemistry (16). Thus, SMART Microreactors™ functionalized as ylide 21 were condensed with the aldehydes 22. The SMART Microreactors 23 were pooled for washing and subsequent deprotection and oxidation to obtain the polymer-bound aldehydes 18. Further sorting and treatment with the dianion of the ketoacids 19 provided the polymer bound carboxylic acids 24 as a mixture of diastereoisomers. Re-sorting and esterification with alcohols 20 afforded metathesis precursors 17. The SMART Microreactors were treated separately with RuCl$_2$(=CHPh)(PCy$_3$)$_2$ catalyst to simultaneously effect cyclization via olefin metathesis, and cleavage of the products, leading to products as mixtures of four 12,13-desoxyepothilones A (25). Each mixture was identified and subjected to preparative thin-layer chromatography to provide pure compounds, which were individually deprotected by treatment with TFA in dichloromethane to obtain compounds 16. The library members (16) displaying good tubulin polymerization activity were then epoxidized to epothilone A analogs 15 for cytotoxicity assays.

Biological evaluation of the epothilone libraries 15 and 16. The epothilone library was screened for induction of tubulin assembly (17) and against an ovarian cancer cell (1A9) and two Taxol™-resistant cell lines which overexpress β_I isotype tubulin (18).

The biological evaluation of the epothilone libraries revealed important information regarding structure-activity relationships for *in vitro* tubulin polymerization and cytotoxicity, and led to several conclusions which are summarized in Figure 4 (15). Inversion of the 3-OH stereochemistry resulted in reduced tubulin polymerization potency. Another clear requirement for tubulin polymerization activity was the (6*R*,7*S*) stereochemistry as revealed by the failure of all (6*S*,7*R*) stereoisomers to induce tubulin polymerization at low concentrations. Interesting, also, was the notable decrease in interaction with tubulin upon inversion of the C8 methyl group and upon introduction of a *gem*-dimethyl group at C8. Thus presumably, the C8 methyl group, as the C4-*gem*-dimethyl group, is important to maintain the active conformation of the epothilone macrocycle. The importance of the natural stereochemistry (12*S*,13*R*) for the epoxide was demonstrated by the general trend of the unnatural 12*R*,13*S* epoxides to exhibit lower activities in inducing tubulin polymerization. Most interestingly, both the *cis* and *trans* olefins corresponding to epothilones A and B were active in the tubulin assembly assays, and the activities of the *cis* olefins were comparable to those of the natural substances. However, we found that the *cis* and, especially, the *trans* olefins were somewhat less cytotoxic than the naturally occurring epoxides. The C12-methyl group consistently bestowed higher potency to all epothilones studied as compared to their C12-*des*-methyl counterparts. Inversion of configuration at C15 led to loss of ability to induce tubulin polymerization. Replacement of the C16-methyl with an ethyl group also reduced activity in the tubulin assay. The epothilone pharmacophore tolerated heterocycle modifications. Most significantly, replacement of the thiazole with a 2-pyridyl moiety was well tolerated. Nevertheless, substitution of the C21-methyl with a phenyl group yielded inactive compounds.

Analysis of these results clearly pointed to two areas of interest requiring further investigation through focused libraries. First, based on the fact that the C26 methyl group provides such marked increase in activity, could elaborations of this methyl group further enhance activity? Second, a pyridyl substitution of the heterocycle was well tolerated, could this be further optimized? These interests motivated the consolidation of various protocols developed in our lab into a highly efficient stereoselective synthesis allowing for diversification of the C26 methyl group (R^1) and the heterocycle (R^2) at a very late stage in the synthesis (Figure 5) (5i, 5p). The heterocycle R^2 would be introduced via a Stille reaction on fully deprotected material (**27**) whereas the C26 modification R^1 would be derived from C26 allylic alcohol in intermediate **29**. Aside from providing a handle on the introduction of diversity, this allylic alcohol should also prove useful in addressing two problems encountered in previous syntheses. First, it was utilized to control the stereo- and regioselectivity of the epoxidation using the Sharpless protocol and second, it played a crucial role in the form of an ester, to control the olefin geometry in the Wittig olefination. Last but not least,

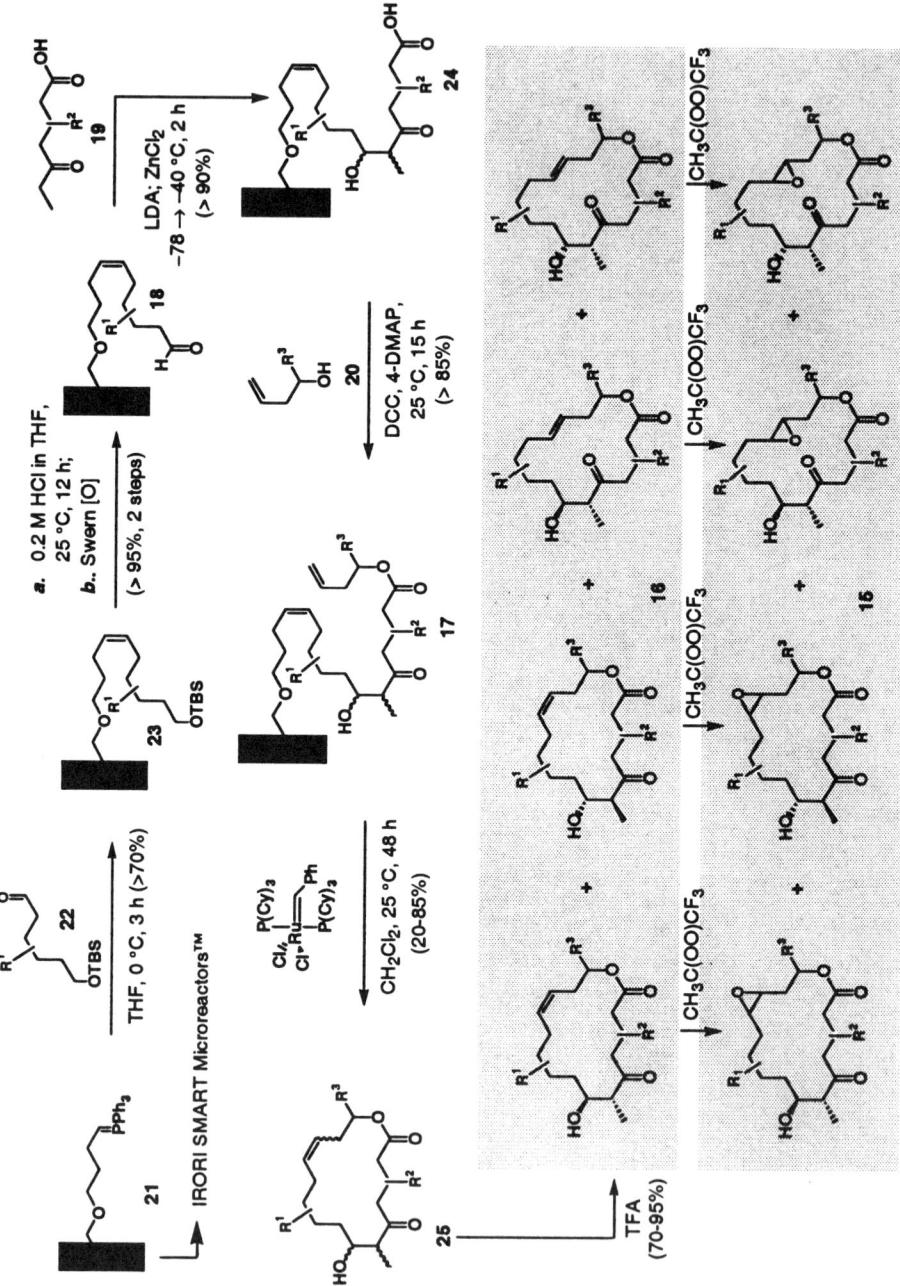

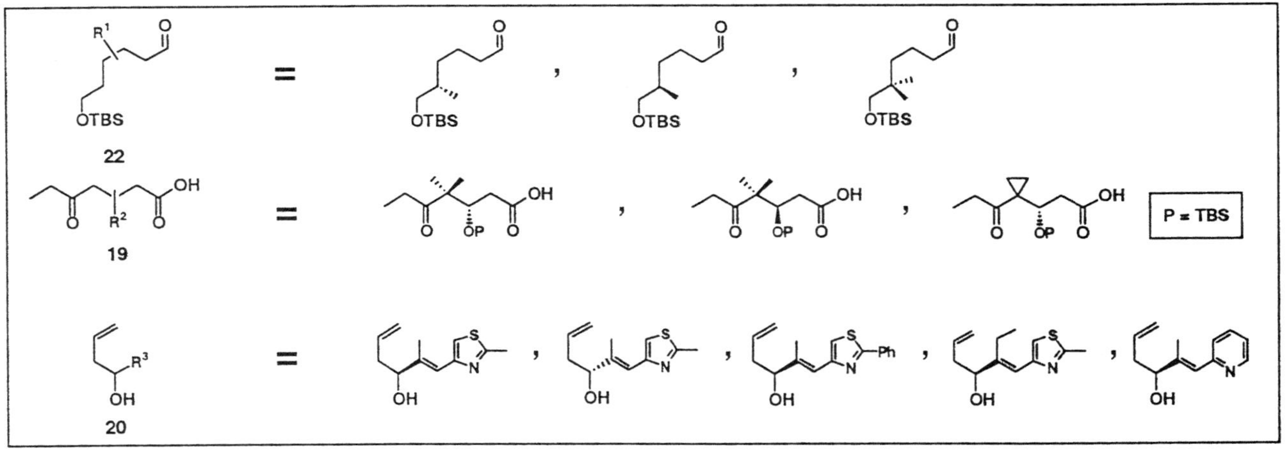

Scheme 1. Solid phase synthesis of epothilone A analogs by a combinatorial approach (14, 15).

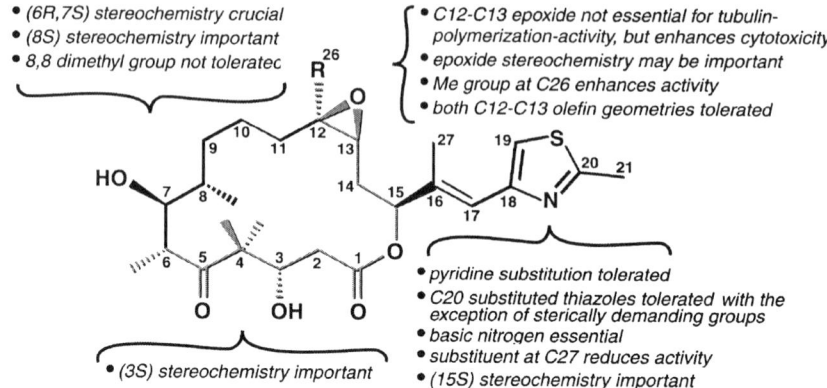

Figure 4. Summary of structure-activity relationships of the epothilone library.

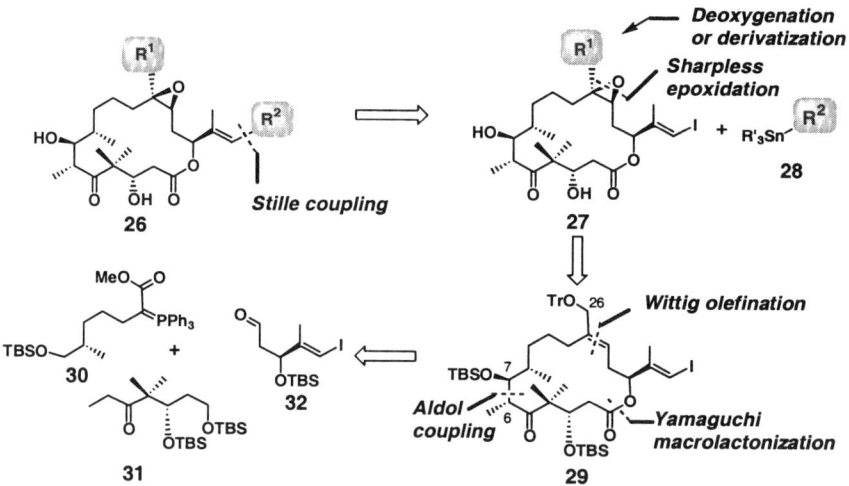

Figure 5. Retrosynthetic analysis of focused libraries of epothilone B analogs (**26**) allowing for diversity of R^1 and using a late stage introduction of heterocycle R^2.

the issue of stereoselectivity in the aldol coupling needed to be addressed as it was clearly determined that the (6S,7R) stereochemistry was important for optimal activity.

Optimized synthesis of epothilone B for designed focused libraries. Intermediate **33** (Scheme 2), obtained in two steps from commercially available materials, was condensed with the commercial butenyl magnesium bromide to provide olefin **34** which was transformed into alcohol **35** *via* ozonolysis and reduction. Alcohol **35** was then converted to the phosphonium salt **36** and the corresponding phosphorane, generated with KHMDS, was condensed with methyl chloroformate to furnish the desired ylide **37**. The crude ylide **37** was then condensed with aldehyde **32** to obtain olefin **38** in excellent selectivity (*E:Z* > 30:1). Aldehyde **39** was reached via DIBAL reduction of the methyl ester and trityl protection of the resulting alcohol, selective silyl deprotection at C7 and subsequent oxidation. We were now in a position to carry out and optimize the aldol coupling. Previous work had shown similar reactions to be somewhat capricious in terms of both yield and stereoselectivity. It was found that excellent results could be obtained by preventing the retroaldol reaction and potential enolate scrambling through a very short reaction time. Silylation of the crude aldol product provided pure tetrasilyl ether **40** from which the primary silyl group was selectively removed and the resulting alcohol was oxidized to the acid. Removal of the C15 hydroxyl-protecting group led to the desired hydroxy-acid **41** which was subjected to macrolactonization under Yamaguchi conditions (19). Global deprotection followed by a highly chemo-, regio- and stereoselective epoxidation of the C12-C13 olefin under Sharpless conditions (20) provided epoxide **42** as a single isomer. This highly advanced intermediate was now the starting point of focused libraries. A series of analogs bearing modifications of the C26 were prepared, however, none of these modifications improved the biological activity (data not shown). Noteworthy though was the substitution of the C26 alcohol for a fluoride which yielded a compound with comparable potency to epothilone B. For the purpose of analogs bearing the epothilone B C26 methyl group, the compound **42** was deoxygenated via a three step sequence involving tosylation, displacement by iodide and reductive deiodination. Finally, the Stille coupling was carried out with a variety of aryl stannanes and proceeded smoothly under our previously developed conditions [$PdCl_2(MeCN)_2$, DMF, 25 °C] (5i) to afford a focused library of epothilone B analogs bearing modifications of the heterocyclic moiety.

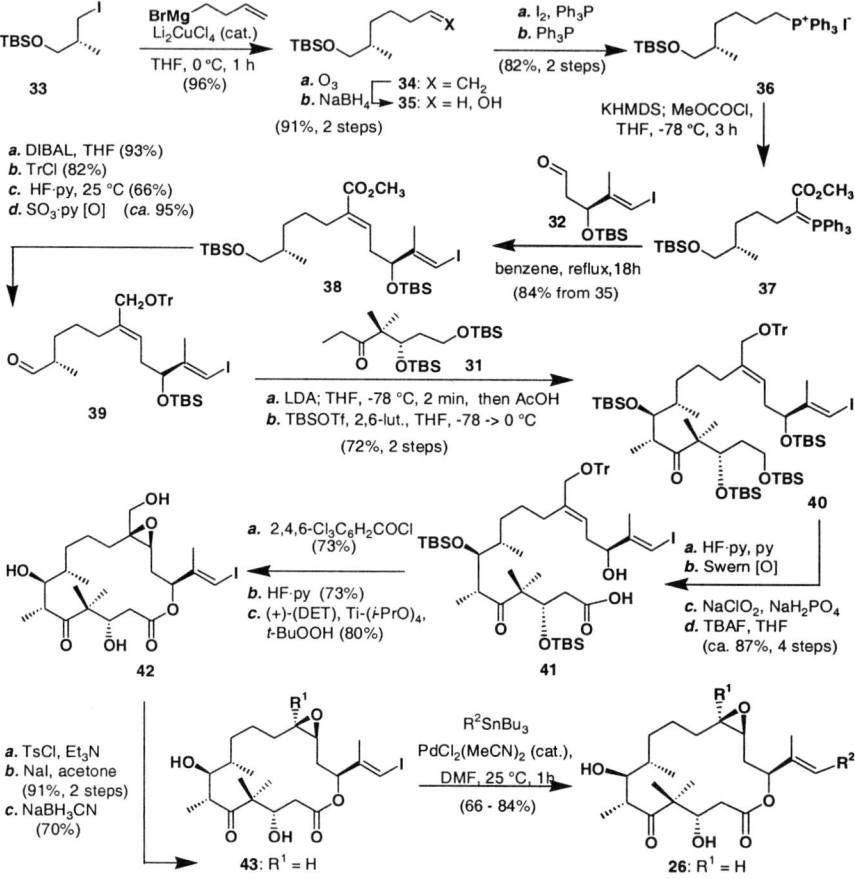

Scheme 2. Stereoselective synthesis of epothilone B analogs (26).

Biological evaluation of the epothilone library 26. In addition to the tubulin assay, the biological properties of this focused library was evaluated in a panel of human cancer cells which included several Taxol™-resistant cell lines characterized by P-glycoprotein overexpression. To our gratification, the activity of the pyridyl analog could indeed be further optimized. The most potent compound of this series (entry D, table I) is more active than epothilone B, especially in a Taxol™-resistant breast tumor cell line (MCF7-ADR). Relocation of the nitrogen away from the ortho position led to severe loss of activity, and, while other basic groups at the ortho position were tolerated, a nitrogen gave the most potency. These results are consistent with molecular docking models based on a common taxane-epothilone binding site (21).

Table I. Biological activities of selected designed epothilone analogs (26).

entry		Tubulin Polymerization	Epidermoid		Lung	Colon		Breast	
			KB-31	KB-8511*	NCI H460	HCT-15	HCT-116*	MCF-7	MCF-7/ADR*
A	4: Epothilone B	85%	0.180	0.179	0.298	0.452	0.313	0.161	3.32
B	1: Taxol™	42%	2.28	545	5.71	136	2.79	1.795	8984
C	26: R^1 = H, R^2 = 2-pyridyl	79.5%	0.304	0.296	0.368	0.455	0.566	0.239	5.29
D	26: R^1 = H, R^2 = 4-methyl-2-pyridyl	89.3%	0.114	0.104	0.175	0.187	0.266	0.118	1.245
E	26: R^1 = H, R^2 = 3-pyridyl	90.3%	0.160	0.163	0.232	0.34	0.367	0.144	3.019

* MDR cell line overexpressing P-glycoprotein.

Sarcodictyins

As with the epothilones, and by sharing the rare and effective mode of action of Taxol™, the sarcodictyins raised the hope that they may become powerful anticancer drugs. Our group took on the total synthesis challenge of these scarce natural products and completed the task within months (22). This accomplishment placed us in the privileged position to begin investigating the structure activity relationships within this class (23). Inspection of sarcodictyins structure reveals a core scaffold decorated with two side chains, the acrylate appended at C8 and the methyl ester at C3 (Figure 6). These groups could certainly be varied in a combinatorial fashion, however, the first task was to develop a linking strategy wherein an advanced intermediate from the synthesis could be tethered so as to allow installation of maximum diversity with minimal manipulation. It was observed during the total synthesis that the hemiketal at C4 participates in a transketalization reaction quite readily by virtue of the stabilization conferred by the two olefins. We asked whether it would be possible to take advantage of this transketalization as a loading and cleavage reaction. An added advantage of such a strategy would be that the transketalization cleavage could be performed in the presence of a variety of nucleophiles and thus the cleavage would be accompanied by the introduction of additional diversity. Despite the reversibility of the transketalization reaction, we thought it could be driven to completion during loading by dehydrating the reaction and likewise, be driven to completion in the cleavage with an excess of nuclophile. Thus, as shown in Figure 6, libraries of sarcodictyin analogs **44** would be obtained from the cleavage of polymer bound intermediate **45** with concomitant introduction of R^3. Intermediate **45** would be derived from polymer bound intermediate **46** bearing an allylic alcohol which would be used to introduce the second element of diversity R^2. It was envisioned that the allylic alcohol could be oxidized to the ester functionality present in the natural product or be directly derivatized or yet converted to an allylic amine for subsequent derivatization. Intermediate **46** would, in turn, stem from hemiketal **48** via **47** through loading and derivatization of the C8 hydroxyl as esters or carbonates.

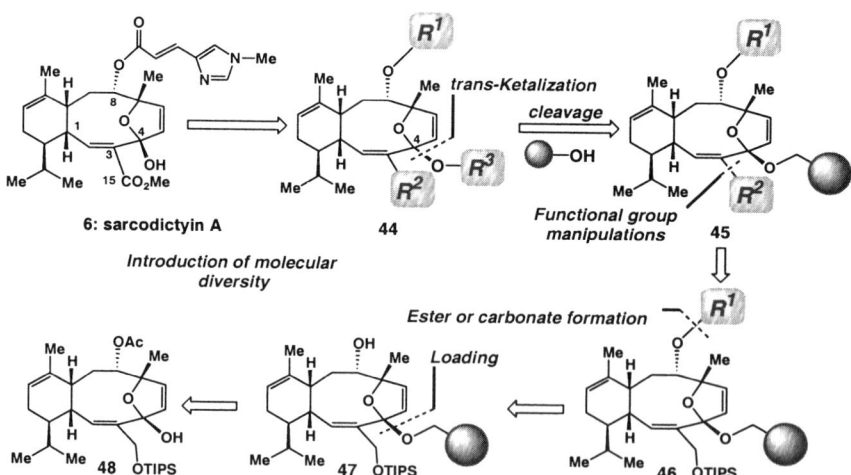

Figure 6. Retrosynthetic analysis of a sarcodictyin library (**44**).

We thus needed access to synthetic intermediate **48**, which, as summarized in Figure 7, could be prepared in gram quantities from (+)-carvone in 23 synthetic steps (22).

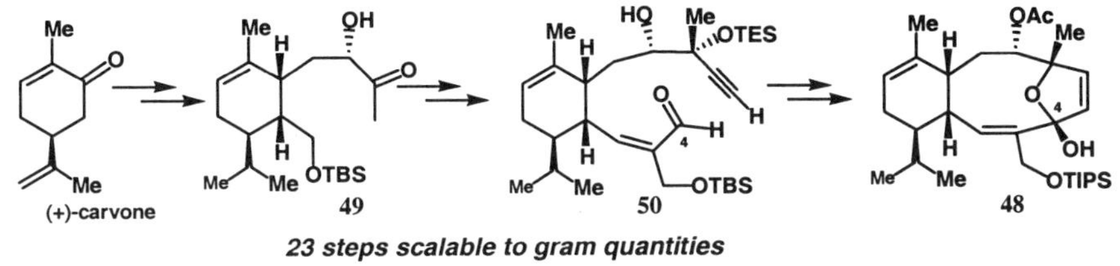

Figure 7. Overview of the synthesis of core sarcodictyin structure **48** (22).

Loading and cleavage of the sarcodictyin scaffold onto a solid phase. Our first efforts to load **48** through a direct transketalization with hydroxy resin **51** (Scheme 3) led to incomplete reaction and low loading despite the use of a variety of dehydrating agents and an excess of the polymer bound alcohol **51**. Nevertheless, the cleavage of the polymer bound material proved to be very efficient. Cleavage of **52** in the presence of MeOH and PPTS afforded the corresponding methyl ketal **54** in 92% yield. To circumvent the loading

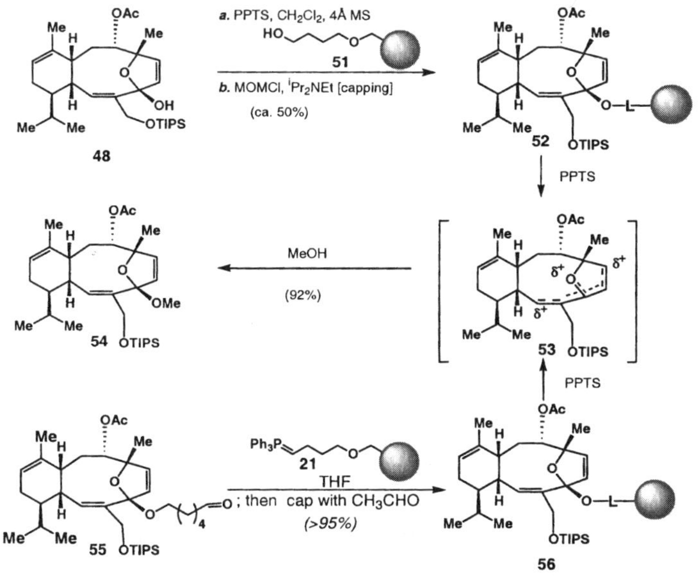

Scheme 3. Loading and cleavage of the sarcodictyin scaffold onto and off the solid phase.

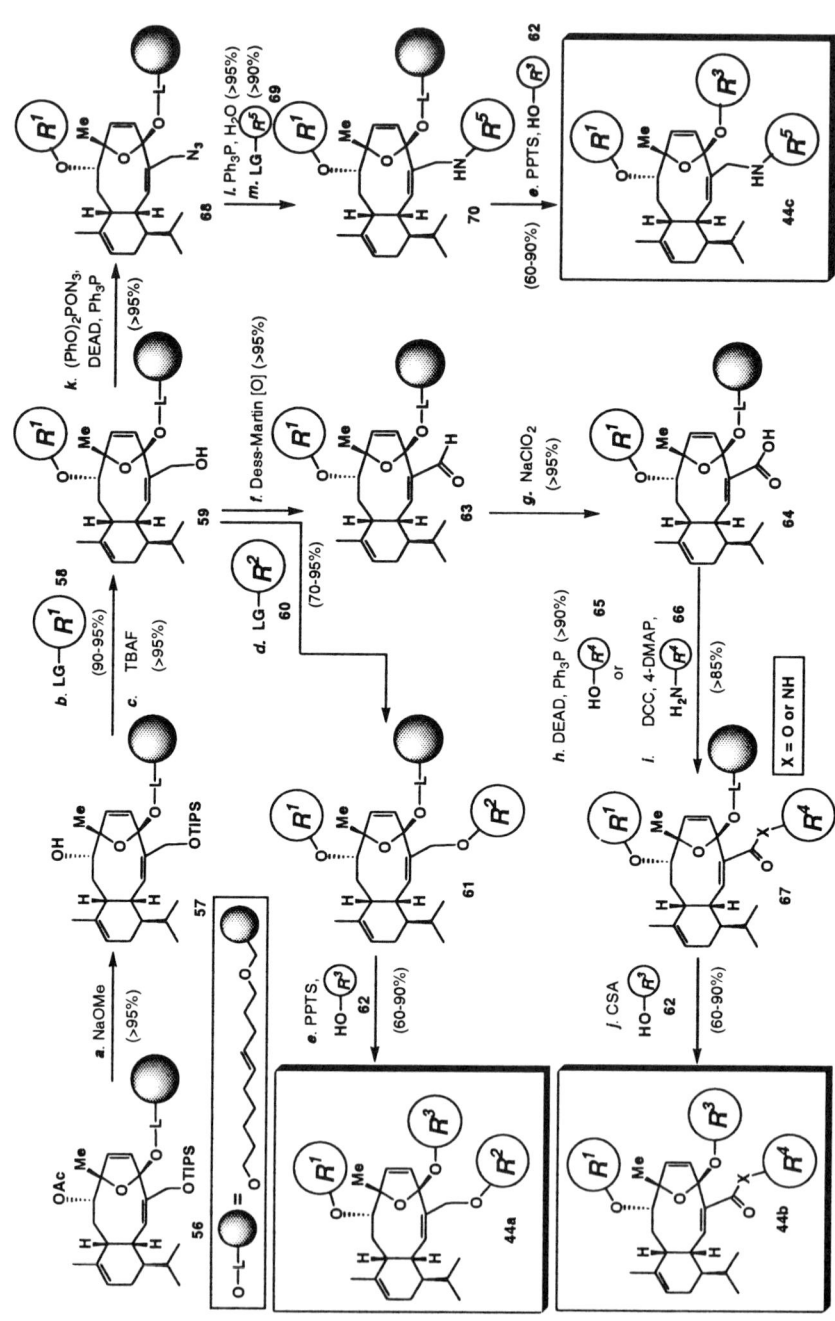

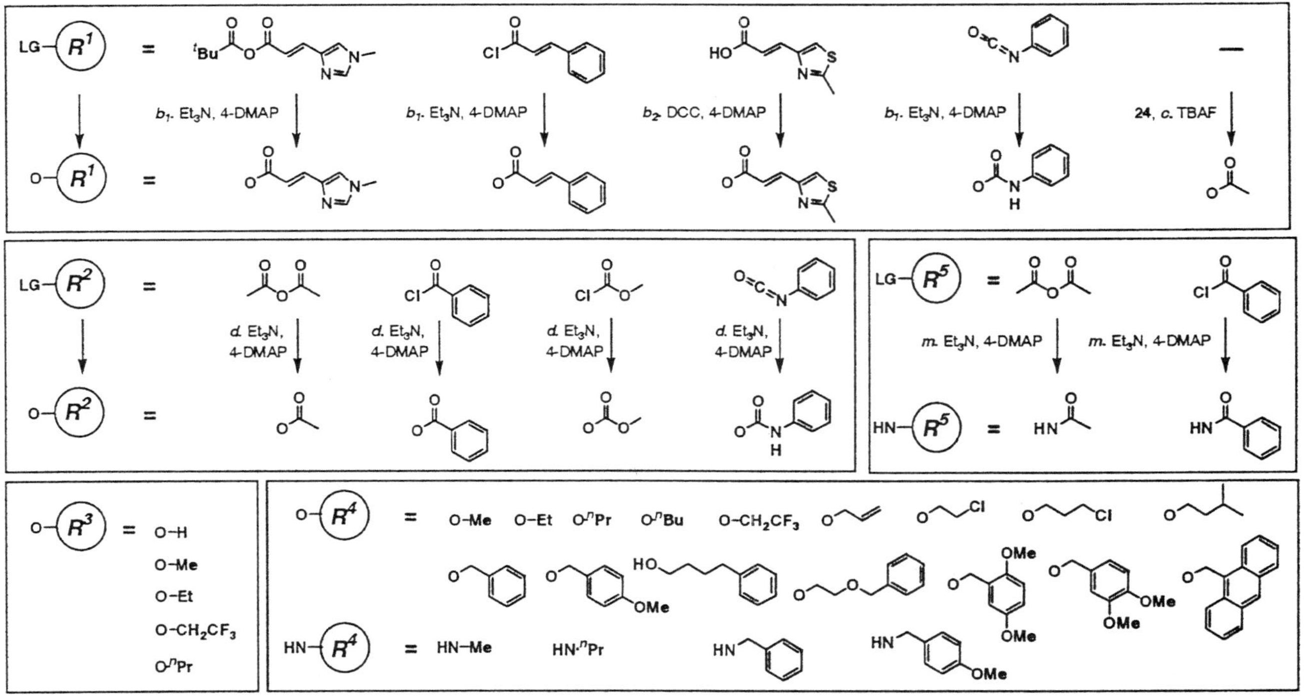

Scheme 4. Synthesis of a library of sarcodictyin analogs 44a - 44c (24).

inefficiency we reasoned that the hemiketal **48** should be transketalized in solution with a tether bearing a functionality which could be smoothly coupled to a polymer. We knew from our work in the epothilone field that a polymer bound ylide could be efficiently condensed with an aldehyde (see Scheme 1), This was successfully achieved with compound **55** (prepared from **48** in 86% yield) bearing a tether functionalized with an aldehyde which was condensed with the polymer bound ylide **21**.

Library synthesis. With an efficient linker-cleavage protocol established, we turned our attention to the construction of a combinatorial library (24). After optimization of the individual chemical transformations to be used in the library, individual flasks with resin **56** in IRORI Microkans™ were utilized to carry out the chemistry summarized in Scheme 4. Thus, **57** (derived by deacetylation of **53**) was reacted with LG-R_1 [acid anhydride, acid chloride, carboxylic acid, or isocyanate] and deprotected with TBAF leading to allylic hydroxy compound **59**. From this point, three divergent sequences were followed. First, a direct derivatization of the allylic alcohol **59** with LG-R_2 [acid anhydride, acid chloride or isocyanate] to afford, after PPTS induced cleavage in the presence of HO-R_3 a series of sarcodictyins **44a** *via* resins **61**. The second sequence involved the oxidation of the allylic alcohol **59** to the carboxylic acid **64** via aldehyde intermediate **63**. The carboxylic acid **64** was subjected to coupling reaction with either alcohols [HO-R_4] under Mitsunobu conditions or amines [H_2N-R_4] under DCC – 4-DMAP conditions. The resulting esters or amides (**67**) were then exposed to acid cleavage conditions (CSA) (25) in HO-R_3 affording sublibrary **44b**. The third sequence involved conversion of allylic alcohol **59** to the corresponding azido group (**68**). Reduction of the azido group and coupling of the corresponding amine with LG-R_5 [anhydride or acid chloride] afforded the amide derivatives **70**. Sublibrary **44c** was then obtained from **70** by exposure to PPTS in HO-R_3 (60-90%). Each library member was obtained in *ca*. 1-5 mg scale (as a single major product) and was purified by silica gel chromatography (flash column or thin layer) or HPLC.

Biological evaluation of the sarcodictyin libraries. The structure-activity relationships derived from these libraries are summarized in Figure 8 (24, 26). The most obvious finding was the importance of the α,β-conjugated imidazole moiety for both tubulin binding and antiproliferative activity. With regards to the C4 hemiketal position, modifications were well tolerated. Remarkably, several of these ketals exhibited marked improvement in antiproliferative properties despite no improvement in tubulin binding activity. In fact, in the ovarian carcinoma cell line (1A9), several of these ketals showed at least a 10-fold improvement in activity. Similar improvements were observed with modification at C15 with small alkyl groups. As shown in Table II, the combined enhancement of activity provided by combined modification of both

C4 ketal and the C15 ester (entry C and F) yielded compounds with 25- to 80-fold improvement in activity over sarcodictyin A and, more importantly, with comparable improvement against the Taxol™-resistant ovarian carcinoma cell line expressing β-tubulin isotypes (1A9PTX10 and 1A9PTX22) (18).

Analysis of the biological data of this library also revealed a subtle, yet intriguing, pattern whereby there is a non-uniform correlation of activity for several of the more active compounds between the tubulin polymerization properties and cytotoxicity (data not shown). A possible explanation of these results could be the fact there are several electrophilic sites on the sarcodictyin core which could make it an alkylating agent (Figure 9).

Conclusion

We have developed efficient synthetic sequences for the rapid exploration of the structure-activity relationships of both the epothilones and the sarcodictyins involving solution and solid phase chemistry. These efforts not only yielded valuable libraries of analogs, but also stimulated the development of new synthetic strategies. For the epothilones, the information gathered from the initial libraries inspired the design of focused libraries yielding compounds significantly more active than epothilone B, the most potent compound in the family of tubulin stabilizing natural products. In the case of the sarcodictyins, the initial libraries contained compounds with activity comparable to epothilone A.

Acknowledgements. We wish to thank our collaborators whose names appear on the referenced publications for their contributions and The Skaggs Institute for Chemical Biology, the National Institutes of Health (USA), Novartis, CaPCURE and the Hewitt Foundation for sponsoring this research.

References

[1] Schiff, P.B.; Fant, J.; Horwitz, S.B. *Nature* **1979**, *277*, 665.

[2] For a reviews, see:, a) Nicolaou, K.C.; Dai, W.-M.; Guy, R.K. *Angew. Chem. Int. Ed. Engl.* **1994**, *33*, 15, b) Rowinsky, E.K. *Annu. Rev. Med.* **1997**, *48*, 353.

[3] For isolation and structure, see: a) Höfle, G.; Bedorf, N.; Gerth, K.; Reichenbach, H. (GBF), DE-4138042, **1993** (*Chem. Abstr.* **1993**, *120*,

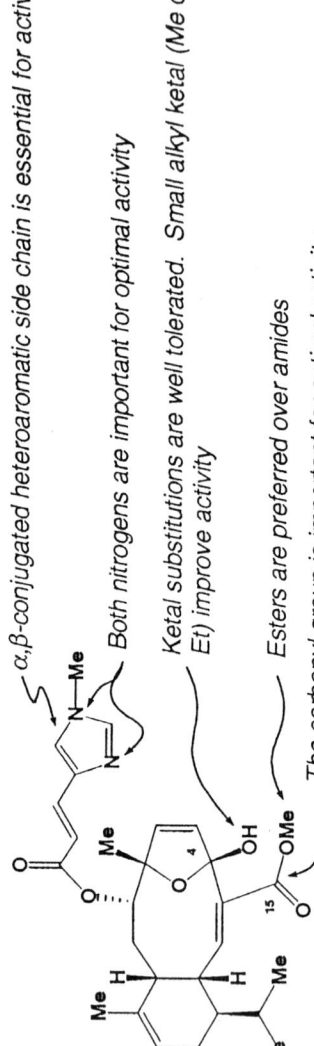

Figure 8. Summary of structure-activity of sarcodictyins (24, 26).

167

Table II. Biological activity of selected sarcodictyin analogs.
Tubulin Polymerization % [IC$_{50}$ 1A9 1A9PTX10 1A9PTX22 nM]

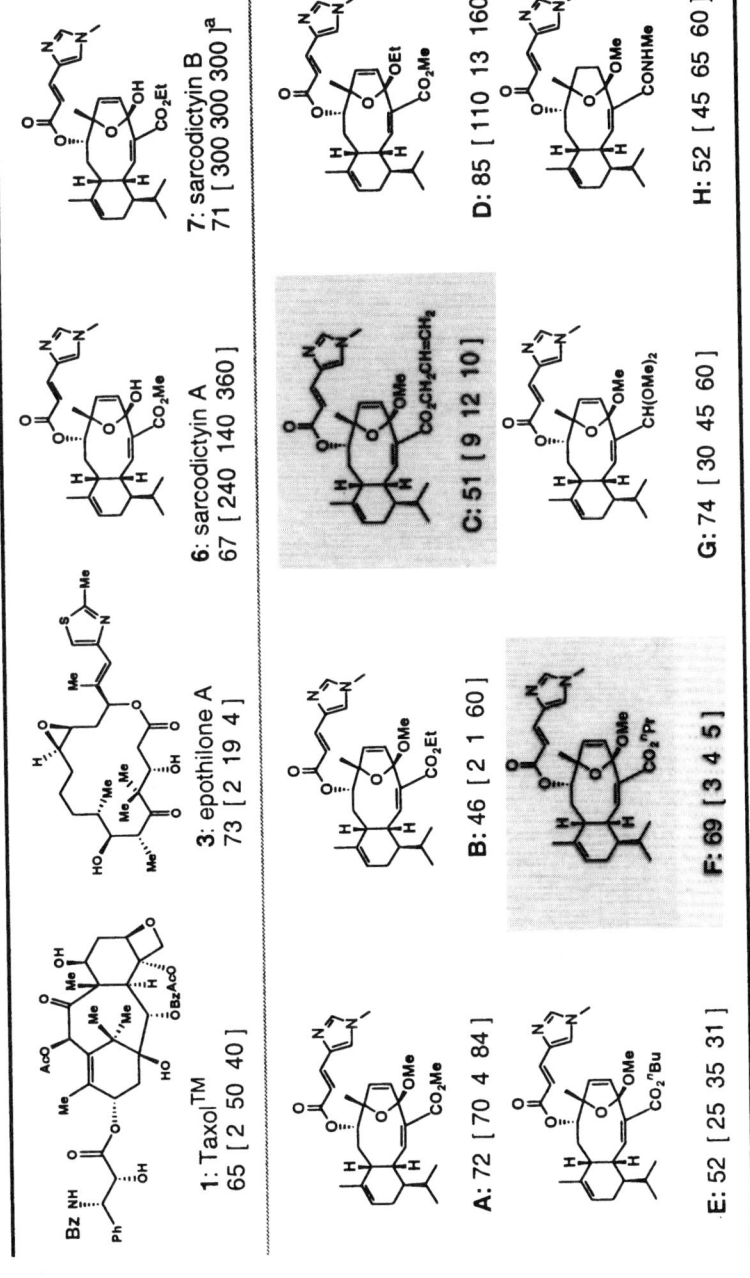

a Ref. 8. For a different IC$_{50}$ values [2, 160, 80] see ref. 24; the previously reported values are probably due to an artifact.

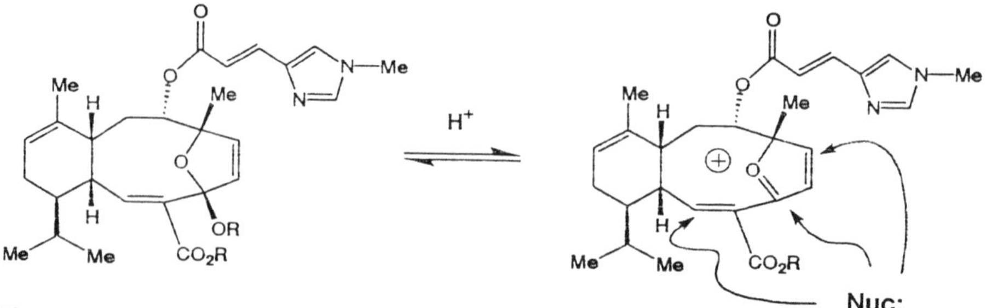

Figure 9. Alkylating potential of the sarcodictyins.

52841); b) Gerth, K.; Bedorf, N.; Höfle, G.; Irschik, H.; Reichenbach, H. *J. Antibiot.* **1996**, *49*, 560; c) Höfle, G.; Bedorf, N.; Steinmetz, H.; Schomburg, D.; Gerth, K.; Reichenbach, H. *Angew. Chem. Int. Ed. Engl.* **1996**, *35*, 1567.

[4] a) Nicolaou, K.C.; Roschangar, F.; Vourloumis, D. *Angew. Chem. Int. Ed. Engl.* **1998**, *37*, 2014; b) Altmann, K.-H.; Wartmann, M.; O'Reilly, T. *Biochim. Biophys. Acta* **2000**, *1470*, M79.

[5] For the initial syntheses, see ref 4a. For recent efforts, see: a) May, S.A.; Grieco, P. *Chem. Commun.* **1998**, 1597; b) Taylor, R.E.; Galvin, G.M.; Hilfiker, K.A.; Chen, Y. *J. Org. Chem.* **1999**, *64*, 7224; c) White, J.D.; Carter, R.G.; Sundermann, K.F. *J. Org. Chem.*, **1999**, *64*, 684; d) Harris, C.R.; Kuduk, S.D.; Balog, A.; Savin, K.; Glunz, P.W.; Danishefsky, S.J. *J. Am. Chem Soc.* **1999**, *121*, 7050; e) Kalesse, M.; Quitschalle, M.; Claus, E.; Gerlach, K.; Pahl, A.; Meyer, H.H. *Eur. J. Org. Chem.* **1999**, 2817; f) Schinzer, D.; Bauer, A.; Schieber, J. *Chem. Eur. J.* **1999**, *5*, 2492; g) White, J.D.; Sundermann, K.F.; Carter, R.G. *Org. Lett.* **1999**, *1*, 1431; h) Sinha, S.C.; Sun, J.S.; Miller, G.; Barbas III, C.F.; Lerner, R.A. *Org. Lett.* **1999**, *1*, 1623; i) Nicolaou, K.C.; King, N.P.; Finlay, M.R.V.; He, Y.; Roschangar, F.; Vourloumis, D.; Vallberg, H.; Sarabia, F.; Ninkovic, S.; Hepworth, D. *Bioorg. Med. Chem.* **1999**, *7*, 665; j) Sawada, D.; Shibasaki, M. *Angew. Chem., Int. Ed.* **2000**, *39*, 209; k) Martin, H.J.; Drescher, M.; Mulzer, J. *Angew. Chem. Int. Ed.* **2000**, *39*, 581; l) Santi, D.V.; Siani, M.A.; Julien, B.; Kupfer, D.; Roe, B. *Gene* **2000**, *247*, 97; m) Johnson, J.; Kim, S.-H.; Bifano, M.; DiMarco, J.; Fairchild, C.; Gougoutas, J.; Lee, F.; Long, B.; Tokarski, J.; Vite, G. *Org. Lett.* **2000**, *2*, 1537; n) Chappell, M.D.; Stachel, S.J.; Lee, C.B.; Danishefsky, S.J. *Org. Lett.* **2000**, *2*, 1633; o) Zhu, B.; Panek, J.S. *Org. Lett.* **2000**, *2*, 2575; p) Nicolaou, K.C.; Hepworth, D.; King, N.P.; Finlay, M.R.V.; Raymond, M.V.; Scarpelli, R.; Pereira, M.M.; Bollbuck, B.; Bigot, A.; Werschkun, B.; Winssinger, N. *Chem. Eur. J.* **2000**, *6*, 2783 and references cited therein.

[6] D'Ambrosio, M.; Guerriero, A.; Pietra, F. *Helv. Chim. Acta* **1987**, *70*, 2019.

[7] a) Ciomei, M.; Albanese, C.; Pastori, W.; Grandi, M.; Pietra, F.; D'Ambrosio, M.; Guerriero, A.; Battistini, C. *Proc. Am. Assoc. Cancer Res.* **1997**, *38*, 5,

Abstract 30; b) Battistini, C.; Ciomei, M.; Pietra, F.; D'Ambrosio, M.; Guerriero, A. Patent application WO 96-EP1688 960423.

[8] Hamel, E.; Sackett, D.L.; Nicolaou, K.C.; Vourloumis, D.; *Biochemistry* **1999**, *38*, 5490.

[9] a) Ojima, I; Chakravarty, S.; Inoue, T.; Lin, S.; He, L.; Horwitz, S.B.; Kuduk, S.D.; Danishefsky, S.J. *Proc. Natl. Acad. Sci. U. S. A.* **1999**, *96*, 4256; b) Giannakakou, P.; Gussio, R.; Nogales, E.; Downing, K. H.; Zaharevitz, D.; Bollbuck, B.; Poy, G.; Sackett, D.; Nicolaou, K.C.; Fojo, Ti. *Proc. Natl. Acad. Sci. U.S.A.* **2000**, *97*, 2904; c) Ojima, I. Book of Abstracts, 219th ACS National Meeting, San Francisco, CA, March 26-30, 2000; d) He, L.; Jagtap, P.G.; Kingston, D.G.I.; Shen, H.-J.; Orr, G.A.; Horwitz, S.B. *Biochemistry* **2000**, *39*, 3972 and citations therein.

[10] Yang, Z.; He, Y.; Vourloumis, D.; Vallberg, H.; Nicolaou, K.C. *Angew. Chem. Int. Ed. Engl.* **1997**, *36*, 166.

[11] Grubbs, R.H.; Miller, S.J.; Fu, G.C. *Acc. Chem. Res.* **1995**, *28*, 446.

[12] a) Nicolaou, K.C.; He, Y.; Vourloumis, D.; Vallberg, H.; Yang, Z. *Angew. Chem. Int. Ed. Engl.* **1997**, *37*, 166; b) Nicolaou, K.C.; He, Y.; Vourloumis, D.; Vallberg, H.; Roschangar, F.; Sarabia, F.; Ninkovic, S.; Yang, Z.; Trujillo, J.I. *J. Am. Chem. Soc.* **1997**, *119*, 7960.

[13] For a general review of cyclorelease reactions, see: van Maarseveen, J.H. *Comb. Chem. High Throughput Screening* **1998**, *1*, 185-214.

[14] Nicolaou, K.C.; Winssinger, N.; Pastor, J.; Ninkovic, S.; Sarabia, F.; He, Y.; Vourloumis, D.; Yang, Z.; Li, T.; Giannakakou, P.; Hamel, E. *Nature* **1997**, *390*, 100.

[15] Nicolaou, K.C.; Vourloumis, D.; Li, T.; Pastor, J.; Winssinger, N.; He, Y.; Ninkovic, S.; Sarabia, F.; Vallberg, H.; Roschangar, F.; King N.P.; Finlay, M.R.V.; Giannakakou, P.; Verdier-Pinard, P.; Hamel, E. *Angew. Chem. Int. Ed. Engl.* **1997**, *36*, 2097.

[16] Nicolaou, K.C.; Xiao, X-.Y.; Parandoosh, Z.; Senyei, A.; Nova, M. P. *Angew. Chem. Int. Ed. Engl.* **1995**, *34*, 2289.

[17] a) Bollag, D.M.; McQueney, P.A.; Zhu, J.; Hensens, O.; Koupal, L.; Liesch, J.; Goetz, M.; Lazarides, E.; Woods, C.M. *Cancer Res.* **1995**, *55*, 2325; b) For an important subsequent study, see: Kowalski, R.J.; Giannakakou, P.; Hamel, E. *J. Biol. Chem.* **1997**, *272*, 2534.

[18] Giannakakou, P.; Sackettt, D.L.; Kang, Y.-K.; Zhan, Z.; Buters, J.T.M.; Fojo, T.; Poruchynsky, M.S. *J. Biol. Chem.* **1997**, *272*, 17118.

[19] Inanaga, J.; Hirata, K.; Saeki, H.; Katsuki, T.; Yamaguchi, M. *Bull. Chem. Soc. Jpn.* **1979**, *52*, 1989.

[20] Katsuki, T.; Sharpless, K.B. *J. Am. Chem. Soc.* **1980**, *102*, 5976.

[21] Nicolaou, K.C.; Scarpelli, R.; Bollbuck, B.; Werschkun, B.; Pereira, M.M.; Wartmann, M.; Altmann, K.-H.; Zaharevitz, D.; Gussio, R.; Giannakakou, P. *Chem. Biol.* **2000**, *7*, 593.

[22] a) Nicolaou, K.C.; Xu, J.-Y.; Kim, S.; Ohshima, T.; Hosokawa, S.; Pfefferkorn, J. *J. Am. Chem. Soc.* **1997**, *119*, 11353; b) Nicolaou, K.C.; Xu, J.Y.; Kim, S.; Pfefferkorn, J.; Ohshima, T.; Vourloumis, D.; Hosokawa, S. *J. Am. Chem. Soc.* **1998**, *120*, 8661.

[23] Nicolaou, K.C.; Kim, S.; Pfefferkorn, J.; Xu, J.-Y.; Ohshima, T.; Hosokawa, S.; Vourloumis, D.; Li, T. *Angew. Chem. Int. Ed.* **1998**, *37*, 1418.

[24] Nicolaou, K.C.; Winssinger, N.; Vourloumis, D.; Ohshima, T.; Kim, S.; Pfefferkorn, J.; Xu, J.Y.; Li, T. *J. Am. Chem. Soc.* **1998**, *120*, 10814.

[25] In the cases where C15 was oxidized to a carbonyl, CSA was required to induce the transketalization cleavage.

[26] Nicolaou, K.C.; Pfefferkorn, J.; Xu, J.Y.; Winssinger, N.; Ohshima, T.; Kim, S.; Hosokawa, S.; Vourloumis, D.; Van Delft, F.; Li T. *Chem. Pharm. Bull.* **1999**, *47*, 1199.

Chapter 10

Synthesis and Structure–Activity Relationship Studies of Cryptophycins: A Novel Class of Potent Antimitotic Antitumor Depsipeptides

Chuan Shih[1], Rima S. Al-Awar[1], Andrew H. Fray[1],
Michael J. Martinelli[1], Eric D. Moher[1], Bryan H. Norman[1],
Vinod F. Patel[1], Richard M. Schultz[1], John E. Toth[1],
David L. Varie[1], Thomas H. Corbett[2], and Richard E. Moore[3]

[1]Lilly Research Laboratories, Eli Lilly and Company, 307 East McCarty Street, Indianapolis, IN 46285
[2]Karmanos Cancer Institute, Wayne State University, Detroit, MI 48202
3Department of Chemistry, University of Hawaii at Manoa, Honolulu, HI 96822

Detailed structure-activity relationship studies of the cryptophycin series of compounds are described. From these extensive SAR efforts, cryptophycin 52 (LY355703) was selected as the first generation cryptophycin for clinical development. The chemical synthesis and pharmacological evaluation of LY355703 and its corresponding chlorohydrin derivative (cryptophycin 55, LY355702) are summarized in this review.

Introduction

Cryptophycins are potent tumor selective depsipeptides isolated first from terrestrial blue-green algae. The cellular mechanism of action of cryptophycins

© 2001 American Chemical Society

is associated with inhibition of mitotic spindle function. This puts cryptophycins into one of the clinically most important classes of anticancer compounds, such as the vinca alkaloids (vincristine, vinblastine, vinorelbine) and the taxanes (taxol and taxotere), that act on the cellular microtubule structures and functions as the primary target for their antitumor activities. Extensive preclinical studies have demonstrated that cryptophycins are highly active against a broad spectrum of murine solid tumors and human tumor xenografts *in vivo*. One very unique pharmacological property of cryptophycins is that they do not serve as substrates for both the Pgp and the MRP multidrug resistant efflux pumps, and are highly active against the resistant tumors that express multiple drug resistant phenotypes both *in vitro* and *in vivo*. This result suggests that cryptophycins may be highly active against human solid tumors that are resistant to taxanes or vincas and may offer alternative efficacy or survival advantages over the current therapies. This review summarizes the collaborative structure-activity relationship studies conducted among the three institutions, Lilly Research Laboratories, University of Hawaii and Wayne State University, in their identification of cryptophycin 52 (LY355703) as the first generation cryptophycin for clinical evaluation.

Discovery of the Antitumor Activity of Cryptophycins

Cryptophycins are peptolides with a 16-membered macrolide ring structure. The peptolide ring is composed of two ester linkages (Fragment A/D and Fragment C/D), two amide linkages (Fragment A/B and Fragment B/C) and 7 asymmetric centers (Figure 1). Cryptophycin (cryptophycin A or cryptophycin 1) was first isolated from the *Nostoc* sp. ATCC 53798 (*1*) by researchers at Merck in the early 1990s and found to be very active against strains of *Cryptococcus*. The development of cryptophycin A as potential antifungal agent was unsuccessful mainly due to the toxicity of the agent. During the same period of time, in an effort to identify novel antitumor agents from the blue-green algae (cyanobacteria), researchers at the University of Hawaii led by Professor Richard Moore discovered that the lipophilic extract of *Nostoc* sp. GSV224 (*2*) was highly cytotoxic to human nasopharyngeal carcinoma (KB) and human colorectal adenocarcinoma (LoVo) cells. More importantly, this lipophilic extract was found to be highly selective toward solid tumor cells versus leukemia cells and normal fibroblasts in the disk diffusion soft agar colony formation assay (*3,4*) developed by Professor Thomas Corbett at Wayne State University. To further explore this activity, a total of 26 naturally occurring secondary metabolites were isolated from the *Nostoc* sp. GSV224

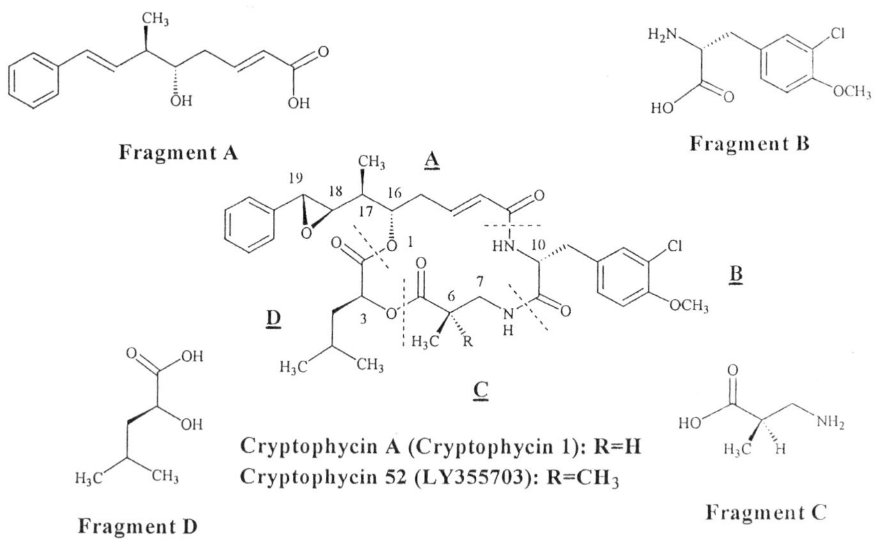

Figure 1. Chemical structure of cryptophycin 1 and its four fragments (A-D).

with cryptophycin 1 accounting for most of the cytotoxic activity in the crude extract. Through rigorous structure elucidation efforts, the University of Hawaii team established the absolute stereochemistry of all of the asymmetric centers of cryptophycin 1 and confirmed that cryptophycin 1 was identical to cryptophycin A reported earlier by the Merck group (5). The antitumor activity of cryptophycin 1 was then rapidly evaluated *in vivo* using various solid tumor models developed either in syngeneic or SCID mice by the Wayne State University group (Professor Thomas Corbett). It was quickly determined that cryptophycin 1 was highly active against a number of murine solid tumors and human tumor xenografts when the compound was given by intravenous (iv) injection. In contrast, the compound was found to be significantly less active when administered by intraperitoneal or oral routes (mainly due to poorer metabolic stability when administered by these routes). Excellent tumor growth inhibition and antitumor activity (6) was observed for cryptophycin 1 in a number of solid tumors. More importantly, it was found that cryptophycin 1 was highly active against the murine Mammary 16/C tumor that is resistant to paclitaxel due to the overexpression of the multiple drug resistant protein Pgp. This makes cryptophycin one of the very few new antitubulin agents that do not serve as substrate for the multiple drug resistant protein, and thus may offer significant advantage for the treatment of resistant human solid tumors.

The discovery of the exciting antitumor activity of cryptophycin 1 prompted the Lilly/Hawaii/Wayne State team to develop an unified strategy not only for producing a large number of analogs for the SAR and the selection of a clinical candidate from the series, but also for developing technologies which could be used for the potential commercial scale production for the cryptophycin class of compounds. Critical analyses revealed that a large scale bioproduction of cryptophycins from cultured blue-green algae was not only technically very challenging but also economically unattractive. For example, cryptophycin 1 can only be obtained with an estimated 0.4 % yield from the dry alga mass after extensive HPLC purification process. In addition to the low yield and technical difficulties in culturing the blue-green algae on large scale, a bioproduction approach also offer limited chemical diversity for structure-activity investigation purposes. The total synthetic approach was eventually selected as the major approach for the chemical investigation of the cryptophycin series of compounds. This latter approach offers several advantages over the bioproduction approach including chemical diversity, synthetic maneuverability, enabling technology, and last but not the least, the speed for the needs of the project. The majority of the structure-activity relationship studies described below were conducted using materials prepared by total synthetic approaches developed jointly between the Lilly Research Laboratories and the University of Hawaii teams. The clinical candidate compound, cryptophycin 52, was prepared by a convergent synthetic scheme that was originally developed by the University of Hawaii team (7) and later modified by the Chemical Process Research and Development group of Lilly Research Laboratories. The 30 step convergent synthetic sequence for cryptophycins illustrates both the complexity as well as the diversity of the approach to this novel chemical series.

Retrosynthetic Analysis of Cryptophycin and the Total Synthesis of Cryptophycin 52 (LY355703)

The retrosynthetic analysis of cryptophycin 52 shown in Figure 1 represents a straightforward disconnection of the depsipeptide into four corresponding Fragments (labeled A-D) through the two ester and two amide linkages. The six asymmetric centers that are associated with cryptophycin 52 can be divided into a group of four contiguous plus the two distal centers around the macrocyclic ring. For the former four, the Sharpless Asymmetric Epoxidation (SAE) was used as the essential step in controlling two of the four asymmetric centers (*16S* and *17S*), and the final diastereoselective epoxidation of the styrene olefin provided the other two asymmetric centers (*18R* and *19R*) in the Fragment A. The two distal asymmetric centers (*3S* and *10R*), on the other hand, were obtained from optically pure amino acids (D-tyrosine for Fragment B, and L-leucine for Fragment D, respectively). In general, the cryptophycin nucleus can be assembled efficiently in a convergent fashion. This was accomplished by first combining the Fragments A and B, and then C and D to give the protected

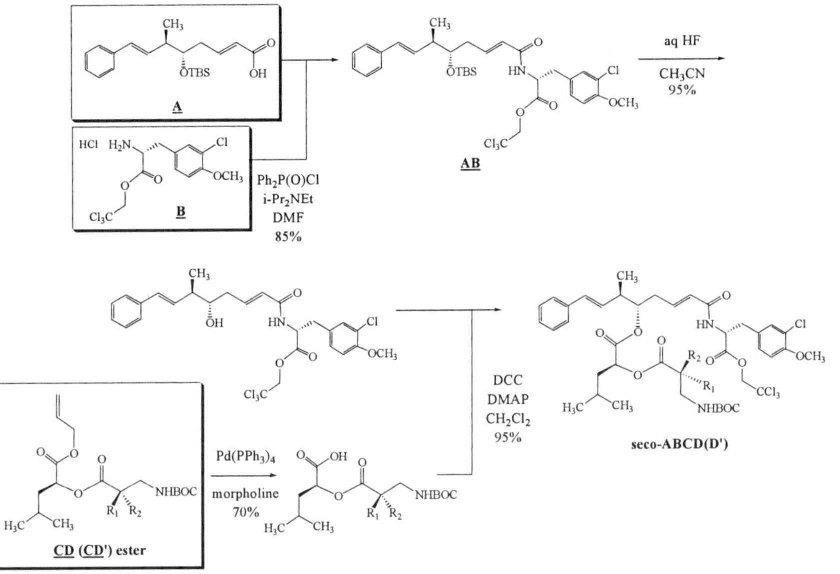

Figure 2. The convergent synthetic approach to the "seco-cryptophycin".

AB and CD fragments. The TBS group of the AB fragment was then removed (aq HF, CH₃CN, 95%) to give the secondary alcohol which was then coupled with the carboxylic acid of the N-BOC CD fragment (DCC, DMAP, CH$_2$Cl$_2$, 95%) to give the fully protected *seco*-ABCD compound (Figure 2). Treatment of the "*seco*" compound with TFA followed by 2-hydroxypyridine effectively catalyzed the macrolactamization to give the cyclized 16-membered depsipeptide in 62 % yield (8). Diastereoselective epoxidation of the styrylolefin (*m*-CPBA, CH$_2$Cl$_2$, 95%) then gave the 2:1 ratio of epoxides with the desirable β-epoxide as the major product. The diastereomeric epoxide mixtures can be separated by HPLC to provide the more active β-epoxide (cryptophycin 52) for biological evaluation. Alternatively, the 2:1 epoxide mixture can be treated with trimethylsilylchloride to give the corresponding chlorohydrins which then can be readily separated by flash silica gel chromatography. Treatment of the pure *18R,19S*-chlorohydrin (cryptophycin 55, LY355702) with potassium carbonate in acetonitrile then converted the chlorohydrin back to the cryptophycin 52 in high yield (Figure 3).

For the synthesis of each of the four fragments of cryptophycins, the fragment A with the four contiguous asymmetric centers clearly represented the most challenging piece among the four fragments. A total 11 step linear synthetic sequence was developed for the preparation of methyl(*5S,6R*)-5-[(tert-

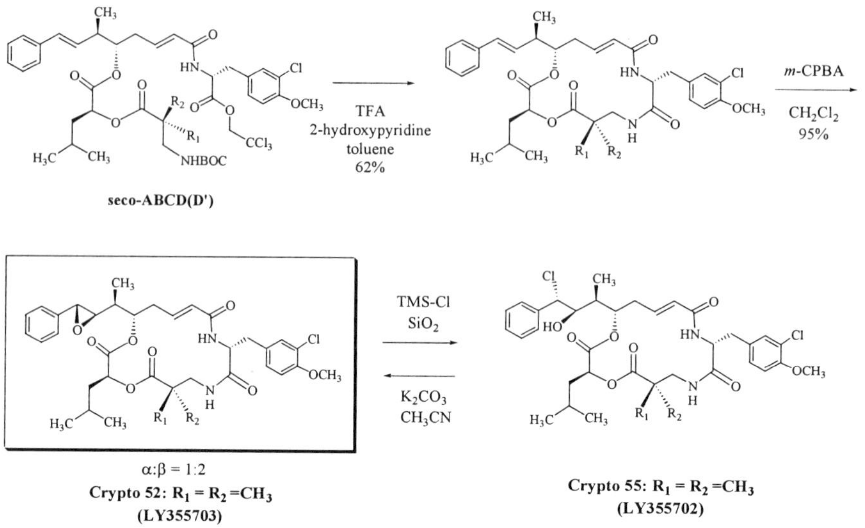

Figure 3. Macrocyclization and the preparation of cryptophycin 52 and its chlorohydrin derivatives.

butyldimethylsilyl)oxy]-6-methyl-8-phenylocta-2(*E*),7(*E*)-dienoate (**1**, 28% overall yield, 95% ee) corresponding to Fragment A. In this sequence, the Sharpless asymmetric epoxidation of the allylic alcohol was used as a key reaction to install two of four asymmetric centers with 95% ee (Figure 4). The other Fragments B, C and D were prepared in a more straightforward fashion. Fragment B was obtained in six steps (40% overall yield, 99 % ee) from D-tyrosine. Fragment C' (for cryptophycin 52) was obtained in three steps (58% overall) from ethylcyanoacetate which was then coupled (CDI in THF, 91%) readily with the allyl-L-leucic acid (obtained in two steps from the L-leucine) to give the corresponding C'-D fragment. The detailed information on the preparation of each of these fragments can be obtained from the original cryptophycin publication (*7*), and thus will not be thoroughly discussed in this review. In summary, although it took more than 30 discrete synthetic operations for the completion of the total synthesis of cryptophycin 52, the convergent nature of the process makes it versatile for creating multiple permutations through the combination of different fragments for SAR purpose, and was practical enough for the preparation of a reasonably large amount of material for the phase I and II clinical evaluations.

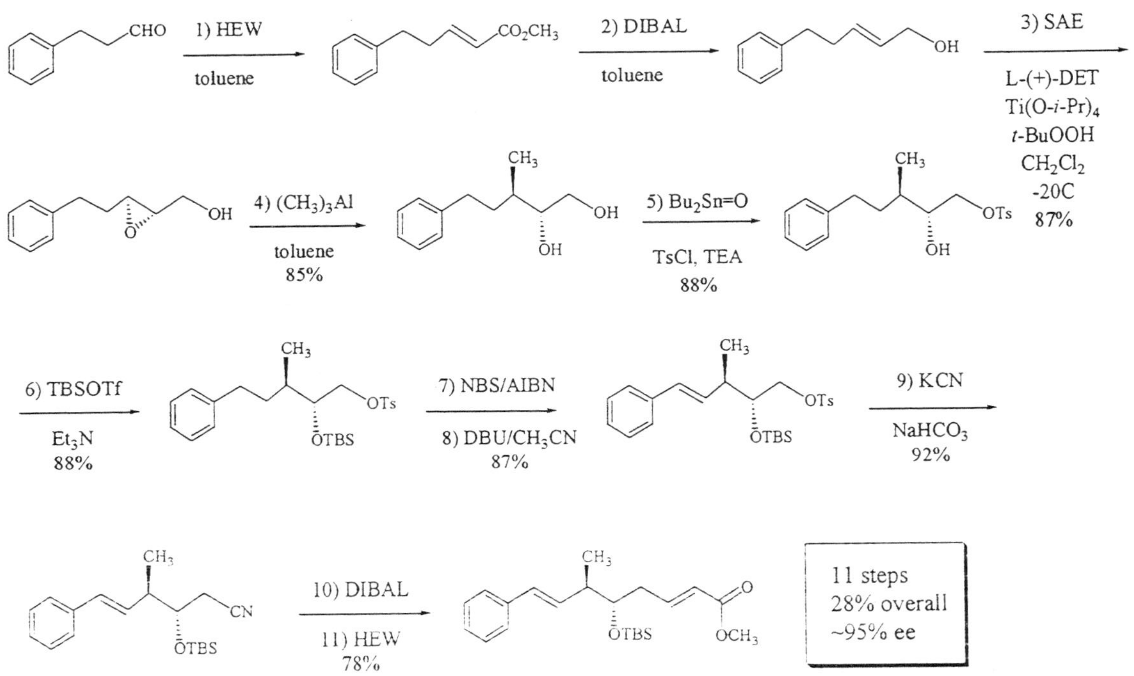

Figure 4. The Sharpless Asymmetric Epoxidation (SAE) route to the preparation of Fragment A.

11 steps
28% overall
~95% ee

Structure-Activity Relationship Studies of Cryptophycins

More than 500 analogs of cryptophycins were prepared by using either the total synthetic approach or semisynthetic modifications of the natural cryptophycin 1 (particularly for the styrylepoxide derivatives) for the SAR studies. Although the molecular target and the mechanism of action of cryptophycin is believed to be cellular tublin and the inhibition of microtubule polymerization, the limited sensitivity of several existing *in vitro* tubulin or microtubule polymerization assays was, in general, not sufficient for structure-activity study purposes (with 50% inhibition concentrations occurring at low to mid micromolar range for most of the cryptophycin analogs). In contrast, cryptophycins exhibited very potent cytotoxicity toward most of the human malignant carcinoma cell lines tested in culture. In 72 h MTT cell-based assays, the IC_{50}s for cryptophycin 1 and 52, for example, were in the range of 20-100 picomolar against several human tumor cell lines (KB, CCRF-CEM, HT-29, SKOV3, LoVo and GC3). These *in vitro* cell based assays provided the required sensitivity and broad range (10pM - >20µM) of activity for cryptophycins, which was essential for differentiating and establishing a firm structure-activity trend for a large number of analogs. In the current studies, most of the cytotoxicity data were collected using three cultured human tumor cell lines: KB (Hawaii), GC3 human colon carcinoma and CCRF-CEM human leukemia (Lilly). The cytotoxicity data cited in the SAR discussion in this
section were mostly derived from GC3 human colon carcinoma cell line unless stated otherwise. The *in vitro* cytotoxicity was then used to prioritize the compounds for the initial *in vivo* antitumor screening using Panc 03 (murine pancreatic adenocarcinoma) model which was then followed by a panel of other murine tumor models and human tumor xenografts (Mam 16C/ADR, Mam 17/ADR, MX-1, GC3, LNCaP, HCT-116 and PC-3) (*9*). Figure 5 summarizes some of the most important SAR findings around the key regions of the cryptophycin molecule. Overall it was noted that: (1) the absolute stereochemistry of all seven asymmetric centers is essential for eliciting the biological activity of cryptophycins; the opposite configuration at these centers led to major losses in activity, (2) the β-epoxides (*R/R*) at the 18- and 19- positions (Fragment A) are consistently much more active than the other isomers (*R/S, S/R and S/S*), (3) substitution of both the amide nitrogen of the macrocyclic ring led to inactive analogs, x-ray and NMR studies suggest that an intra-peptolide hydrogen bond may exist between the NH(8) and CO(12) groups in order for the macrocyclic ring to adopt the bioactive conformations, (4) amide replacement of the ester linkage may be tolerated between the Fragment C/D but not between the A/B fragments, (5) small substituents on the phenyl ring of fragment A can be tolerated, however substitutions and modifications of the chloromethyl tyrosine unit of Fragment B are more stringent and limited to several options.

The SAR described in this review will be primarily focused on four regions: (A) the C/D ester bond and C6, C7-positions of Fragment C, (B) the epoxide of

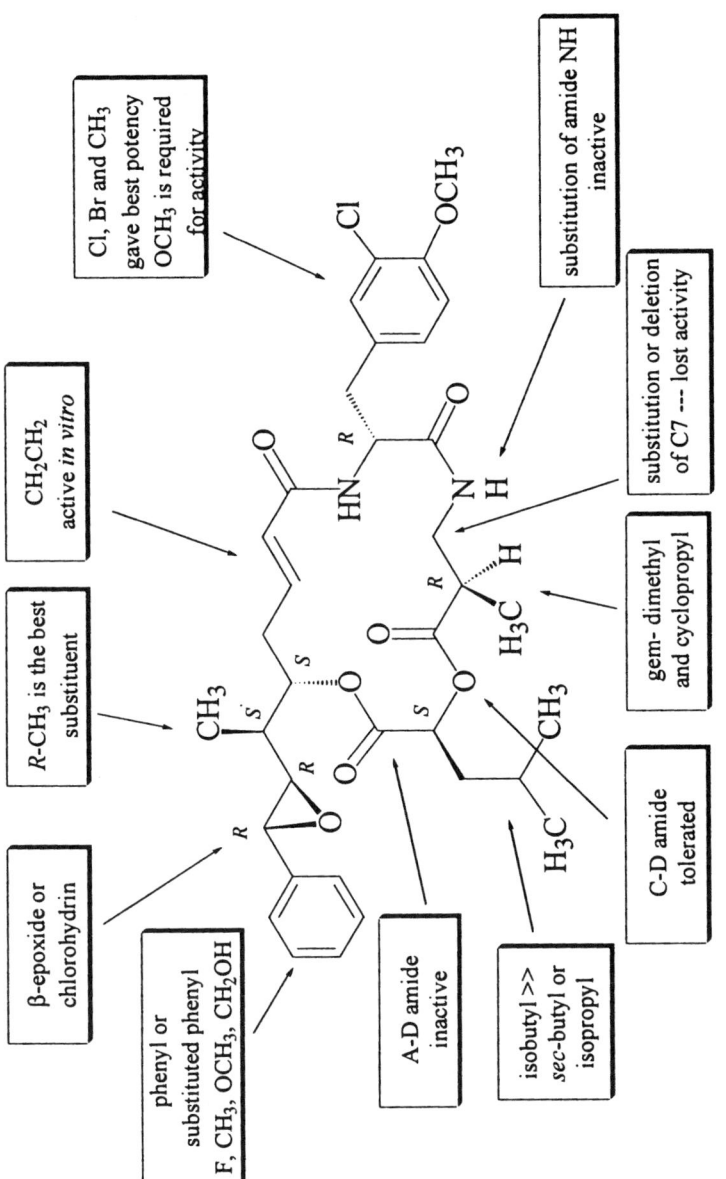

Figure 5. Key SAR findings of cryptophycin.

Fragment A, (C) substitution of the phenyl group on Fragment A, and finally (D) modification of the D-chloromethyltyrosine unit of Fragment B.

SAR of the C/D Ester Bond and the C6, C7- Positions

Some of the early studies of cryptophycin 1 suggested that the ester bond between fragments C and D of the macrocyclic ring may be metabolically labile due to nucleophilic attack at the carbonyl group of the ester bond. This is supported by the observation that arenastatin (a closely related natural product isolated from Japanese sponge *Dysidea Arenaria*) is completely devoid of activity *in vivo*, albeit it has potent activity in culture (*10*). Two approaches were then investigated for the stabilization of the C-D ester bond: (a) geminal substitution α- to the carbonyl group and (b) isosteric replacement of the ester bond with other functionalities such as an amide. Both approaches produced very active derivatives. In particular, introduction of gem-dimethyl group at the C6 position gave an analog (cryptophycin 52, LY355703) with a very much improved *in vivo* antitumor efficacy profile compared to cryptophycin 1. This geminal disubstitution effect in protecting the carbonyl group from nucleophilic attack was also successfully extended to the spirocyclopropyl group (**2**). However, it was found that this effect rapidly diminished when larger groups (gem-diethyl, gem-di-*n*-propyl, spirocyclopentyl, and spirocyclohexyl) were introduced into the C6 position (Figure 6, reference *11*). The amide isosteric placement of the C-D ester bond also gave very active analogs **3** and **4**, regardless of whether there was a monomethyl or a geminal dimethyl group at the C6 position (*12*). Both agents exhibited potent activity in culture ($IC_{50}=$ 0.015 nM); however it was found that the *in vivo* potency for the gem-dimethyl amide analog **4** was significantly lower than that of cryptophycin 52 when compared head-to-head in the Panc 03 model. Replacement of the C-D ester with the corresponding ether linkage (**5**) caused a major loss of activity in cell culture (240-fold) (*12*). As part of the SAR studies of Fragment C, it was also found that moving the substituent (R= CH_3, **6**) from the C6 to the C7 position retains the activity *in vitro*. However, similar to what was observed for arenastatin, absence of substituents at the C6 position caused the compound to be completely inactive *in vivo* presumably due to the metabolic instability of the C-D ester bond. Larger substituents at the C7 position (R= phenyl, **7**) did not improve the activity of the compound either *in vitro* or *in vivo*. Excising the C6 carbon atom of peptolide ring was also attempted and the result was disappointing. The smaller macrocyclic ring (15-membered) analogs **8** and **9** were significantly less active in cell culture (1,400-fold) than the natural cryptophycins (*13*).

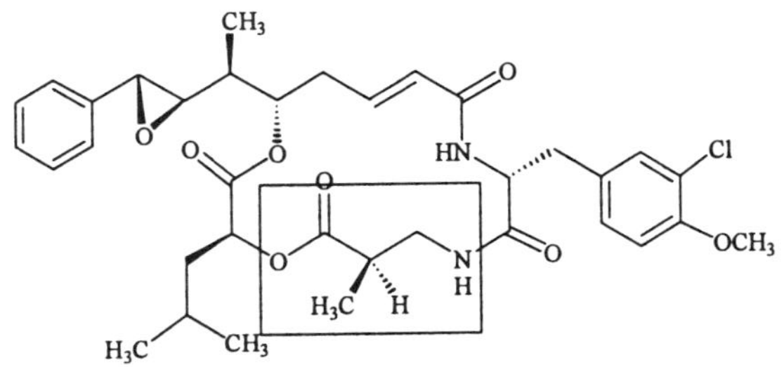

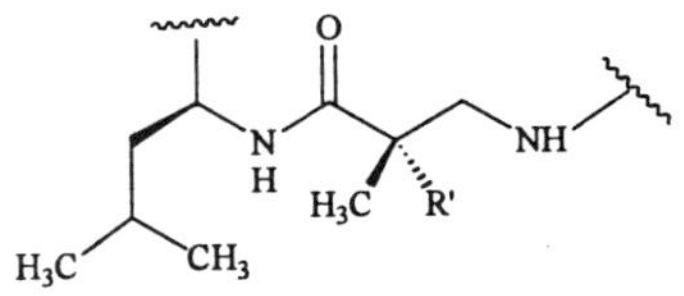

3 R'= H, 0.015 nM
4 R'= CH_3, 0.014 nM
5 C-D ether analog
(IC_{50}= 3.6nM)

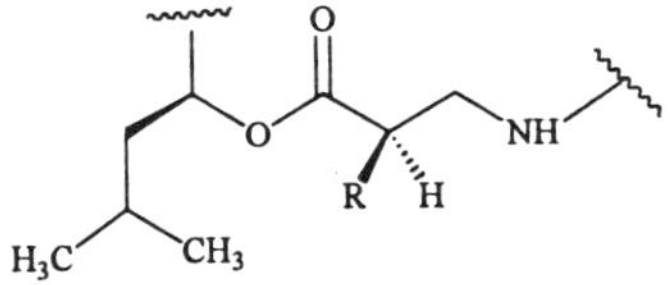

R= H, Arenastatin
(IC_{50}= 0.017 nM)
inactive in vivo

R= CH_3, Cryptophycin 1
(IC_{50} = 0.009 nM)

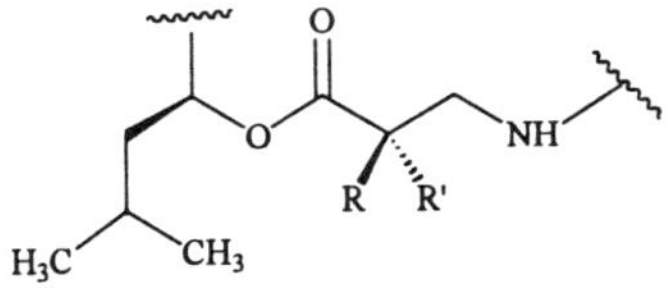

Cryptophycin 52 R= R'= CH_3,
(IC_{50}= 0.038 nM)

2 R= R'= cyclopropyl
(IC_{50}= 0.010 nM)

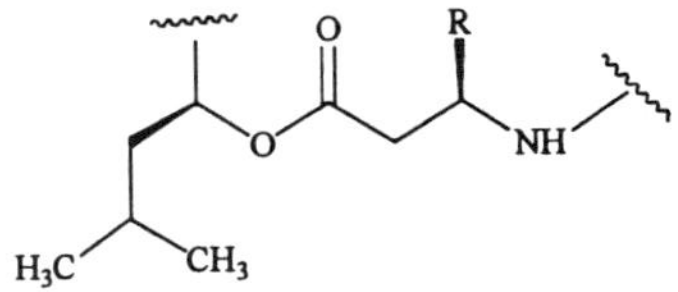

6 R= CH_3, 0.4 nM
7 R= Ph, 1.0 nM

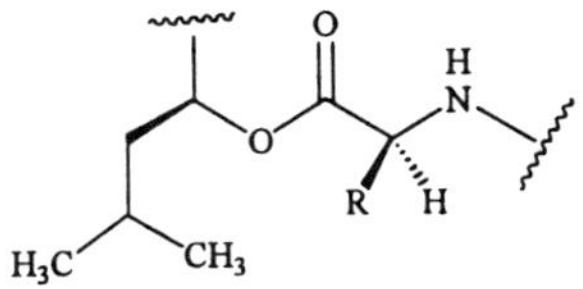

8 R= CH_3, 13 nM
9 R= isopropyl, 306 nM

Figure 6. SAR of CD ester bond and C6, C7- positions.

SAR of β-Epoxide Region of Fragment A

One very important structure-activity feature of the cryptophycins is centered on the configuration of the styrylepoxide portion of the molecule. Through extensive structure-activity studies, it was established that only the β-epoxide or its masked equivalents (such as the corresponding *18R,19S*-chlorohydrin, *18R,19S*-bromohydrin and the reversed *18R,19S*-bromohydrin) possessed the potent cytotoxicity (IC_{50}= 0.004 –0.027 nM) in cell culture assays (Figure 7). Epoxides with other configurations (*R/S, S/R and S/S*) were much less active. Displacement of the oxygen atom of the epoxide with sulfur (episulfide, **10**) or nitrogen (aziridine, **11**) led to major loss of cytotoxicity (*14*). Removal of the oxygen functionality at C18/C19 also led to inactive phenethyl (**12**), styryl (**13**) and chlorobenzyl (**14**) derivatives of cryptophycin. The corresponding diol and the fluorohydrin (**15**) derivatives of β-epoxide were also much less active in the *in vitro* assays. In summary, the SAR in the C18 and C19 region of cryptophycin clearly delineate the need for a highly reactive electrophilic center such as an epoxide or functionality that can serve as a masked β-epoxide. For example, pharmacokinetic data (*vide infra*) clearly suggested that the *18R,19S*-chlorohydrin was rapidly converted to the corresponding β-epoxide *in vivo* in rodents and could account for the majority of the antitumor activity seen in murine tumors and human tumor xenografts.

SAR of the Phenyl Group of Fragment A

It was observed that the activity of cryptophycins can be modulated by introducing either small electron-donating or electron-withdrawing groups onto the phenyl ring of the styrene oxide region (*15*). Introduction of electron-donating groups such as methyl, hydroxymethyl, methoxy, amino, and substituted amino groups to the phenyl ring usually led to highly potent cryptophycin derivatives *in vitro* (*m*-methoxy, IC_{50}= 0.015 nM; *p*-hydroxymethyl, IC_{50}= 0.004 nM; compared with IC_{50}= 0.019 nM for cryptophycin 52, measured in CCRF-CEM cells). Excellent antitumor activity with a significant increase of potency was also observed for these analogs *in vivo*. For example, an 8-fold increase of *in vivo* potency was observed for the *p*-methyl (IC_{50}= 0.036 nM) derivatives of cryptophycin 52 in Panc 03 model.

Introduction of an electron-withdrawing group such as nitro, cyano, carboxylic acid (and its methyl ester) and trifluoromethyl groups to the phenyl ring also led to highly cytotoxic derivatives in cell culture (*p*-nitro, IC_{50}= 0.4 nM; *p*-carboxymethyl, IC_{50}= 0.36 nM; cryptophycin 52, IC_{50}= 0.1 nM). Along the same trend, introduction of halogen atom(s) (chlorine or fluorine) at either the *ortho*- or *para*- position relative to the oxirane ring of the styrene oxide also led to derivatives with potent cytotoxicity *in vitro* (*o*-fluoro, IC_{50}= 0.039 nM; *p*-fluoro, IC_{50}= 0.005 nM; *p*-chloro, IC_{50}= 0.020 nM) and good antitumor activity *in vivo* (*p*-chloro, %T/C= 0 for Panc 03 tumor model).

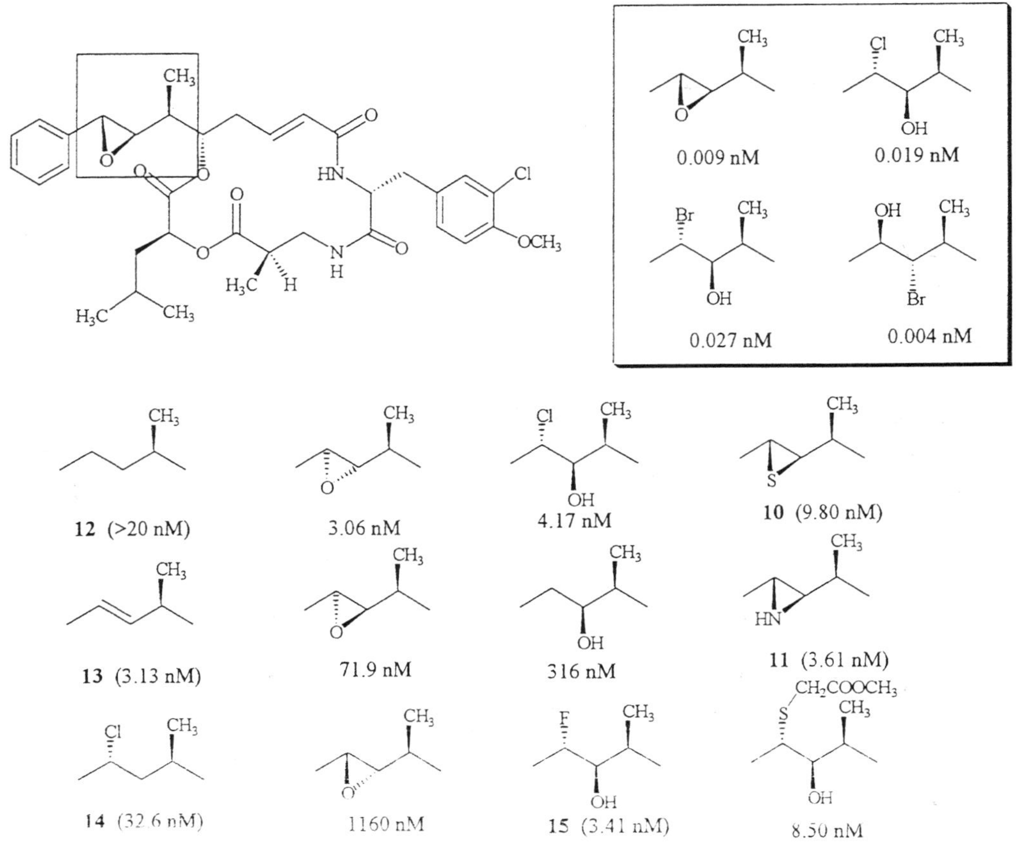

Figure 7. SAR of β-epoxide of Fragment A.

Replacing the phenyl group of the styrene oxide with heteroatom-containing aromatics such as pyridine, thiophene, furan or with naphthalene, however, all led to a major loss of cytotoxicity in cell culture. Replacement of the phenyl group with simple alkyl groups (methyl, ethyl, *n*-propyl, methoxy, formyl and *t*-butyl ester) also led to less active cryptophycin compounds in cell culture studies and inactivity in corresponding animal studies.

Using radiolabeled cryptophycin 52 (OCT3 in the chloro-D-tyrosine unit), Professor Lester Wilson's group at the University of California at Santa Barbara was able to clearly demonstrate that the binding of cryptophycin to tubulin was not an irreversible process (*16*). The data suggested that no covalent interaction had taken place between the protein (tubulin) and the epoxide functionality of cryptophycins. This ruled out the earlier hypothesis of a possible SN2 nucleophilic opening of the styrene-β-oxide. The SAR showed that both electron-donating and -withdrawing groups can enhance antiproliferative effects of cryptophycins which also suggests that a transition state involving a fully or partially developed benzylic carbonium ion (SN1 process) is unlikely. In conclusion, the SAR studies around the β-styrene oxide region indicated that the reactivity of this portion of the molecule toward tubulin may have little dependency on the electronic nature of phenyl ring. The binding pocket of tubulin and microtubules which can accommodate the β-styrene oxide region may be hydrophobic in nature. The steric fitness of the substituted styrene oxide and the shape of the binding pocket may have a more profound role in determining the binding affinity of the new analogs to the target.

SAR of the D-Chlorotyrosine of Fragment B

The SAR in fragment B indicated that this is probably the region that is least amenable to structure modification. Various structural modifications of this region of the molecule all led to significant loss of cytotoxicity (*17*). These include replacing the chlorine atom with fluorine, introduction of additional halogen atoms onto the phenyl ring, modification of the methoxy ether, replacing the chloromethoxyphenyl with other aromatic or non-aromatic groups and replacing chloromethyltyrosine with other simple natural and unnatural amino acids (such as glycine, valine, proline, *p*-aminophenylalanine, *p*-hydroxyphenylglycine, phenylethylglycine, etc.). Methylation of the asymmetric methine carbon at C10 also gave inactive analogs. The only tolerable modification in fragment B was the replacement of the chlorine with a bromine or a methyl group. This for example led to the corresponding methyl D-tyrosine derivative that had only a 2-fold decrease of cytotoxicity. (IC_{50}= 0.04 nM; cryptophycin 52, IC_{50}= 0.019 nM, measured in CCRF-CEM cells)

Comparison of the Epoxide and the Chlorohydrin Derivatives of Cryptophycin

As described earlier in the SAR of Fragment A, the chlorohydrin derivatives of the β-epoxides are highly active both in cell culture and *in vivo* tumor models. For example, cryptophycin 55 (LY355702, chlorohydrin derivative of cryptophycin 52) was found to be consistently more active *in vivo* than cryptophycin 52 when both agents were dosed to the maximum tolerated doses (MTD), although the MTD for the chlorohydrin was usually 4-6 times higher than that of the epoxide (*18*). This initial excitement, however, was later dampened by the discovery that the chlorohydrin derivatives were unstable, and were converted to the more potent epoxides under various *in vitro* as well as *in vivo* conditions. It was found that the chlorohydrin derivatives could be converted to the epoxides in the dosing vehicle containing 5% ethanol and 5% cremophor. This conversion was also documented in plasma (*in vitro*) and in animal pharmacokinetic studies. For example, in mice dosed intravenously with 80 mg/kg of cryptophycin 55, the plasma levels of cryptophycin 52 (converted from cryptophycin 55) were almost identical to those found after administration of 20 mg/kg of cryptophycin 52 itself (a dose relationship that was consistent with the findings of the efficacy studies). This observation suggested that the chlorohydrin derivatives cryptophycin 55 may act as prodrugs of the β-epoxides of cryptophycin 52. Finally, formulation stability studies indicated that cryptophycin 52 was stable up to three months in the ethanol/cremophor concentrate while the chlorohydrin derivative LY355702 was converted quite extensively to the epoxides under the same condition limiting the shelf life of the compound.

In conclusion, the SAR described above has covered a large number of cryptophycins with a broad array of structural diversity. The SAR pinpointed some critically important areas for structure-activity optimization. These include the styrene β-oxide region (chlorohydrins versus epoxides), the C/D ester bond region and the phenyl region of Fragment A. From assessment of the collective data of *in vitro/in vivo* pharmacology, biopharmaceutical properties, ADME/pharmacokinetic evaluation and toxicological findings, cryptophycin 52 (LY355703) was selected as a first generation cryptophycin for clinical evaluation.

The *In Vivo* Antitumor Activity of Cryptophycin 52 (LY355703)

The mode of action of cryptophycin 52 is associated with inhibition of mitotic spindle function. On the biochemical level, cryptophycin 52 inhibited microtubule polymerization in a concentration-dependent manner. Cryptophycin 52 also powerfully suppressed microtubule dynamics that are critical for mitotic progression. Cryptophcin's effects on microtubule dynamics and mitotic spindle functions have been extensively studied by Wilson *et al* (20, 21). Similarly, detailed investigation of the cellular pharmacology of cryptophycin 52, including cytotoxicity, cell cycle effect and its sensitivity toward the multidrug resistance proteins (Pgp and MRP), have also been investigated by Williams *et al* (19) and thus will not be discussed in this review.

Table I. Antitumor Activity (Log Kill*) of LY355703 and Other Oncolytics Against Human Tumor Xenografts

Agent	HCT-116 Colon	H125 Colon	LNCaP Prostate	TSU Prostate	PC-3 Prostate
Cryptophycin 52	+++	++	++++	++	+
Paclitaxel	+++	+	++++	++++	++
Doxorubicin		-	-	-	-
Etoposide	±	+	+	±	-
VLB/VCR	+	+++	++++	++++	++
Cytoxan	-	-	++++	++++	-
CisPlatin	-	-	-	-	-
5-FU	-	-	-	-	-

* Log Kill = T-C/3.2XTd (*6*). Log Kill activity rating: ++++ (>2.8), +++ (2.0-2.8), ++ (1.3-1.9), + (0.7-1.2), - (inactive, <0.7)

The antitumor activity of cryptophycin 52 *in vivo*, however, is summarized here to illustrate the broad spectrum activity in a variety of human tumor xenografts. Due to the lipophilic nature of the cryptophycins, cryptophycin 52 was formulated in a mixture of 5% Cremophor and 5% ethanol in 90% saline, a formulation that is used clinically for paclitaxel. The treatment schedule varied but was generally consistent with at least five repeated intermittent intravenous injections, usually every other day (q2d), and the doses used in the xenograft studies were usually in the range of 6-10 mg/kg/injection/day for cryptophycin 52. The MTD for cryptophycin 52 in nude mice is 10 mg/kg/injection and a total dose of ca. 50-60 mg/kg. Table 1 summarizes the antitumor activity (expressed in Log Kill) of cryptophycin 52 against five human tumor xenografts (HCT-116 colon, H125 lung, LNCaP prostate, TSU prostate and PC-3 prostate carcinomas). In comparison, the antitumor activity of a group of commonly used oncolytic agents (paclitaxel, doxorubicin, etoposide, vincristine/vinblastine, cytoxan, cisplatin and 5-flurouracil) is also included (*18, 6b*). It is clear from these studies that cryptophycin 52 is a very active anticancer agent with superior antitumor activity than many of the commonly used oncolytic drugs. In addition, cryptophycin 52 also exhibited excellent activity against tumors that are resistant to adriamycin (Mamm16C/Adr, Mamm17/Adr, both are syngeneic murine tumors) and vinblastine (UCLA-P3, human lung xenograft), whereas paclitaxel, doxorubicin, etoposide and vinblastine were completely inactive in these multidrug resistant tumors.

Conclusions

Cryptophycins represent a new macrocyclic chemical platform with a novel mode of action that targets mitotic spindles and have excellent antitumor activity in preclinical models. The activity of cryptophycins against resistant tumors that express the multidrug resistant pump suggests that cryptophycins may be highly active against human solid tumors that are resistant to anthracyclines, taxanes and vinca alkaloids, and thus may offer alternative efficacy or survival advantages over the current therapies. The successful development of a very efficient convergent total synthesis for the cryptophycin class of depsipeptides facilitated the rapid SAR studies for this novel series of antimitotic agents, and the selection of cryptophycin 52 as the first generation of cryptophycin for clinical trial. Three phase I dose escalation studies have been successfully completed (*22*) for cryptophycin 52 and the compound is currently being evaluated in phase II trials against various tumor types. It will be interesting to see if this novel class of antimitotic agents can demonstrate significant efficacy against human solid tumors and provide additional benefits to patients who have failed chemotherapy due to the resistant nature of their diseases.

References

1. Schwartz, R. E.; Hirsch, C. F.; Sesin, D. F.; Flor, J. E.; Chartrain, M.; Fromcling, R. E.; Harris, G. H.; Salvatore, M. J.; Liesch, J. M.; Yudin, K.; *J. Ind. Microbiol.* **1990**, 5, 118-126.
2. Patterson, G. M.; Baldwin, C. L.; Bolis, C. M.; Caplan, F. R.; Karuso, H.; Larsen, L. K.; Lwevine, I. A.; Moore, R. E.; Nelson, C. S.; Tschappat, K. D.; Tuang, G. D.; Furusawa, S.; Norton, T. R.; Raybourne, R. B.; *J. Phycol.* **1991**, 27, 530-536.

3. Corbett, T. H.; Valeriote, F. A.; Polin, L.; Panchapor, C.; Pugh, S.; White, K.; Lowichik, N.; Knight, J.; Bissery, M. C.; Wozniak, A.; LoRusso, P.; Biernat, L.; Polin, D.; Knight, L.; Biggar, S.; Looney, D.; Demchik, L.; Jones, J.; Jones, L.; Blair, S.; Palmer, K.; Essenmacher, S.; Lisow, L.; Mattes, K. C.; Cavanaugh, P.F.; Rake, J. B.; Baker, L.; *Cytoxic Anticancer Drugs: Models and Concepts for Drug Discovery and Development*; Valeriote, F. A.; Corbett, T. H.; Baker, L. H.; Eds.; Kluwer Academic Publishers, **1992**, 35-87.
4. Valeroite, F. A.; Moore, R. E.; Patterson, G. M. L.; Paul, V. J.; Scheuer, P. J.; Corbett, T. H.; *Discovery and Development of Anticancer Agents*, Valeriote, F. A.; Corbett, T. H.; Baker, L. H.; Eds.; Kluwer Academic Publishers, **1997**, 67-93.
5. (a) Golakoti, T.; Ohtani, I.; Patterson, G. M .L; Moore, R. E.; Corbett, T. H.; Valeriote, F. A.; Demchik, L.; *J. Am. Chem. Soc.* **1994**, 116, 4729-4737. (b) Golakoti, T.; Ogino, J.; Heltzel, C. E.; Le Husebo, T.; Jensen, C. M.; Larsen, L. K.; Patterson, G. M. L.; Moore, R. E.; Mooberrr, S. L.; Corbett, T. H.; Valeriote, F. A.; *J. Am. Chem. Soc.* **1995**, 117, 12030-12049.
6. (a) Corbett, T. H.; Valeriote F. A.; LoRusso, P.; Polin, L.; Panchapor, C.; Pugh, S.; White, K.; Knight, J.; Demchik, L.; Jones, J.; Jones, L.; Lisow, L.; *Anticancer Drug Development Guide*, Teicher, B. A.; Eds.; Humans Press, **1997**, 75-99. (b) Polin, L.; Valeriote, F.; White, K.; Panchapor, C.; Pugh, S.; Knight, J.; LoRusso, P.; Hussain, M.; Liversidge, E.; Peltier, N.; Golakoti, T.; Patterson, G.; Moore, R.; Corbett, T. H.; *Invest. New Drugs,* **1997**, 15, 99-108.
7. (a) Barrow, R. A.; Hemscheidt, T.; Liang, J.; Paik, S.; Moore, R. E.; Tius, M. A.; *J. Am. Chem. Soc.* **1995**, 117, 2479-2490. (b)Martinelli, M. J.; Nayyar, N. K.; Moher, E. D.; Dhokte, U. P.; Pawlak, J. M.; Vaidyanathan, R.; *Org. Lett.* **1999**, 1(3), 447-450. (c) Martinelli, M. J.; Vaidyanathan, R.; Van Khau, V.; *Tetrahedron Lett.* **2000**, 41(20), 3773-3776.
8. Fray. A. H.; *Tetrahedron Asymmetry*, **1998**, 9(16), 2777-2781.
9. Moore, R. E.; Corbett, T. H.; Patterson, G. M. L.; Valeriote, F. A.; *Current Pharmaceutical Design*, 1996; 2(3), 317-330.
10. (a) Kobayashi, M.; Aoki, S.; Ohyabu, N.; Kurosu, M.; Wang, W.; Kitagawa, I.; *Tetrahedron Lett.* **1994**, 35, 7969-7972. (b) Kobayashi, M.; Kurosu, M.; Ohyabu, N.; Wang, W.; Fujii, S.; Kitagawa, I.; *Chem. Pharm. Bull.* **1994**, 42, 2196-2207.
11. Varie, E.; Shih, C.; Hay, D.; Andis, S.; Corbett, T.; Gossett, L.; Janisse, S.; Martinelli, M.; Moher, E.; Schultz R.; *Bioorg. Med. Chem. Lett.* **1999**, 9, 369-374.
12. Norman, B. H.; Hemscheidt, T.; Schultz, R. M.; Andis, S. L.; *J. Org. Chem.* **1998**, 63(15), 5288-5294.

13. Shih, C.; Gossett, L. S.; Gruber, J. M.; Grossman, C. S.; Andis, S. L.; Schultz, R. M.; Worzalla, J. F.; Corbett, T. H.; Metz, J. T.; *Bioorg. Med. Chem. Lett.* **1999**, 9, 69-74.
14. Moore, R. E.; *J. Ind. Microbiol.* **1996**, 16(2), 134-143.
15. Al-Awar, R. S., personal communication, Lilly Research Laboratories.
16. Panda, D.; DeLuca, K.; Williams, D.; Jordan, M. A.; Wilson, L.; *Proc. Natl. Acad. Sci. USA* **1998**, 95(16), 9313-9318.
17. Patel, V. F.; Andis, S. L.; Kennedy, J. H.; Ray, J. E.; Schultz, R. M.; *J. Med. Chem.* **1999**, 42(14), 2588-2603.
18. (a) Worzalla, J. F.; Cao, J.; Ehlhardt, W. J.; Harrison, S. D.; Law, K. L.; Martinelli, M. J.; Self, T. D.; Starling, J. J.; Shih, C.; Thoebald, K. S.; Toth, J. E.; Zimmermann, J. L.; Corbett, T. H.; *Proc. Amer. Assoc. Can. Res.* **1997**, 38, 225 (Abs. 1516). (b) Polin, L.; Valeriote, F.; Moore, R.; Tius, M.; Barrow, R.; Hemscheidt, T.; Liang, J.; Paik, S.; White, K.; Harrison, S.; Shih, C.; Martinelli, M.; Corbett, T.; *Proc. Amer. Assoc. Can. Res.* **1997**, 38, 225, (Abs. 1514).
19. (a) Wagner, M. M.; Paul, D. C.; Shih, C.; Jordan, M. A.; Wilson, L.; Williams, D. C.; *Cancer Chemotherapy and Pharmacology,* **1999**, 43(2), 115-125. (b) Williams, D. C.; Wagner, M. M.; Paul, D. C.; Jordan, M. A.; Wilson, L.; Shih, C.; *Annals of Oncology,* **1998**, 9, 342-348.
20. (a) Jordan, M. A.; Wilson, L.; *Curr. Opin. Cell. Biol.* **1998**, 10(1), 123-130. (b) Panda, D.; Jordan, M. A.; DeLuca, K.; Wilson, L.; *Molecular Biology of the Cell*, **1998**, 9 (Suppl), 273A. (c) Wilson, L.; Jordan, M. A.; *Chem. Biol.* **1995**, 2, 569-573. Panda, D.; Williams, D.; Shih, J.; Jordan, M. A.; Wilson, L.; *Annals of Oncology*, **1998**, 9, 341-351.
21. Panda, D.; Williams, D.; Shih, J.; Jordan, M. A.; Wilson, L.; *Annals of Oncology*, **1998**, 9, 341-351.
22. Stevenson, J. P.; Gallagher, M.; Vaughn, D.; Schucter, L.; Haller, D.; Fox, K.; Algazy, K.; Aviles, V.; Hahn, S.; O'Dwyer, P. J.; *Proc. Amer. Assoc. Can. Res.* **1999**, 40, 92.

Chapter 11

Farnesyltransferase Inhibitors as Potential Anticancer Agents

J. B. Gibbs[1], N. J. Anthony[2], I. Bell[2], C. A. Buser[1], J. P. Davide[1],
S. J. deSolms[2], C. Dinsmore[2], S. L. Graham[2], G. D. Hartman[2],
D. C. Heimbrook[1], H. Huber[1], K. S. Koblan[1], N. E. Kohl[1],
R. B. Lobell[1], and T. M. Williams[2]

[1]Cancer Research and [2]Medicinal Chemistry, Merck Research Laboraotries, Sunnytown Pike, West Point, PA 19486

Localization of the Ras oncoproteins to the inner surface of the plasma membrane is essential for their biological activity. This observation suggested that the enzyme that mediates the membrane localization, farnesyl-protein transferase (FTase), would be a target for the development of novel anti-cancer agents. We have developed potent, cell-active inhibitors of FTase that exhibit antiproliferative activity in cell culture and block the morphological alterations associated with Ras-induced transformation of mammalian cells in monolayer cultures. In vivo, these compounds block the growth of *ras*-transformed fibroblasts in a nude mouse xenograft model and block the growth and in some cases cause regression of mammary and salivary tumors in several strains of *ras* transgenic mice in the absence of any detectable side effects. The results of our preclinical studies and those of others suggest that FTIs may have utility against a variety of human cancers, a hypothesis that is currently being tested in the clinic.

Introduction

The Ras proteins serve as GTP binding molecular switches that interface transmembrane tyrosine kinase growth factor receptors with intracellular signaling cascades [1]. Of the 3 alleles of *ras* (Ha, Ki, N-*ras*), the Ki-*ras* gene is found mutated in many common human cancers such as colon, lung, and pancreatic carcinomas. Ras protein functions in the plasma membrane, and post-translational modification of the Ras C-terminal domain (the so-called CaaX sequence) is required for this localization [2]. The first step is farnesylation of the CaaX cysteine residue; the second step is proteolytic cleavage of the aaX tripeptide, and the final step is methylation of the cysteine carboxylate. The farnesyl group is derived from a metabolite of the mevalonate biosynthetic pathway. Inhibitors of HMG-CoA reductase, the rate limiting enzyme of this pathway, were shown to inhibit Ras farnesylation and also inhibit Ras membrane localization. However, with a view towards developing a possible cancer therapeutic, efforts were initiated in 1989 to identify the enzyme which was responsible for placing a farnesyl group onto Ras.

Farnesyl-protein transferase (FTase) modifies specific CaaX sequences such as that found in Ras proteins using farnesyl diphosphate as substrate in a reaction that requires magnesium and zinc ions [2]. The identification of FTase attracted attention within the pharmaceutical industry for two main reasons. First, farnesylation of Ras is essential to both its membrane localization and cell transforming activity. This was shown either by mutating the CaaX sequence so that Ras could no longer serve as substrate for FTase or, in yeast, by deleting the gene encoding a unique subunit of FTase. Second, FTase afforded an enzyme target, and enzymes have historically been amenable to inhibition by designed small molecules antagonists. Several recent reviews provide more details on the background of this field as well as the breadth of pharmaceutical effort to develop FTIs [2-6]. In this review, we will focus on some of our recent work on FTIs.

Biology of farnesyltransferase inhibitors in vitro

An early example of an FTase inhibitor (FTI) was L-744,832 (Fig 1). This compound serves as a mimetic of the CaaX sequence and indeed inhibits FTase by a mechanism that is competitive with respect to protein substrate [7]. Although L-744,832 inhibits FTase selectively over a related enzyme, geranylgeranyl-protein transferase type I (GGTase-I), this compound is not specific for inhibition of Ras substrate alone. This was demonstrated by incubating cells with [^{3}H]mevalonate acid (to radiolabel all farnesylated and geranylgeranylated proteins) either in the presence or absence of L-744,832, followed by 2-dimensional gel electrophoresis and autoradiography. As shown in Fig. 2, a number of spots observed in the vehicle control group are not present in the FTI-treated group. The spots that remain represent geranylgeranylated proteins exemplifying the enzyme specificity of this compound *in vivo*. This result indicates that FTIs inhibit the farnesylation of several proteins in addition to Ras. Example of farnesylated proteins include nuclear lamins and kinesins, tyrosine and inositol phosphatases, and the ras related proteins RhoB, Rheb, and Rap1, as well as several other proteins. Inhibition of the farnesylation and function of these proteins may contribute to the biology of FTIs [2,4,5,8].

In spite of the fact that FTIs are not specific to Ras function, they nevertheless display an antiproliferative effect against tumor cells in a manner distinct from cytotoxic agents like doxorubicin and paclitaxel. In a soft agar assay, Ha-*ras* transformed Rat1 fibroblast cells and PSN-1 human pancreatic tumor cells are more sensitive to an FTI than Rat1 cells transformed by the Raf oncogene protein which is not dependent on farnesylation for function [9]. In contrast, paclitaxel or doxorubicin was equally inhibitory against all 3 cells lines. Using a panel of human breast cancer cell lines, Rosen, Kohl, and Sepp-Lorenzino demonstrated that L-744,832 had greater anti-proliferative activity when used in combination with cytotoxic agents such as doxorubicin, cisplatin, vincristine, paclitaxel, and an epothilone [10]. It was interesting that the combination of FTI with either of the two agents that enhance microtubule polymerization, paclitaxel and epothilone, appeared to have the most inhibitory effect in these experiments.

Biology of farnesyltransferase inhibitors in animals

Several animal tumor models were used to evaluate the activity of FTIs. We focused on a transgenic mouse v-Ha-*ras* breast cancer model because it was readily available [7]. L-744,832 was capable of inducing regression of pre-existing tumors in a manner that appeared to be non-toxic to the animal. This anti-tumor activity was superior to doxorubicin both in terms of efficacy as well as safety.

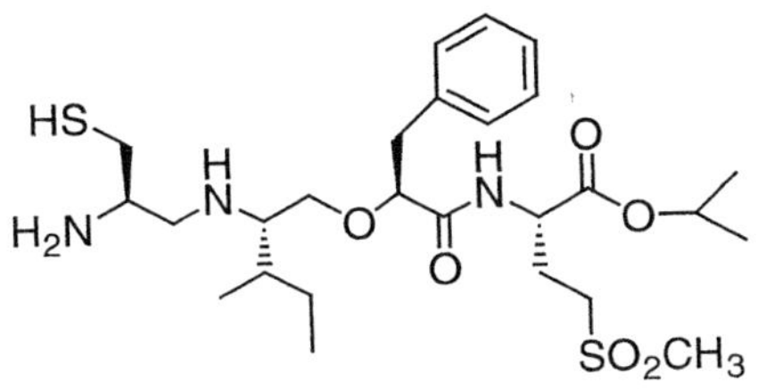

Figure 1. Structure of the peptidomimetic FTI L-744,832

2D Gel Analysis:
[^{3}H-mevalonate]-labelled Rat1/v-ras proteins

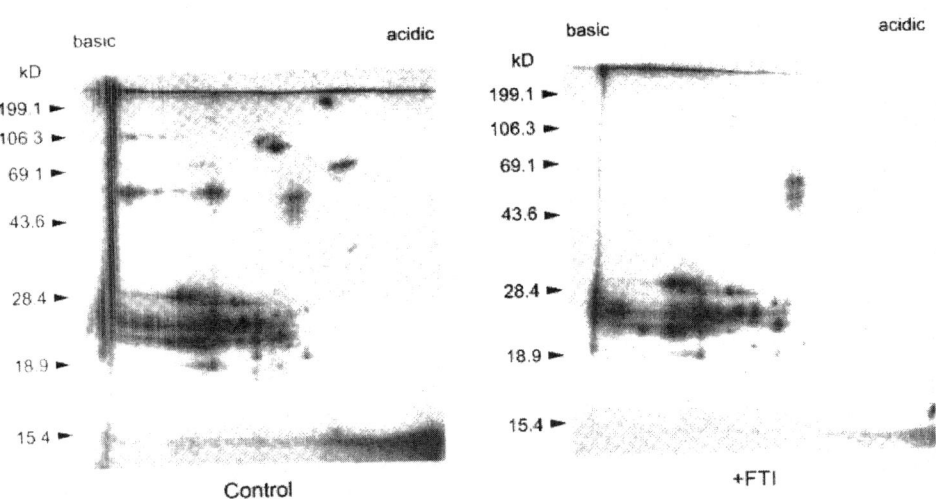

Figure. 2: v-H-ras transformed Rat 1 cells were treated with FTI for 4 hours prior to addition of 100 µCi/ml ^{3}H-mevalonolactone (60 Ci/mmol, American Radiolabeled Chemicals). Incubation in the FTI and tracer continued for an additional 20 hrs. Cells were harvested, lysed and focused on IEF gels for a total of 9kV. The first dimension was fixed atop of a 14% SDS polyacrylamide gel and electrophoresed at 40 mAmps until the dye front reached the bottom. Gels were fixed, enhanced with Enlightening (NEN), dried and set under film for 5 to 30 days.

The mechanism whereby L-744,832 caused tumor regression was by induction of apoptosis [11]. The apoptotic mechanism was independent of wild-type p53 function (a key regulator of apoptosis) because mice having the genotype v-Ha-ras/p53$^{-/-}$ also demonstrated tumor regression by an apoptotic mechanism when treated with L-744,832 [11]. Interestingly, the tumor regression observed was not a complete cure. There was histological evidence for residual tumor, and if treatment with L-744,832 was stopped, tumors would regrow [7]. Overt systemic toxicity was not observed in L-744,832 treated v-Ha-*ras* animals in contrast to the doxorubicin-treated animals which displayed weight loss when treated at the maximally tolerated dose.

While the experiments with v-Ha-*ras* transgenic mice were encouraging, this allele of *ras* is not as clinically relevant as Ki-*ras* which is found mutated in many solid tumors and in some leukemias. In cells, Ki-*ras* appears to have similar biological properties to Ha-*ras*, but the biochemistry of the proteins encoded by these genes differs. Due to different sequences in the CaaX box, Ha-Ras is a substrate for FTase only whereas Ki-ras is a substrate for both FTase and GGTase-I [2,4]. The hypothesis that this result generated was that tumors having mutant Ki-ras would be resistant to the action of FTIs. For this reason, an MMTV-[Val12]Ki4B-*ras* mouse was created and tested with L-744.832 [12]. Inhibition of tumor growth was noted, although frank tumor regression was not as prominent as that seen in the MMTV-v-Ha-*ras* mice. In tumor tissue of animals treated with L-744,832, the processing of Ki-Ras was not inhibited, perhaps because of GGTase-I activity. The observed inhibition of Ki-Ras tumor formation in this model in the absence of inhibiting Ki-Ras protein processing supports models of FTI biological function in which proteins in addition to Ras contribute to the anti-tumor activity of FTIs.

Nonpeptidomimetic inhibitors

Medicinal chemistry efforts were able to convert peptidomimetics such as L-744,832 into novel nonpeptidomimetics such as those shown in Fig. 3 [13]. These FTIs lack a thiol group, peptide bonds, and do not have labile esters as does L-744,832. As shown in Fig. 3, compounds 1 and 2 are potent against FTase, but display differential specificities with respect to GGTase-I. Compound 1 is highly specific for FTase whereas compound 2 has relaxed specificity. The differences in the *in vitro* biochemical specificities was also manifest in a cell-based assay. Whereas, both compound 1 and 2 were able to inhibit the processing of a farnesylated substrate (not shown), only compound 2 was able to inhibit the processing of the geranylgeranylated protein Rap1.

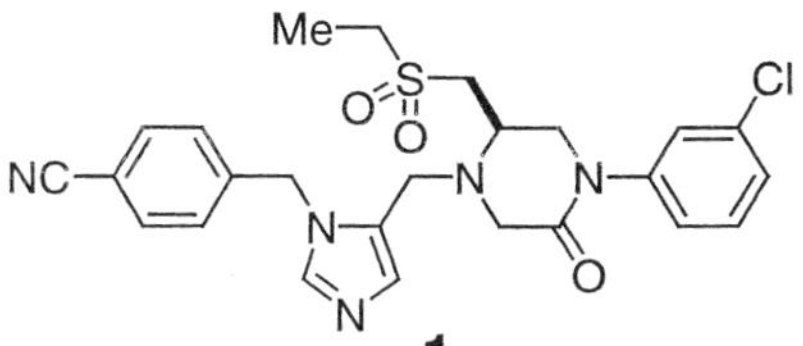

	FTase (IC$_{50}$, nM)	GGTase-I (5 mM ATP) (IC$_{50}$, nM)	Rap1a Processing (MIC, nM)
1	0.2	3,800	>10,000
2	2	98	1,000

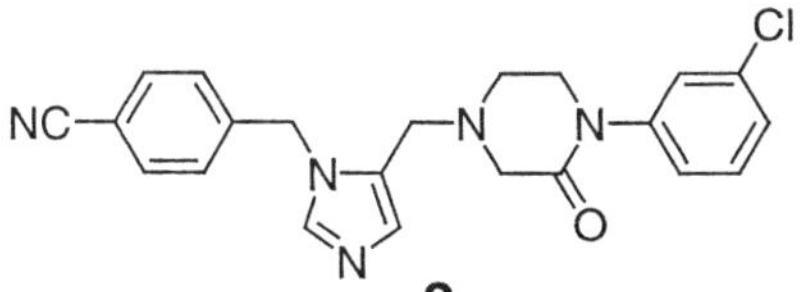

Figure 3. Structures of non-peptide FTIs and biochemical properties. Compounds were assayed with human FTase or with human GGTase-I in the presence of 5 mM ATP. Activity in cells was also tested by monitoring the processing of Rap1 which is a substrate for GGTase-I (MIC, minimal inhibitory concentration).

Compound 2 was specifically tested for inhibition of Ki-Ras processing. Buser et al. developed an LC-MS methodology to accurately identify the specific modification of the Ki-Ras C-terminus. In untreated cells, Ki-Ras is modified only by a farnesyl group as previously reported by others using radioisotope methodologies. Treatment of cells with compound 1 revealed Ki-Ras modified by both farnesyl and geranylgeranyl. However, in cells treated with compound 2, fully unmodified Ki-Ras was detected in addition to farnesylated Ki-Ras. This indicates that the GGTase-I inhibitory activity of compound 2 was sufficient to block the alternative prenylation of Ki-Ras.

Both compounds 1 and 2 were active in animal tumor models (xenografts and transgenic mice) containing mutant Ki-*ras*. However, higher doses were needed to see anti-tumor efficacy in these models than in corresponding ones containing mutant Ha-*ras*. Furthermore, at the highest doses tested, some evidence of toxicity was noted as monitored by the histology of bone marrow and the gastrointestinal tract. Ki-Ras processing was not inhibited in these experiments.

Monitoring clinical development

Clinical trials were initiated with compound 2. It was administered during phase I into cancer patients by continuous infusion over 7 days [14]. The toxicities noted were neutropenia and prolonged QTc interval in electrocardiograms. Both of these toxicities were observed in preclinical studies and were reversible. No responses were reported in phase I when compound 2 was administered as monotherapy.

To assess whether plasma concentrations sufficient to inhibit protein prenylation were achieved, Lobell et al. developed an assay that could be adapted to white blood cells collected from patients [15]. In this assay, isolated white blood cells were lysed and then protein was applied to an SDS-PAGE to separate proteins. The processing of the heat shock protein hDJ (which is farnesylated) was monitored by Western blotting using mobility shift as an indicator. This protein has a robust signal and a large dynamic range of processed/unprocessed for ease of detection. Indeed, inhibition of hDJ processing was detected as a function of increasing plasma concentrations of compound 2. This inhibition was maximal when compound 2 began reaching plasma concentrations >10 uM. This suggests that sufficient concentrations of compound 2 were achieved to inhibit protein farnesylation in a circulating blood cell type. Similar pharmacodynamic assays have been reported with the FTI SCH 66336 which is currently in phase II clinical trials [16].

Conclusions

FTIs were developed based upon knowledge of the ras oncogene functions, but it is clear now that other proteins are likely involved with the anti-tumor activity of these compounds. As such, toxicity to some normal proliferating cells should be expected. Clinical trials were initiated by several pharmaceutical companies including Janssen, Schering-Plough, Merck, Abbott, Bristol-Myers Squibb, and Pfizer [2]. The more advanced trials are also evaluating FTIs in combination with other therapeutic modalities. The results of these trials are eagerly awaited as a test of the hypothesis that targeting FTase may have a beneficial anti-tumor effect in human cancer patients.

References

1) Lowy, D. R.; Willumsen, B. M. Annu. Rev. Biochem. **1993**, 62, 851-891.
2) Oliff, A. Biochim Biophys Acta **1999**, 1423, C19-30.
3) Sebti, S.; Hamilton, A. D. Curr Opin Oncol **1997**, 9, 557-61.
4) Gibbs, J. B.; Oliff, A. Annu Rev Pharmacol Toxicol **1997**, 37, 143-66.
5) Cox, A. D.; Der, C. J. Biochim Biophys Acta **1997**, 1333, F51-71.
6) Williams, T. M. EXPERT OPINION ON THERAPEUTIC PATENTS **1999**, 9, 1263-1280.
7) Kohl, N. E.; Omer, C. A.; Conner, M. W.; Anthony, N. J.; Davide, J. P.; deSolms, S. J.; Giuliani, E. A.; Gomez, R. P.; Graham, S. L.; Hamilton, K.; Handt, L. K.; Hartmen, G. D.; Koblan, K. S.; Kral, A. M.; Miller, P. J.; Mosser, S. D.; O'Neill, T. J.; Rands, E.; Schaber, M. D.; Gibbs, J. B.; Oliff, A. Nature Medicine **1995**, 1, 792 - 797.
8) Prendergast, G. C.; Du, W. DRUG RESISTANCE UPDATES **1999**, 2, 81-84.
9) Kohl, N. E.; Wilson, F. R.; Mosser, S. D.; Giuliani, E. A.; deSolms, S. J.; Conner, M. W.; Anthony, N. J.; Holtz, W. J.; Gomez, R. P.; Lee, T.-J.; Smith, R. L.; Graham, S. L.; Hartmen, G. D.; Gibbs, J. B.; Oliff, A. Proceedings of the National Academy of Sciences USA **1994**, 91, 9141-9145.

10) Moasser, M. M.; Sepp Lorenzino, L.; Kohl, N. E.; Oliff, A.; Balog, A.; Su, D. S.; Danishefsky, S. J.; Rosen, N. Proc Natl Acad Sci U S A **1998**, 95, 1369-74.

11) Barrington, R. E.; Subler, M. A.; Rands, E.; Omer, C. A.; Miller, P. J.; Hundley, J. E.; Koester, S. K.; Troyer, D. A.; Bearss, D. J.; Conner, M. W.; Gibbs, J. B.; Hamilton, K.; Koblan, K. S.; Mosser, S. D.; O'Neill, T. J.; Schaber, M. D.; Senderak, E. T.; Windle, J. J.; Oliff, A.; Kohl, N. E. Mol Cell Biol **1998**, 18, 85-92.

12) Omer, C. A.; Chen, Z.; Diehl, R. E.; Conner, M. W.; Chen, H. Y.; Trumbauer, M. E.; Gopal-Truter, S.; Seeburger, G.; Bhemnathwala, H.; Abrams, M. T.; Davide, J. P.; Ellis, M. S.; Gibbs, J. B.; Greenberg, I.; *Hamilton, K.; Koblan, K. S.; Kral, A. S.; Liu, D.; Lobell, R. B.; Miller, P. J.;* Mosser, S. D.; O'Neill, T. J.; Rands, E.; Schaber, M. D.; Senderak, E. T.; Oliff, A.; Kohl, N. E. Cancer Res. **2000**, 60, 2680-2688.

13) Williams, T. M.; Bergman, J. M.; Brashear, K.; Breslin, M. J.; Dinsmore, C. J.; Hutchinson, J. H.; MacTough, S. C.; Stump, C. A.; Wei, D. D.; Zartman, C. B.; Bogusky, M. J.; Culberson, J. C.; Buser Doepner, C.; Davide, J.; Greenberg, I. B.; Hamilton, K. A.; Koblan, K. S.; Kohl, N. E.; Liu, D. M.; Lobell, R. B.; Mosser, S. D.; TJ, O. N.; Rands, E.; Schaber, M. D.; Wilson, F.; Senderak, E.; Motzel, S. L.; Gibbs, J. B.; Graham, S. L.; Heimbrook, D. C.; Hartman, G. D.; Oliff, A. I.; Huff, J. R. JOURNAL OF MEDICINAL CHEMISTRY **1999**, 42, 3779-3784.

14) Britten, C. D.; Rowinsky, E.; Yao, S.-L.; S. Soignet; Rosen, N.; S. G. Eckhardt; Drengler, R.; Hammond, L.; Siu, L. L.; Smith, L.; McCreery, H.; Pezzulli, S.; Lee, Y.; Lobell, R.; Deutsch, P.; Hoff, D. V.; Spriggs, D. Proceedings of the American Society of Clinical Oncology **1999**, 17, #597.

15) Soignet, S.; Yao, S. L.; Britten, C.; Spriggs, D.; Pezzulli, S.; McCreery, H.; Mazina, K.; Deutsch, P.; Lee, Y.; Lobell, R.; Rosen, N.; Rowinsky, E. Proceedings of the American Association for Cancer Research Annual Meeting. March **1999**, 40, 517.

16) Adjei, A. A.; Erlichman, C.; Davis, J. N.; Cutler, D. L.; Sloan, J. A.; Marks, R. S.; Hanson, L. J.; Svingen, P. A.; Atherton, P.; Bishop, W. R.; Kirschmeier, P.; Kaufmann, S. H. Cancer Res **2000**, 60, 1871-7.

Chapter 12

Farnesyltransferase Inhibitors: From Squalene Synthase Inhibitors to the Clinical Agent BMS-214662

John T. Hunt

Department of Oncology Chemistry, Bristol-Myers Squibb Pharmaceutical Research Institute, Route 206 and Province Line Road, Princeton, NJ 08543–4000

We have discovered a number of distinct classes of inhibitors of the enzyme farnesyltransferase, an attractive new therapeutic target in oncology. The inhibitor classes include ones that are farnesylpyrophosphate-based, bisubstrates, thiol- and imidazole-based tetrapeptides, imidazole-based dipeptides and ultimately imidazole-based tetrahydrobenzodiazepines. The latter class yielded (R)-7-cyano-2,3,4,5-tetrahydro-1-(1H-imidazol-4-ylmethyl)-3-(phenylmethyl)-4-(2-thienylsulfonyl)-1H-1,4-benzodiazepine (BMS-214662), which has entered clinical trials.

Introduction

Ras proteins are a family of small GTPases which play a central role in signaling processes from the cell membrane to the nucleus (1). Signals from extracellular growth factors are transmitted through growth factor receptors, leading to the assembly at the interior of the cell membrane of signaling complexes which include the Ras protein. Formation of the signaling complex leads to the catalyzed exchange of GDP, from the signaling-incompetent Ras-GDP complex, for GTP, to form the signaling-competent Ras-GTP complex. Signals are then transmitted through downstream protein kinase cascades, resulting in gene activation. Ultimately, signaling is turned off by the intrinsic GTPase activity of the Ras protein, which is facilitated by the GTPase activating protein (GAP).

In the early 1980's, oncogenic forms of Ras were found (2) to be present in a variety of human tumors, including those of the pancreas (90%), colon (50%) and lung (30%). The mutated forms of the Ras protein were found to remain inordinately long in their GTP-bound signaling forms because of resistance to GAP-assisted GTP hydrolysis. Downregulation of this uncontrolled signaling pathway came to be viewed as a novel mechanism to potentially achieve an antitumor effect.

In order to mediate signaling, both normal and oncogenic forms of Ras must be membrane associated. Membrane localization of the Ras protein is the result of a three-step sequence of post-translational modifications of cytosolic Ras. Farnesylation of a cysteine located four residues from the C-terminus is followed by tripeptide hydrolysis and methyl esterification of the new C-terminal farnesylcysteine. Of these steps, farnesylation is most critical to membrane localization (3) and therefore signaling. This fact has made the enzyme catalyzing this transformation, protein farnesyltransferase (FT), the primary target of therapeutic intervention (1).

FPP- and Bisubstrate-Based Inhibitors

FT is a heterodimeric enzyme which catalyzes the transfer of a farnesyl group from farnesyl pyrophosphate (FPP) to the sulfur of a cysteine residue four amino acids from the C-terminus of the unprocessed Ras protein (Figure 1). The enzyme shows a requirement for the divalent cations Mg^{+2} and Zn^{+2} (4), with the Mg^{+2} believed to complex the pyrophosphate group to facilitate its leaving group ability, and the Zn^{+2} believed to play a role in catalysis by coordinating the cysteine thiol and activating it for attack at C-1 of FPP (5). In terms of rational drug design, inhibitors can be envisioned to be derived from FPP, from the protein substrate, or from combinations of the two, namely bisubstrate inhibitors.

Some of our earliest efforts in the design of FT inhibitors were based on an FPP-based approach to take advantage of our previous experience in the design of such inhibitors of the FPP-utilizing enzyme squalene synthase. Building in components which take advantage of the hydrophobic binding site available to the farnesyl side chain, as well as anionic groups to mimic the pyrophosphate, compounds such as **1** (IC_{50} = 50 nM) were obtained (6). Such compounds join other inhibitors based on FPP, both designed (7) and natural product-derived (8).

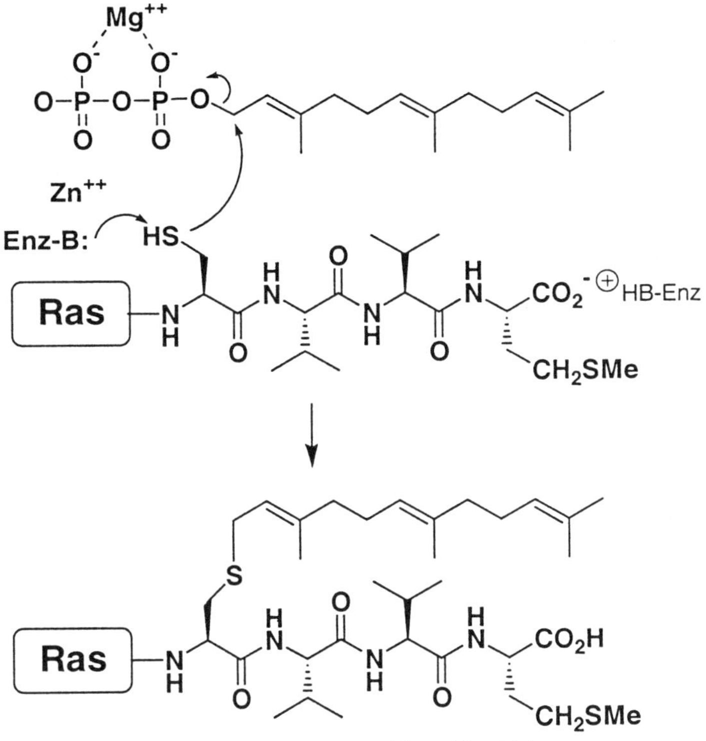

Figure 1. Reaction Catalyzed by FT.

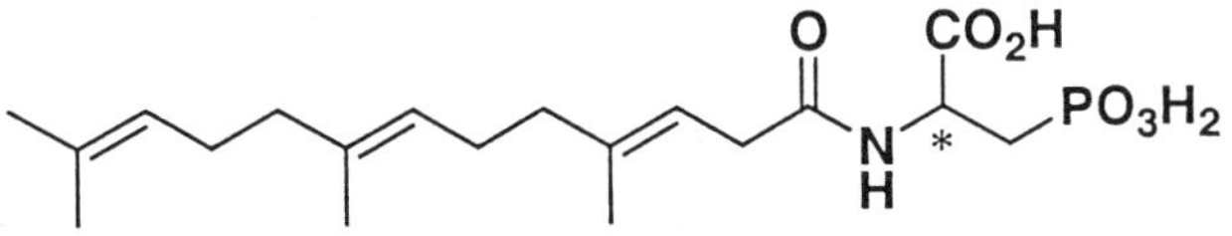

    Because there are a variety of enzymes which use farnesylpyrophosphate as a substrate, bisubstrate inhibitors were investigated with the expectation that more selective FT inhibitors would result. Although the second natural substrate of FT is the Ras protein, studies of peptide fragments of Ras had shown that a C-terminal tetrapeptide sequence of the protein, termed a C-A-A-X (Cys-Aliphatic1-Aliphatic2-Xaa) box, contained essentially all of the binding determinants necessary for substrate behavior (9). We therefore prepared tetrapeptide analogs in which the cysteine residue was substituted with a linker which would allow attachment onto a VIM or VVM tripeptide of an aliphatic chain, intended to mimic the farnesyl group, and a charged moiety, intended to mimic the pyrophospate. These efforts resulted in the preparation of carboxylic acid based inhibitors such as **2** (IC_{50} = 33 nM) (*10*) and phosphinic acid based inhibitors such as **3** (IC_{50} = 6 nM) (*11*). These inhibitors demonstrated poor effects in cellular assays, presumably due to their polyanionic nature. For example, **3** was unable to inhibit the transformation of oncogenic *H-ras* transfected NIH-3T3 cells at concentrations as high as 100 µM, while the corresponding phosphinic acid methyl ester, presumably acting as a cellular prodrug of the phosphinic acid, caused about 75% inhibition of transformation at this concentration.

Although the bisubstrates were likely to lead to FT inhibitors of greatly increased specificity, their physicochemical properties were poorly conducive to their development as drugs. For example, using Lipinski's rules (*12*), both **2** and **3** are flagged for unacceptably high molecular weight, and **2** is also flagged for containing an unacceptably high number of N's and O's (13, where >10 results in a flag) as well as NH's and OH's (7, where >5 results in a flag).

Tetrapeptide-Based Inhibitors

We next turned our attention toward CAAX-based peptidic inhibitors. As noted above, aliphatic residues in the A1 and A2 positions afford good substrates of FT. Contemporaneous with our own scanning studies of the sequence specificity of CAAX-based tetrapeptide FT substrates, it was reported (*13*) that substitution of Phe at the A2 position provided relatively high affinity inhibitors (**4**, CVFM, IC$_{50}$ = 37 nM (*14*)). Spurred by this discovery, we discovered that constraining the phenylalanine onto it's α-amine to form the tetrahydro-isoquinoline carbonyl (Tic) analog **5** led to an extremely potent inhibitor (CV-Tic-M, IC$_{50}$ = 1 nM) (*15*). The high potency of **5** may be a result of constraining the phenylalanine side chain in a preferred conformation for enzyme binding. Alternatively, early NMR experiments suggested that peptide substrates bind to FT in a turn conformation (*16*), and the Tic residue may favor such a turn conformation in the tetrapeptide inhibitor.

5, X = O; R$_1$ = H; R$_2$ = SMe
6, X = H,H; R$_1$ = H; R$_2$ = SMe
7, X = H,H; R$_1$ = Me; R$_2$ = CONH$_2$

Although **5** was poorly active in cellular assays, the reduced analog **6** (IC$_{50}$ = 0.6 nM) was active in both *ras* transformation inhibition and soft agar growth assays. Further optimization led to **7** (IC$_{50}$ = 2.8 nM), which displayed antitumor activity when administered *ip* in an *ip* Rat-1 tumor model (*15*).

Despite this interesting biological activity, we considered the thiol of these tetrapeptide inhibitors to be a potential liability due to facile oxidation to the disulfide. Operating on the assumption that the thiol was functioning as a ligand to the active site zinc of FT, we began a search for alternative zinc ligands which could afford potent inhibitors in the tetrapeptide series. Tripeptides containing N-terminal groups such as phenol (*17*), thiophenol, alkanol, hydroxamic acids and sulfonamides all afforded only modest inhibitors at best (unpublished results).

Enzymes generally ligate zinc atoms using either thiol, thioether, carboxyl or imidazole-containing amino acid side chains. Potent thiol- and carboxyl-containing zinc-metalloprotease inhibitors are well known, but imidazole-containing inhibitors are less effective (*18*). This apparent anomaly appeared to hold with FT as well, where CVFM is a 200-fold more potent inhibitor than HVFM (*14*). Nevertheless, we were led to further investigate imidazole-containing tetrapeptide FT inhibitors, driven by the hypothesis that conformational factors would be more critical in imidazole tetrapeptide binding than in thiol tetrapeptide binding. That is, the directional lone pair projecting from the planar imidazole would have more severe demands for accessing the ligand sphere of the zinc than would the spherical thiol attached to a series of rotatable single bonds. Indeed, a tetrapeptide analog in which imidazole was linked to the tripeptide via an ethyl linker afforded an extremely potent inhibitor (**8**, IC_{50} = 0.79 nM) (*19*). Surprisingly, a variety of other linkers also provided relatively potent inhibitors. Most interestingly, while the histidine analog **10** (IC_{50} = 520 nM) afforded a modest inhibitor, the des-α-amino-histidine analog **9** (IC_{50} = 4.4 nM) was a potent inhibitor. A possible explanation is that the *tau*-nitrogen of the imidazole is involved in ligating the zinc of FT, and the presence of the α-ammonium ion provides an internal H-bond donor, shifting the imidazole tautomer such that the *tau*-nitrogen is protonated. Thus, the reason for the poor inhibitory potency of HVFM was primarily the result of the presence of the α-amine, rather than an inappropriate linker, as we had hypothesized.

While compounds such as **8** were potent FT inhibitors, the absence of the oxidizable thiol did not lead to dramatic improvements in cellular activity. For example, the IC_{50} for inhibition of H-Ras processing in oncogenic H-Ras overexpressing cells was only 5 μM, as determined by the differences in electrophoretic mobility between the H-Ras protein before and after the 3 step processing sequence (*19*). Data such as these suggested that the peptidic nature of inhibitors such as **8**, and in particular the presence of the C-terminal carboxylate, would have to be addressed to obtain highly permeable inhibitors.

Dipeptide-Based Inhibitors

During our earlier exploration of the SAR of tetrapeptide thiol-based inhibitors, we observed that deletion of either the C-terminal carboxylate (compare **6** and **11**) or truncation of the side chain (compare CV-Tic-M, IC_{50} = 1 nM and CV-Tic-A, IC_{50} = 6.4 µM) produced dramatic reductions in FT inhibitory potency. Complete removal of the C-terminal amino acid also led to poor inhibitors (e.g. **12**). Further truncation of the tripeptide was not expected to offer the promise of improvements in potency. Nevertheless, just such an exercise appears to have been performed by deSolms et al. (*20*), who showed that simple dipeptide amides of cysteine restored FT inhibition to sub-micromolar levels. An example from our group of this effect in a tetrahydroisoquinoline-based inhibitor is **13**.

6, R = iPr, IC_{50} = 0.6 nM

11, R = iPr, IC_{50} = 1.3 µM

12, R = tBu, IC_{50} = 2 µM

13, R = iBu, IC_{50} = 0.58 µM

Attracted to this finding, but with the aim of avoiding the presence of the thiol, we prepared imidazole-based dipeptides and amino aicd amides. Disappointingly, the best of these compounds were low micromolar inhibitors. However, such compounds appeared to be very interesting starting points for drug design, since they were of low MW, uncharged, and highly cell permeable.

In general, the concentrations needed for inhibition of FT were only 10-fold lower than the concentrations required for cellular effects (e.g., **14**, IC_{50} = 8.9 µM, IC_{50} for inhibition of Ras processing ~100 µM).

Tetrahydrobenzodiazepine Inhibitors

In the thiol-dipeptide series, others had shown that constraining the dipeptide into a piperazine ring *(21)* provided improvements in FT inhibition, and further optimization led to thiol-containing compounds such as **15** with low nanomolar potency. In our hands, the direct application of this template to imidazole-based compounds, e.g. **16**, did not afford potent inhibitors. Nevertheless, such compounds fit a pharmacophore hypothesis for FT inhibitors which had developed in our group over the course of our program. Namely, high FT inhibitory potency required not only a zinc ligand, but also two appropriately positioned hydrophobic groups and an appropriately positioned H-bond acceptor. One variation of the piperazine template which we explored involved cyclization of the methoxyethyl side chain onto the piperazine (a), and concomitant aromatization of the resulting ring, effectively replacing the alkyl hydrophobe with an aryl hydrophobe. Although the tetrahydroquinazoline compound which resulted (not shown) did not show improved FT inhibitory potency (IC_{50} = 2.2 µM), homologation of the piperazine to a diazepine ring afford the submicromolar tetrahydrobenzodiazepine **17** *(22)*.

Encouraged by the promising inhibitory potency of **17**, we embarked on an intensive SAR investigation of this lead compound. Initial studies focused on positions 1 and 4 of the diazepine ring as well as the most synthetically accessible positions on the aromatic ring, namely positions 7 and 8. At position 1, there was some tolerance for alterations in the linker length and its point of attachment onto the imidazole ring, with the imidazol-4-ylethyl group providing an equipotent inhibitor. At position 4, there appeared to be tolerance for both other H-bond acceptors such as sulfonyl as well as tolerance for a variety of other hydrophobic groups. Increased potency was achieved with a variety of substituents at position 7 such as a pyridyl, phenyl or cyclohexyl ring or a bromo group. Most of the exploration of 8-substituents was based on the 8-amino analog, and a variety of amides, carbamates and ureas which contained hydrophobic appendages provided potent inhibitors *(22)*.

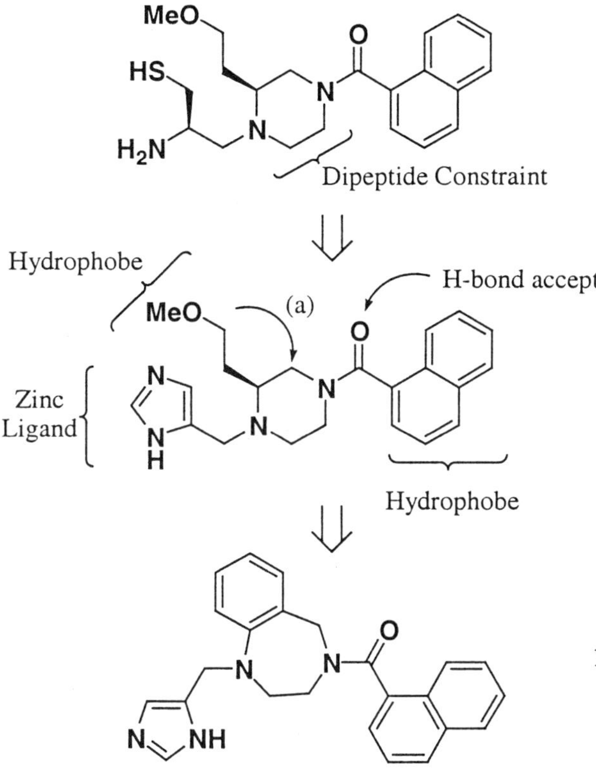

15, IC$_{50}$ = 3 nM

Dipeptide Constraint

Hydrophobe

H-bond acceptor

Zinc Ligand

16, IC$_{50}$ = 1.3 μM

Hydrophobe

17, IC$_{50}$ = 0.46 μM

7- and 8-substituents independently improve potency
7: 4-pyridyl>phenyl>3-pyridyl>c-hexyl>Br>H
8: NHCO-cHx>NHCOphenyl>NHCOMe>Cl=H

Some variability tolerated in linker length and point of imidazole attachment

Other H-bond acceptors tolerated such as CONH, SO$_2$, SO$_2$NH

A variety of hydrophobic groups tolerated

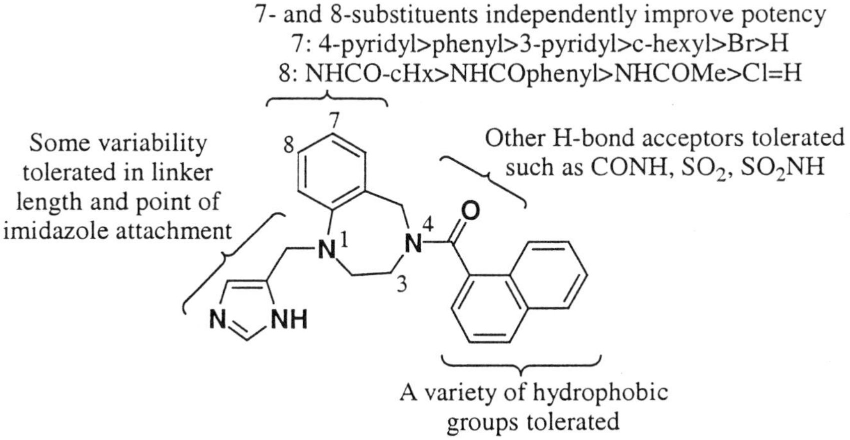

Probe studies of positions 2 and 3 did not immediately identify any potency enhancing substitution patterns. However, we were intrigued by the idea that the critical N-4 hydrophobic group could be accessed alternatively by a 3-substituent. Gratifyingly, the racemic 3-phenylmethyl analog **18**, which lacked the critical N-4 hydrophobic group, was roughly equipotent to **17** (*23*). Further optimization, including identification of the R-enantiomer as the preferred one, led to the potent inhibitor **19**.

18, $IC_{50} = 0.84 \, \mu M$ **19**, $IC_{50} = 11 \, nM$

Continued optimization of the 3-(R)-phenylmethyl series showed that, in general, previously investigated SAR trends continued to hold. At N-4, a variety of side chains attached to a carbonyl or sulfonyl group provided potent inhibitors. Some of the most potent compounds in cellular assays contained hydrophobic side chains, and the overall lipophilicity of the molecules was becoming higher than optimal. During an effort to reduce the overall lipophilicity of the analogs, it was discovered that the small, hydrophilic cyano group could be incorporated at position 7 to afford highly potent inhibitors. In addition, cellular potency was also increased in these analogs. It is not clear why the nitrile functionality should be an exceptional replacement for the relatively lipophilic 7-substituents which we had previously identified to be important for inhibitory potency. However, it is interesting to note that a 4-cyanophenylmethyl substituent in several different series (*24, 25*) has been independently found to provide potent FT inhibitors.

Final optimization of the 3-(R)-phenylmethyl-7-CN series resulted in a number of potent analogs containing N-4 side chains of differing character. For example, the cationic side chain of **20** produced an inhibitor of equal potency to the analogs with alkyl and aryl side chains. The highest cellular potencies were realized with hydrophobic side chains, with analogs **21-23** producing 60-90% phenotypic reversion of *H-ras* transformed Rat-1 cells at 100 nM.

20 IC$_{50}$ = 2.5 nM — ⁓⁓⁓NMe$_2$

21 IC$_{50}$ = 1.8 nM — ⁓⁓⁓

22 IC$_{50}$ = 1.8 nM — ⁓Ph

23 IC$_{50}$ = 1.4 nM — ⁓(2-thienyl)

As SAR optimization of the tetrahydrobenzodiazepine series progressed, we concomitantly evaluated compounds for pharmacokinetic properties and antitumor efficacy. In addition to the *H-ras* transformed Rat-1 tumor model described above, we made extensive use of a human colon tumor xenograft carrying a *K-ras* mutation, namely the HCT-116 model. At their optimal dose, a number of compounds were active against staged subcutaneous HCT-116 tumors when given once a day orally for 10 days. For example, **19** (700 mpk) and **20** (600 mpk) each achieved 1 LCK (log cell kill, defined as tumor growth inhibition equal to 3.3 tumor doubling times). Interestingly, several compounds showed curative efficacy at their optimal dose under these conditions (**21**, 7/7 cures at 600 mpk; **22**, 7/8 cures at 600 mpk; **23**, 8/8 cures at 600 mpk). Impressive efficacy was also evident at suboptimal doses. For example, as shown in Figure 2, **23** produced 3.5 LCK at 300 mpk, half of the optimal dose.

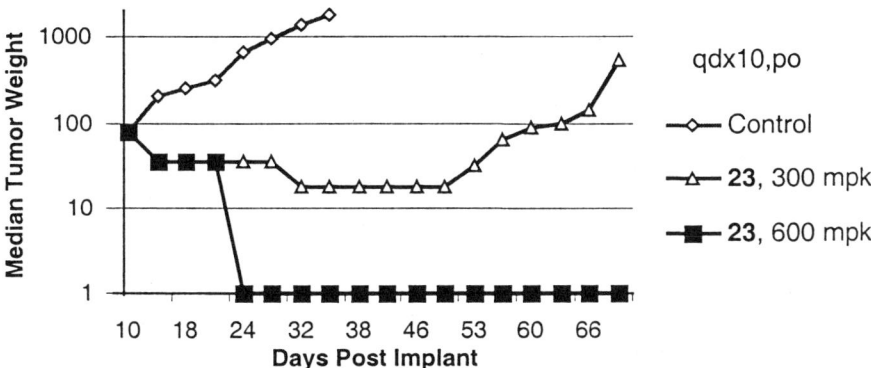

*Figure 2. In vivo antitumor activity of **23** in the HCT-116 xenograft model.*

For a number of the potent compounds, the pharmacokinetics underlying the good in vivo efficacy were analyzed. For example, as shown in Figure 3, analysis of the pharmacokinetics of **23** in rats after an intravenous (*iv*) dose revealed a relatively slow clearance (5.7 mL/min/kg), a moderate volume of distribution (1.4 L/kg) and a moderate half life ($t_{1/2}$ = 2.4 hr). Upon *po* administration of an equal dose, the compound was found to give a high C_{max} value (20 μM) and an oral bioavailability of 56%.

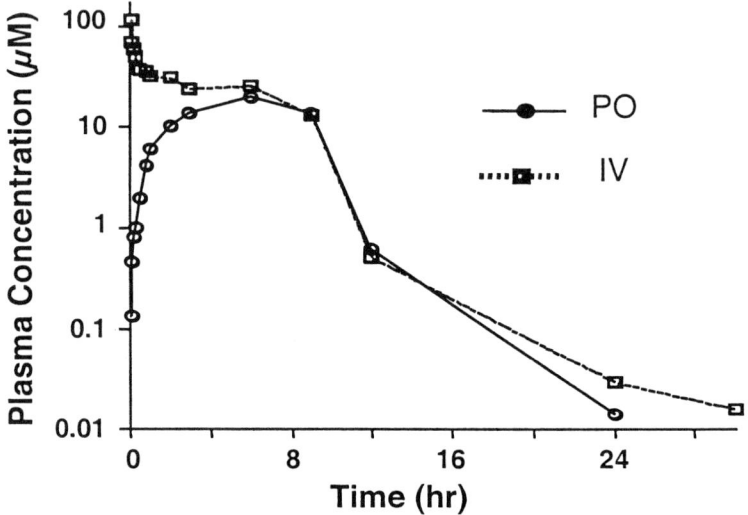

Figure 3. Pharmacokinetics of 23 in rats.

Based upon its high FT inhibitory potency, spectrum of excellent antitumor efficacy (data not shown), good pharmacokinetics and good physical properties, **23** (BMS-214662; (R)-7-cyano-2,3,4,5-tetrahydro-1-(1H-imidazol-4-ylmethyl)-3-(phenylmethyl)-4-(2-thienylsulfonyl)-1H-1,4-benzodiazepine) was selected for advancement into human clinical trials (*23*). Another feature of **23** which is advantageous for clinical and commercial development is its straightforward chemical synthesis. The synthesis utilized by our discovery group utilized bromoisatoic anhydride and (R)-phenylalanine ethyl ester hydrochloride as the chiral starting material. Condensation to form the tetrahydrobenzodiazepine-dione (**24**) was followed by reduction to form the tetrahydrobenzodiazepine (**25**). Cyanide replacement of the bromine, followed by sulfonylation with 2-thienylsulfonyl chloride and reductive alkylation to introduce the imidazolylmethyl group afforded **23** in 5 steps in 50% overall yield with no chromatography. The basic imidazole was then used to prepare a variety of acid salts for optimization of pharmaceutical properties.

24, X = O; R = Br
25, X = H, H; R = Br
26, X = H, H; R = CN

23

In summary, the farnesyltransferase inhibitor program at Bristol-Myers Squibb was an evolving one based continuously on rational drug design. The initial tack focused on FPP-based inhibitors developed as an offshoot of a squalene synthase program. The direction shifted toward bisubstrate-based inhibitors in order to gain additional selectivity, but these inhibitors were limited by disadvantageous physical properties. Tetrapeptide thiol inhibitors were developed which allowed for both cellular and *in vivo* proof of concept. However, both the thiol and the peptidic nature of these inhibitors were considered liabilities for drug development. The novel use of imidazole as a putative zinc ligand was pioneered in the tetrapeptide series, and clues to accessing non-peptide frameworks were gained from the literature. Finally, the melding of these concepts reached fruition with the discovery of the imidazolylmethyltetrahydrobenzodiazepine FT inhibitors, culminating in the discovery of BMS-214662. This compound is a potent inhibitor with excellent cellular activity, superior in vivo antitumor activity, good pharmacokinetics and good physical properties. BMS-214662 (**23**) is currently in human clinical trials, and joins R115777 and Sch66336 as FT inhibitors which should finally answer the question of the therapeutic potential of this novel class of agents.

References

1. Rowinsky, E. K.; Windle, J. J.; Von Hoff, D. D. *J. Clin. Oncology* **1999**, *17*, 3631-3652.
2. Barbacid, M. *ras Genes*; Richardson, C., Ed.; Annual Reviews Inc.: Palo Alto, CA, 1987; Vol. 56, pp 779-827.

3. Kato, K.; Cox, A. D.; Hisaka, M. M.; Graham, S. M.; Buss, J. E.; Der, C. J. *Proc. Natl. Acad. Sci.* **1992**, *89*, 6403-6407.
4. Reiss, Y.; Brown, M. S.; Goldstein, J. L. *J. Biol. Chem.* **1992**, *267*, 6403-6408.
5. Huang, C.-C.; Casey, P. J.; Fierke, C. A. *J. Biol. Chem.* **1997**, *272*, 20-23.
6. Patel, D. V.; Schmidt, R. J.; Biller, S. A.; Gordon, E. M.; Robinson, S. S.; Manne, V. *J. Med. Chem.* **1995**, *38*, 2906-2921.
7. Lamothe, M.; Perrin, D.; Blotieres, D.; Leborgne, M.; Gras, S.; Bonnet, D.; Hill, B. T.; Halazy, S. *Bioorg. Med. Chem. Lett.* **1996**, *6*, 1291-1296.
8. Singh, S. S.; Zink, D. L.; Liesch, J. M.; Goetz, M. A.; Jenkins, R. G.; Nallin-Omstead, M.; silverman, K. C.; Bills, G. F.; Mosley, R. T.; Gibbs, J. B.; Albers-Schonberg, G.; Lingham, R. B. *Tetrahedron* **1993**, *27*, 5917-5926.
9. Reiss, Y.; Goldstein, J. L.; Seabra, M. C.; Casey, P. J.; Brown, M. S. *Cell* **1990**, *62*, 81-88.
10. Bhide, R. S.; Patel, D. V.; Patel, M. M.; Robinson, S. P.; Hunihan, L. W.; Gordon, E. M. *Bioorg. Med. Chem. Lett.* **1994**, *4*, 2107-2112.
11. Patel, D. V.; Gordon, E. M.; Schmidt, R. J.; Weller, H. N.; Young, M. G.; Zahler, R.; Barbacid, M.; Carboni, J. M.; Gullo-Brown, J. L.; Hunihan, L.; Ricca, C.; Robinson, S.; Seizinger, B. R.; Tuomari, A. V.; Manne, V. *J. Med. Chem.* **1995**, *38*, 435-442.
12. Lipinski, C. A.; Lombardo, F.; B, W., Dominy; Feeney, P. J. *Advanced Drug Delivery Reviews* **1997**, *23*, 3-25.
13. Goldstein, J. L.; Brown, M. S.; Stradley, S. J.; Reiss, Y.; Gierasch, L. M. *J. Biol. Chem.* **1991**, *266*, 15575-15578.
14. Leftheris, K.; Kline, T.; Natarajan, S.; DeVirgilio, M. K.; Cho, Y. H.; Pluscec, J.; Ricca, C.; Robinson, S.; Seizinger, B. R.; Manne, V.; Meyers, C. A. *Bioorg. Med. Chem. Lett.* **1994**, *4*, 887-892.
15. Leftheris, K.; Kline, T.; Vite, G.; Cho, Y. H.; Bhide, R. S.; Patel, D. V.; Patel, M. M.; Schmidt, R.; Weller, H. N.; Andahazy, M. L.; Carboni, J. M.; Gullo-Brown, J. L.; Lee, F. Y. F.; Ricca, C.; Rose, W. C.; Yan, N.; Barbacid, M.; Hunt, J. T.; Meyers, C. A.; Seizinger, B. R.; Zahler, R.; Manne, V. *J. Med. Chem.* **1996**, *39*, 224-236.
16. Stradley, S. J.; Rizo, J.; Gierasch, L. M. *Biochemistry* **1993**, *32*, 12586-12590.
17. Patel, D. V.; Patel, M. M.; Robinson, S. S.; Gordon, E. M. *Bioorg. Med. Chem. Lett.* **1994**, *4*, 1883-1888.
18. Cecchi, R.; Ciabatti, R.; Favara, D.; Barone, D.; Baldoli, E. *Il Farm., Ed. Sci.* **1985**, *40*, 541-554.
19. Hunt, J. T.; Lee, V. G.; Leftheris, K.; Seizinger, B. R.; Carboni, J. M.; Mabus, J.; Ricca, C.; Yan, N.; Manne, V. *J. Med. Chem.* **1996**, *39*, 353-358.
20. deSolms, S. J.; Deana, A. A.; Giuliani, E. A.; Graham, S. L.; Kohl, N. E.; Mosser, S. D.; Oliff, A. I.; Pompliano, D. L.; Rands, E.; Scholz, T. H.;

Wiggins, J. M.; Gibbs, J. B.; Smith, R. L. *J. Med. Chem.* **1995**, *38*, 3967-3971.
21. Williams, T. M.; Ciccarone, T. M.; MacTough, S. C.; Bock, R. L.; Conner, M. W.; Davide, J. P.; Hamilton, K.; Koblan, K. S.; Kohl, N. E.; Kral, A. M.; Mosser, S. D.; Omer, C. A.; Pompliano, D. L.; Rands, E.; Schaber, M. D.; Shah, D.; Wilson, F. R.; Gibbs, J. B.; Graham, S. L.; Hartman, G. D.; Oliff, A. I.; Smith, R. L. *J. Med. Chem.* **1996**, *39*, 1345-1348.
22. Ding, C. Z.; Batorsky, R.; Bhide, R.; Chao, H. J.; Cho, Y.; Chong, S.; Gullo-Brown, J.; Guo, P.; Kim, S. H.; Patel, M.; Penhallow, B. A.; Ricca, C.; Rose, W. C.; Schmidt, R.; Slusarchyk, W. A.; Vite, G.; Yan, N.; Manne, V.; Hunt, J. T. *J. Med. Chem.* **1999**, *42*, 5241-5253.
23. Hunt, J. T.; Ding, C. Z.; Batorsky, R.; Bednarz, M.; Bhide, R.; Cho, Y.; Chong, S.; Chao, S.; Gullo-Brown, J.; Guo, P.; Kim, S. H.; Lee, F. Y. F.; Leftheris, K.; Miller, A.; Mitt, T.; Patel, M.; Penhallow, B. A.; Ricca, C.; Rose, W. C.; Schmidt, R.; Slusarchyk, W. A.; Vite, G.; Manne, V. *J. Med. Chem.*, **2000**, *43*, 3587-3595.
24. Breslin, M. J.; deSolms, S. J.; Giuliani, E. A.; Stocker, G. E.; Graham, S. L.; Pompliano, D. L.; Mosser, S. D.; Hamilton, K. A.; Hutchinson, J. H. *Bioorg. Med. Chem. Lett.* **1998**, *8*, 3311-3316.
25. Williams, T. M.; Bergman, J. M.; Brashear, K.; Breslin, M. J.; Dinsmore, C. J.; Hutchinson, J. H.; MacTough, S. C.; Stump, C. A.; Wei, D. D.; Zartman, C. B.; Bogusky, M. J.; Culberson, J. C.; Buser-Doepner, C.; Davide, J.; Greenberg, I. B.; Hamilton, K. A.; Koblan, K. S.; Kohl, N. E.; Liu, D.; Lobell, R. B.; Mosser, S. D.; O'Neill, T. J.; Rands, E.; Schaber, M. D.; Wilson, F.; Senderak, E.; Motzel, S. L.; Gibbs, J. B.; Graham, S. L.; Heimbrook, D. C.; Hartman, G. D.; Oliff, A. I.; Huff, J. R. *J. Med. Chem.* **1999**, *42*, 3779-3784.

Chapter 13

Inhibiting Fornesyl Protein Transferase with Sch-66336: Potentially a Selective, Noncytotoxic Therapy for Human Cancer

A. G. Taveras, F. G. Njoroge, R. J. Doll, J. Kelly, S. Remiszewski,
A. K. Mallams, C. S. Alvarez, J. del Rosario, R. R. Rossman,
B. Vibulbhan, P. Pinto, J. Deskus, M. Connolly, J. Wang, J. Desai,
R. Wolin, A. Afonso, A. B. Cooper, D. F. Rane, Y.-T. Liu, C. J. Aki,
J. Chao, C. Strickland, P. Weber, M. Liu, M. S. Bryant,
A. A. Nomeir, R. Patton, L. Wang, L. James, D. Carr,
P. Kirschmeier, W. R. Bishop, V. Girijavallabhan,
and A. K. Ganguly

Schering-Plough Research Institute, 2015 Galloping Hill Road, Kenilworth, NJ 07033

The discovery of Farnesyl Protein Transferase (FPT) and its role in signal transduction and cellular proliferation has inspired the identification of inhibitors of this process as potential cancer therapy. Although thousands of FPT inhibitors (FTIs) have been prepared, only a select few have advanced to human clinical trials. **Sch-66336**, a potent FTI with demonstrated *in vivo* efficacy in mice, is currently in Phase II clinical trials for the treatment of human cancers. Structure Activity Relationship studies, which led to the discovery of **Sch-66336**, are described and related to the observed differences in interactions made with FPT as determined by x-ray crystal structure analysis. Its potency against a panel of tumor cell lines and its efficacy in mouse models as mono or combination therapy is reviewed.

Introduction

The treatment of human cancers is often limited due to side effects and poor response rates associated with these treatments. The mechanism of action of many chemotherapeutics is not specific, and therefore cytotoxic, relying solely on the narrow difference in growth rates between tumor and normal cells. It is widely believed that advances in technology can lead to a better understanding of the biochemical processes that lead to cellular proliferation. Understanding how these processes differ between normal and tumor cells may eventually lead to more efficacious and selective treatments for cancers without associated side effects.

Nearly 30% of human tumors, more commonly those of pancreatic, colon and lung, have been identified as having Ras mutations (*1*). Moreover, recent delineations of the signal transduction pathway and its role in cellular proliferation have identified Ras as an important mediator of this process (*1-6*). Ras, a 21-Kd GTP-binding protein, is found in its inactive state in the cytosol of cells and can only mediate cell signaling upon structural modification and subsequent association with the cell membrane (*7*). The carboxy terminus of Ras consists of a four-amino acid sequence commonly known as the CaaX motif. Association of this CaaX region with the prenyl transfer protein farnesyl protein transferase (FPT) results in alkylation of the cysteine residue of Ras by a farnesyl moiety from farnesyl pyrophosphate (FPP). This prenylated Ras protein, following post-translational modification, can actively associate with the cell's signaling machinery (*8-10*).

An x-ray crystal structure determination of FPT (*11*) shows a dimeric association between helical-rich α- and β- subunits. Their interaction creates a cavity of approximately 15Å in diameter in which is bound a zinc cation ligated by three amino acid residues from the β-subunit. Lipophilic interactions between the farnesyl moiety of FPP and residues in the β-subunit non-covalently bind FPP to FPT (*12*). Although an x-ray crystal structure determination of Ras has been obtained, its rapid farnesylation when bound to FPT prohibits delineation of the specific association in the Ras-FPT complex. However, an x-ray crystal structure determination (*13*) of a ternary complex consisting of FPT, the stabilized FPP mimic α-hydroxyfarnesylphosphonic acid (αHFP) and the CaaX tetrapeptide Cys-Val-Ile-Met shows the interaction of this CaaX motif with FPT.

Competitive inhibitors of Ras binding can effectively inhibit farnesylation and downstream signaling *in vitro* and *in vivo* (*14*). Yet, while the inhibition of H-Ras provided the initial rationale for the development of farnesyl protein transferase inhibitors (FTIs), the observation that two structurally related Ras proteins, K-Ras and N-Ras, are differentially prenylated suggests that other farnesylated proteins mediate the effects of FTIs (15). Recent evidence suggests these farnesylated protein candidates might include RhoB (17) and nuclear centromere associated proteins (18, 19). These proteins have C-terminal CaaX sequences and are substrates for FPT.

Due to the size of the active site cavity of FPT, it is not surprising that many structurally diverse FTIs can be accommodated (14). Some of the early FTIs have been designed to mimic and compete with the CaaX motif of Ras (20). However, the progression of these tetra-peptide or peptido-mimetic antagonists have been hindered by poor cellular penetration and poor metabolic stability (20). Subsequent structural modifications incorporating an imidazole moiety, a reported replacement for thiol as a zinc-binding ligand, led to second-generation, non-peptidic FTIs such as Merck's cyanobenzylimidazoles (21, 22) and Brystol Myers Squibb's diazepines (23). Other non-peptidomometic FTIs were discovered by conventional screening paradigms. For example, in-house screening of Janssen's anti-fungal libraries led to the development of imidazole-containing dihydroquinolone FTIs (24, 25) while screening of compound libraries at Schering-Plough Research Institute (SPRI) and subsequent structural optimization led to the identification of **Sch-66336** as a potent representative of the novel benzocycloheptapyridyl class of FTIs (26). **Sch-66336** has been found to potently inhibit the growth of human tumor cells *in vitro* and *in vivo* (27) and is currently completing Phase II human clinical trials for its use in the treatment of human cancers. Herein, we report the details of the discovery of **Sch-66336** and discuss in depth the drug-protein interactions between benzocycloheptapyridine FTIs and FPT.

Sch-66336
FPT $IC_{50} = 0.002$ μM

The Discovery of Benzocycloheptapyridine FTIs

Early Structure Activity Relationship Studies

Screening of compound libraries at SPRI led to the identification of **Sch-37370** as a relatively weak inhibitor of farnesylation of Ras in an enzyme assay (IC$_{50}$ = 26.9 µM). This compound has been reported (*28*) to also antagonize the binding of histamine (IC$_{50}$ = 0.32 µM) and PAF (IC$_{50}$ = 0.61 µM). Thus, structural modifications which enhanced both potency and selectivity of this benzocycloheptapyridine lead for FPT were sought. Of the many modifications made of the amide moiety in **Sch-37370** (*29 - 31*), the best potencies were achieved with arylalkyl and heteroarylalkyl derivatives (i.e., **2**). 4-Pyridylmethyl amide **2** was found to be optimal while isomeric 3- and 2-pyridyl analogs were less potent. The preferred spacing between the amide carbonyl and the heteroaryl ring was determined to be one methylene unit (*32*).

Sch-37370
FPT IC$_{50}$ = 26.9 µM

1, n = 0, FPT IC$_{50}$ = 4.7 µM
2, n = 1, FPT IC$_{50}$ = 0.25 µM

The x-ray crystal structure of **2** complexed with FPT (Figure 1) lends insight into the binding orientation of 8-chlorobenzocycloheptapyridine FTIs (*33*). The tricyclic moiety is seen to span the α- and β-subunits of FPT as well as associate with the lipophilic region of FPP. The pyridylmethyl amide substituent of **2** interacts with the tyrosine moiety in a lipophilic binding site comprised of Asp359β, Tyr93β and Cys95β. A more focussed view of this interaction is shown in Figure 2. Extension of the pyridyl ring with additional methylene spacers would weaken its aryl-aryl interaction with Tyr93β while placement of the pyridyl nitrogen either *ortho* or *meta* would position the lone pair of electrons of the pyridyl nitrogen atom away from solvent and toward a more lipophilic region of this pocket. The decreased activity of benzyl derivatives of **2** (i.e., where the pyridylmethyl moiety has been replaced by a benzyl group, $IC_{50} = 0.8$ μM) is consistent with a poorer aryl-aryl interaction of the more electron rich phenyl ring and Tyr93β compared to pyridyl.

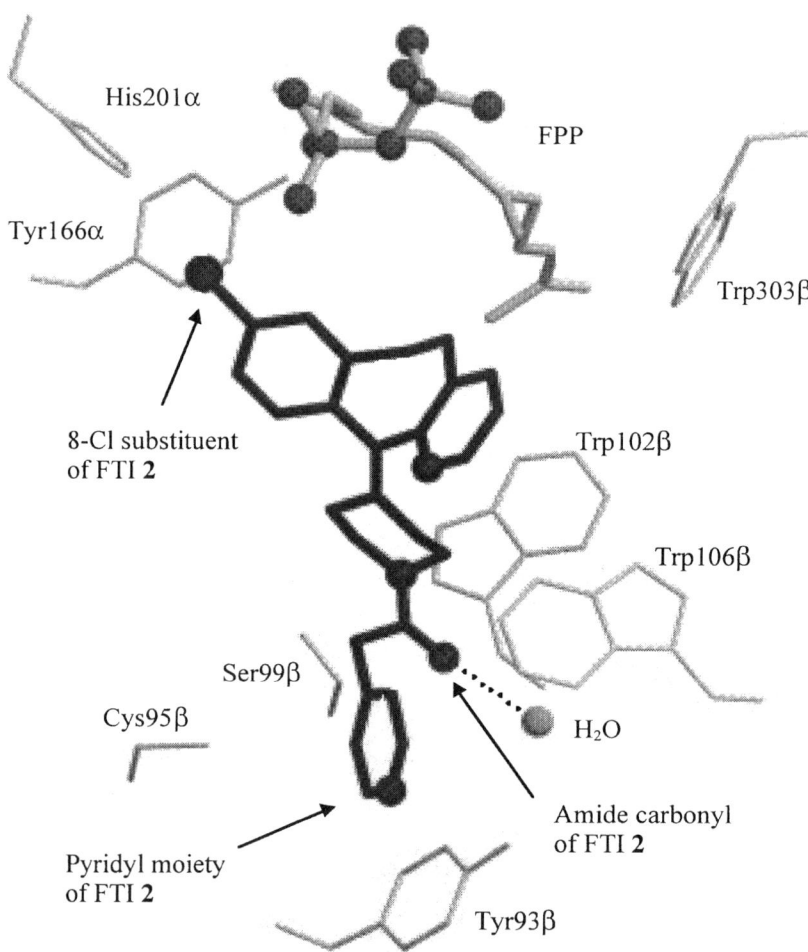

Figure 1. X-ray crystal structure of 2-FPT.

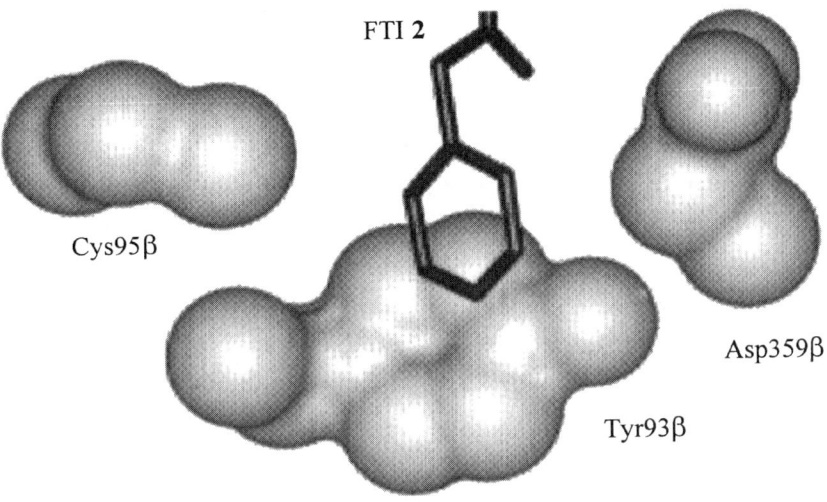

*Figure 2. X-ray crystal structure of **2**-FPT. Pyridylacetyl moiety interacts with Asp359β, Tyr93β and Cys95β of FPT.*

2
FPT IC_{50} = 0.25 μM

3
FPT IC_{50} = 15.8 μM

The position of the amide carbonyl of **2** is also critical (*31*) as the inhibitory potency of the isomeric ketone **3** is 63 times less active (IC$_{50}$ = 15.8 µM). The x-ray crystal structure of the 2-FPT complex (Figure 1) shows the carbonyl oxygen atom bound to a water molecule which is engaged in hydrogen bonding contacts with Phe360β (*33*).

Most of the potency enhancements achieved with benzocycloheptapyridine FTIs were derived from modifications to the tricyclic moiety (*34*). This finding is explainable based on the significant interactions made between the tricyclic group and residues from both the α- and β-subunits of FPT as well as with FPP as seen from the x-ray crystal structure of FTI-FPT complexes (*33*). Removal of the C(5)-C(6) bridge, as in **6**, causes significant loss in potency presumably due to entropic as well as enthalpic factors (*30*).

4, IC$_{50}$ = 0.17 µM (X=C, Y=N)
5, IC$_{50}$ = 0.42 µM (X=N, Y=C)

6, IC$_{50}$ = 38 µM

Substitution of the pyridyl ring of **2** at C(2) produced compounds of inferior potency (i.e., CH$_3$, IC$_{50}$ = 0.93 µM; Cl, IC$_{50}$ = 1.2 µM; OH, IC$_{50}$ = 6.0 µM) while substitution at C(4) gave mixed results (i.e., Cl, IC$_{50}$ = 0.19 µM; OCH$_2$CH$_2$N(Me)$_2$, inactive). Quite surprisingly, substitution at this site with a benzotriazolyloxy moiety did not alter the FPT inhibitory potency drastically (IC$_{50}$ = 0.6 µM) suggesting some ability to accommodate larger moieties in this α-3 pocket (*33*). Substitution at C(3) with small, lipophilic groups such as chloro, bromo or methyl gave FTIs demonstrating a ten-fold enhancement (0.04 - 0.07 µM) in potency (*29, 30*). Substitution at this site with more bulky moieties (i.e., t-butyl or phenyl) caused significant decreases (IC$_{50}$ > 4 µM) in their ability to inhibit Ras processing (*29*).

Figure 3 illustrates the binding interactions of 3-bromo FTI **7** (IC$_{50}$ = 0.05 µM) with several binding pockets in FPT as determined by x-ray crystal structure analysis (*33*) of the 7-FPT complex.

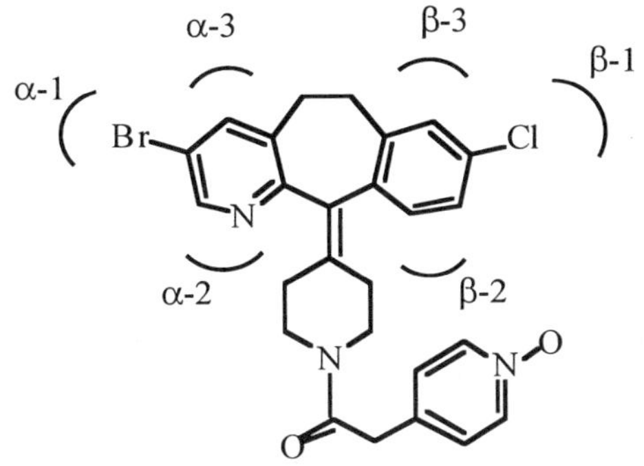

Figure 3. FPT binding pockets surround tricyclic structure of 7.

The x-ray structure (Figure 4) shows the C(3)-bromine substituent of **7** occupying the α-1 pocket of FPT *(33)*. Interestingly, in the absence of a C(3)-bromine substituent (i.e., **2**), the binding orientation of 8-chlorobenzocycloheptapyridine FTIs is such that the C(8)-chlorine atom occupies the α-1 pocket (Figure 1). This pocket, consisting of amino acid residues Lys164α, His201α and Tyr166α, anchors FTI **7** more effectively thereby accounting for the improved potency of this class of molecules (**7**, IC$_{50}$ = 0.05 μM). The x-ray structure in Figure 4 also highlights the interaction between the pyridyl nitrogen of **7** and an ordered water molecule.

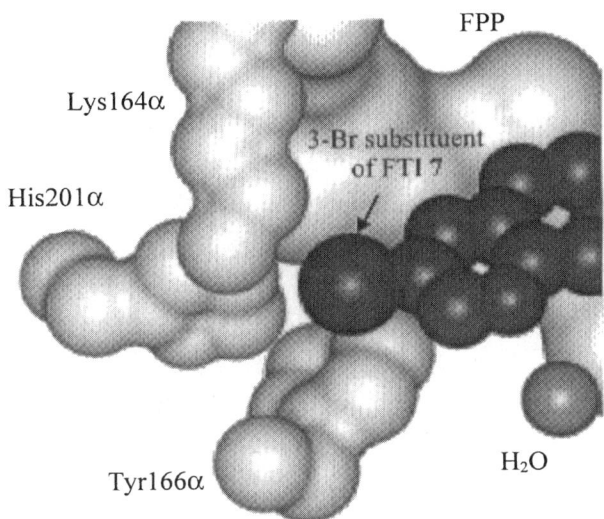

Figure 4. X-ray Crystal Structure of 7-FPT. C(3)-Br-Substituent fills α-1 Pocket of FPT; Water molecule H-bonds to pyridyl nitrogen of 7.

The binding interactions incurred by the chlorophenyl ring of dihalo FTI **7** and FPT are lipophilic with FPP, and both lipophilic and electronic with the amino acid residues comprising the β-1 pocket of the binding site (*33*). Tyr361β, Trp106β and Trp102β are seen to engage in edge-to-face and π-aromatic interactions with this phenyl ring (Figure 5). Replacement of the chlorine atom at C(8) with hydrogen, esters or amides renders the analogs less active ($IC_{50} > 3$ μM).

As was the case with halogen modifications of the pyridyl ring of the tricyclic moiety in **2** ($IC_{50} = 0.25$ μM), halogen substitution at either C(7) or at C(10) of the phenyl ring provided di-halo derivatives (*26*) with significantly improved potency ($IC_{50} = 0.01 - 0.06$ μM). Combining the modifications which gave the best potency, i.e., C(3)-Br with either C(7)- or C(10) bromination of the existing chlorophenyl ring, yielded tri-halobenzocycloheptapyridyl FTIs with exceptional activity (*26*). Of the two enantiomeric FTIs isolated from derivatization at C(7), only *S* isomer **9** inhibited FPT-mediated farnesylation of Ras with an IC_{50} of 2 nM. The opposite enatiomer (**8**) was significantly less potent. Interestingly, bromine modification at C(10) of benzocycloheptapyridine FTIs yielded two enatiomers whereby the *R* isomer (**10**) was found to be the most potent with an $IC_{50} = 2$ nM.

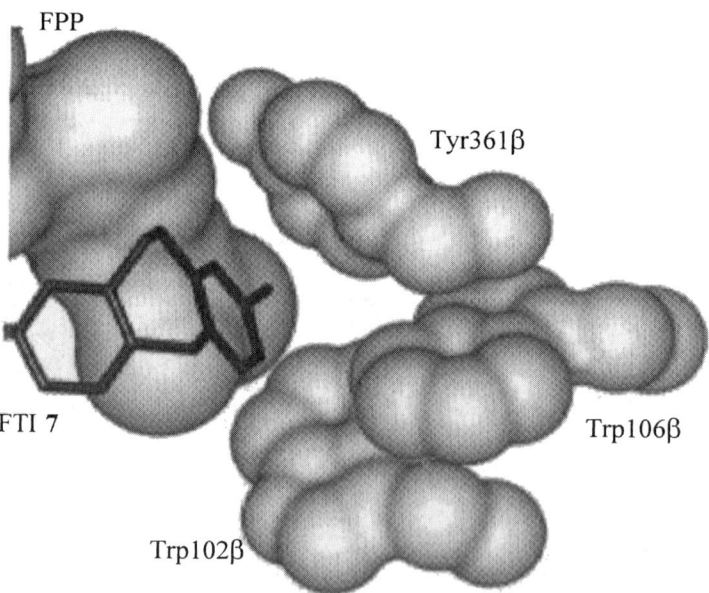

Figure 5. X-ray crystal structure of 7-FPT. C(7)-Cl Substituent fills β-1 pocket.

The relationship of binding orientation and C(11) chirality is readily understood from evaluation of the x-ray crystal structures (Figure 6) of the complexes created from 3,7,8-trihalo- and 3,8,10-trihalobenzocycloheptapyridine FTIs with FPT (*33*). Accordingly, the benzocycloheptapyridine moiety of **9** binds with the pyridyl ring closer to the β-subunit and the dihalophenyl ring interacting with the α-subunit of FPT. This binding mode is contrary to what is usually observed with dihalobenzocycloheptapyridines (i.e., **7**) and is perhaps a consequence of a more shallow β-3 pocket relative to the α-3 pocket and the relief of steric interactions created upon binding of 3,7,8-trihalobenzocycloheptapyridine FTIs in this reversed manner.

R (+) **8**, FPT IC$_{50}$ >> 0.10 μM
S (-) **9**, FPT IC$_{50}$ = 0.002 μM

R (+) **10**, FPT IC$_{50}$ = 0.002 μM
S (-) **11**, FPT IC$_{50}$ >> 0.10 μM

The binding situation for 3,8,10-trihalobenzocycloheptapyridine FTIs is opposite to that seen with 3,7,8-trihalobenzocycloheptapyridine FTIs (Figure 6). In this case, the 10-bromine substituent of **10** is seen to occupy the β-2 binding pocket interacting with lipophilic residues surrounding this pocket. An ordered water molecule normally bound to the Ser99β hydroxyl group is displaced upon binding of 10-bromobenzocycloheptapyridine FTIs (i.e., **10**) suggesting potency enhancement in part due to entropic benefit. Additionally, the dihalophenyl ring is perhaps more tightly engaged in aromatic-aromatic interactions as a result of the enhanced electron deficiency of the C(10)-halogenated system.

The bromine substituent at C(10) may play a more fundamental role apart from its interaction with residues of the β-2 pocket. We propose the presence of an intramolecular constraint exerted by the 10-bromine substituent in **10** whereby steric interference restricts the mobility of the pendant C(11) piperidyl moiety. The resultant decrease in entropy would translate to enhanced potency for 10-bromobenzocycloheptapyridine FTIs. This concept has inspired the design and synthesis of indolocycloheptapyridines (i.e., **13** and **14**) which were found to potently inhibit the farnesylation of Ras in a manner dependent on the degree of substitution of the β-carbon of the indole framework (*35*). Thus, β-

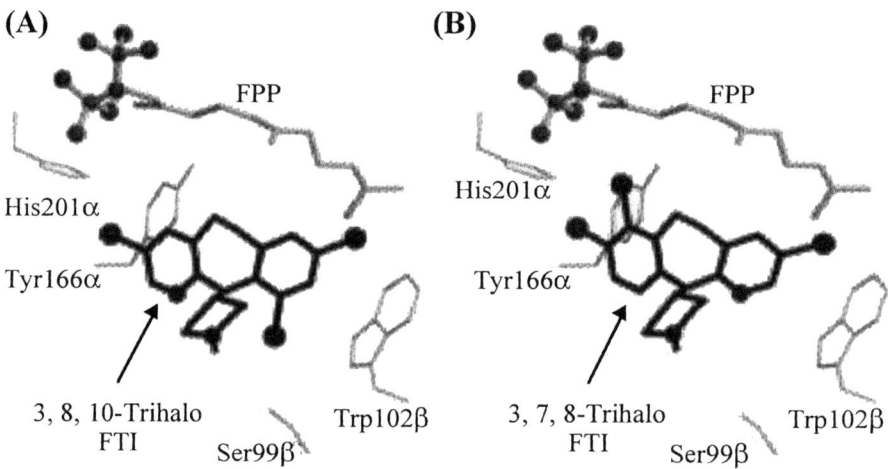

*Figure 6. X-ray crystal structure (R)-3,8,10- **(A)** and (S)-3,7, 8- **(B)** trihalobenzocycloheptapyridine FTIs complexed with FPT-FPP(34).*

12, IC$_{50}$ = 0.019 μM

13, R = Me, IC$_{50}$ = 0.007 μM
14, R = H, IC$_{50}$ = 0.042 μM

methyl indole **13** was found to be a more potent FTI than β–des-methyl **14**. The x-ray crystal structure of the **13**-FPT complex has been determined and will be reported shortly (*35*).

Optimizing Cellular and Pharmacokinetic Properties Leads to Sch-66336

Pharmacokinetic studies with pyridyl N-oxide **10** showed it to have poor serum concentrations in orally treated mice (*26*). Piperidylacetyl analogs of mono- and dihalo- benzocycloheptapyridine FTIs have previously been identified and found to be equi-potent with pyridyl N-oxide FTIs similar to **7** while demonstrating superior pharmacokinetic stability (*30*). Thus, significant modifications (Table I) of the terminal portion of 3,8,10-trihalobenzocycloheptapyridine FTIs were undertaken to identify structures with improved pharmacokinetic profiles while maintaining potencies in enzyme and cellular assays (*36*). Of the many analogs prepared, nicotinic N-oxides **17** and **18** demonstrated good AUCs when orally administered to mice (Table I). Sulfonamide analogs (i.e., **15**) were rarely bioavailable while urea **Sch-66336** and oxalamide **16** demonstrated good serum concentrations in mice when dosed orally. Pharmacokinetic evaluation of **Sch-66336** showed (*26, 36*) serum concentrations of 1.9 μM (Cmax) and an AUC of 22 μM.hr in orally-treated cynomolgus monkeys (10 mpk).

Although trihalobenzocycloheptapyridines demonstrated single digit nanomolar potency against FPT, their activity in cellular assays was not predictable (*26*). Typically, 3,7,8-trihalobenzocycloheptapyridine FTIs (i.e., **9**) poorly inhibited the growth of tumor colonies on agar media while 3,8,10-trihalobenzocycloheptapyridine FTIs (i.e., **10**) proved to be more effective.

Table I. FPT Potency and Pharmacokinetic Profile of Piperidylacetyl FTIs

Compound	R	IC$_{50}$ (μM)	AUC (μM.hr)[a]
Sch-66336	CONH$_2$	0.0019	24
15	SO$_2$Me	0.002	1
16	C(O)CH$_2$NH$_2$	0.0027	45
17	C(O)-(pyridyl N-oxide, 4-)	0.0025	16
18	C(O)-(pyridyl N-oxide, 3-)	0.0021	70

a. nude mice, 25 mpk p.o.

Tumor colony formation (NIH-H cells) in the soft agar assay was inhibited at 75 nM (IC$_{50}$) by **Sch-66336** while it affected ras processing in a Cos cell assay (IC$_{50}$ = 10 nM). In a panel of tumor cell lines representing human lung, colon, pancreatic and breast cancers, **Sch-66336** inhibited their growth in agar media at low concentrations (26, 37). The cell lines evaluated in this panel included NCI-H146 (IC$_{50}$ = 50 nM), HCT-116 (IC$_{50}$ = 70 nM), Mia Paca (IC$_{50}$ = 250 nM) and MCF-7 (IC$_{50}$ = 50 nM).

With a good *in vitro* and pharmacokinetic profile, **Sch-66336** was selected for further development as a recommendation candidate for the treatment of human cancers (38). **Sch-66336** was found to inhibit the farnesylation of Ras with an IC$_{50}$ = 1.9 nM without affecting the function of a structurally related prenyl transfer protein GGPT (IC$_{50}$ > 50 μM). *In vivo* evaluation of **Sch-66336** has shown promising results. In Wap-ras transgenic mice treated with **Sch-66336** (p. o.), tumor regression has been recorded while mice implanted with tumor cells (xenograft model) and dosed with **Sch-66336** (p. o.) show a dose dependent decrease in tumor volume (27). Thus, **Sch-66336** inhibited the growth of HCT-116 (human colon carcinoma) cells in nude mice by 84% at 40 mpk (qid). A similar effect was seen in human prostate carcinoma (DU-145)-bearing mice (86% @ 40 mpk, qid) and in mice exposed to HTB-177 cells (lung carcinoma, 83% @ 40 mpk, qid).

A synergistic effect was observed in mice treated with **Sch-66336** in combination with other anticancer agents (*27, 39*). Thus, **Sch-66336** (40 mpk, p. o.) in combination with vincristine (1 mpk, i. p.) inhibited tumor growth (HTB-177) by 80% in a mouse xenograft model. As a single agent, vincristine causes only a 7% decrease in the growth rate of these same tumors. Similarly, Cytoxan (200 mpk, i. p.) combined with **Sch-66336** (40 mpk, qid) slowed the growth of tumors by 81% compared to 9% and 60%, respectively, for these anticancer agents administered individually. The implications of these synergies are promising. With the potential for reducing the dosage of known cytotoxic agents such as vincristine or Cytoxan by their combined administration with **Sch-66336**, the associated side effects normally seen with much higher, efficacious doses may be limited.

The specific mechanism by which tumor growth inhibition results is, in part, due to the inhibition of farnesylation of Ras (*18, 19*). In tumor cells containing the more sensitive H-ras mutation, this is indeed the case (*37*). In cell lines containing K-ras or N-ras mutations, their sensitivities to FTIs are much less understood (*19*). What is clear is that exposure of cells to FTIs affects cell function (*27, 37*). In Wap-ras transgenic mice treated orally with **Sch-66336** (80 mpk, 5 days), an increase in apoptotic cells and a decrease in proliferating cells are detected upon histopathological examination of the excised mammary adenocarcinomas. Undoubtedly, the combination of these two factors, cellular proliferation inhibition and induction of apoptosis, led to regression of tumors in transgenic mice.

Conclusion and Prospects

Sch-66336 has been selected as a potent, selective benzocycloheptapyridine FTI for evaluation of its potential use in the treatment of human cancers. This drug candidate has demonstrated oral *in vivo* efficacy inhibiting the growth of tumors in mouse xenograft models while affecting tumor regression in transgenic models. Phase I clinical studies are complete and show inhibition of farnesylation in humans with some patients exhibiting stable disease over multiple treatment cycles (*38*). Combination therapy may offer the greatest opportunities for cancer therapy with FTIs coupling the unique and novel mechanism of FPT inihibition with reduced concentrations of conventional chemotherapeutics.

References

1. Barbacid, M. In *Annu. Rev. Biochem.* **1987**, *56*, 779-827.
2. Der, C.; Cox, A. D. *Cancer Cells* **1991**, *3*, 331-340.
3. Willumsen, B. M.; Norris, K.; Papageorge, A. G.; Hubbert, N. L.; Lowry, D. R. *EMBO J.* **1984**, 3, 2581-2585.
4. Lowry, D. R.; Willumsen, B. M. *Nature(London)* **1989**, *341*, 384-385.

5. Gibbs, J. B. *Cell* **1991**, *65*, 1-4.
6. Khosravi-Far, R.; Cox, A. D.; Kato, K.; Der, C. J. *Cell Growth and Differen.* **1992**, *3*, 461-469.
7. Barbacid, M. *Eur. J. Clin. Invest.* **1990**, *20*, 225-235.
8. Casey, P.J.; Solski, P. A.; Der, C.I.; Buss, J. E. *Proc. Natl. Acad. Sci USA* **1989**, *86*, 8323-8327.
9. Reiss, Y.; Goldstein, J. L.; Seabra, M. C.; Casey, P. J.; Brown, M. S. *Cell* **1990**, *62*, 81-88.
10. Hancock, J. F.; Magee, A. I.; Childs, J. E.; Marshall, C. J. *Cell* **1989**, *57*, 1176-77.
11. Park, H. W.; Boduluri, S. R.; Moomaw, J. F.; Casey, P. J.; Beese, L. S. *Science* **1997**, *275*, 1800-1804.
12. Long, S. B.; Casey, P. J.; Beese, L. S. *Biochemistry* **1998**, *37*, 9612-9618.
13. Strickland, C. L.; Windsor, W. T.; Syto, R.; Wang, L.; Bond, R.; Wu, Z.; Schwartz, J.; Le, H. V.; Beese, L. S.; Weber, P. C. *Biochemistry* **1998**, *37*, 16601-16611.
14. Rotella, D. P. *Curr. Opin. Drug Disc. & Devel.* **1998**, *1*, 165-174.
15. Cox, A. D. and Der, C. J. *Biochimica et Biophysica Acta* **1997**, *1333*, F51-F71.
16. Whyte, D. B.; Kirschmeier, P.; Hockenberry, T. N.; Nunez-Olivia, I.; James, L.; Catino, J. J.; Bishop, W. R.; Pai, J. K. *J. Biological Chemistry* **1997**, *272*, 14459-14464.
17. Prendergast, G. C. *Curr. Opin. Cell Bio.* **2000**, *12*, 166-173.
18. Ashar, H. R.; Armstrong, L.; James, L. J.; Carr, D. M.; Gray, K.; Taveras, A. G.; Doll, R. J.; Bishop, W. R. and Kirschmeier, P. T. *Chemical Research in Toxicology* **2000**, *13*, 949-952.
19. Ashar, H. R.; James, L.; Gray, K.; Carr, D.; Black, S.; Armstrong, L.; Bishop, W. R. and Kirschmeier, P. *J. Biological Chemistry* **2000**, *275*, 30451-30457.
20. Gibbs, J. B.; Olif, A. *Ann. Rev. Pharmacol. Toxicol.* **1997**, *37*, 143-166.
21. Britten, C. D.; Rowinski, E.; Yao, S.-L.; Rosen, N.; Eckhardt, S. G.; Drengler, R.; Hammond, L.; Siu, L. L.; Smith, L.; McCreery, H.; Pezzulli, S.; Lee, Y.; Lobell, R.; Deutsch, P.; Von Hoff, D.; Spriggs, D. *Proc. Am. Soc. Clinical Oncol.* **1999**, 597.
22. Gibbs, J. B.; Anthony, N. J.; Buser, C. A.; de Solms, S. J.; Graham, S. L.; Hartman, G. D.; Heimbrook, D. C.; Lobell, R. B.; Koblan, K. S.; Kohl, N. E.; Williams, T. M. *Abstracts 219th Am. Chem. Soc. Natl. Meeting* **2000**, Med. Chem., 292.
23. Hunt, J. T. *Abstracts 219th Am. Chem. Soc. Natl. Meeting* **2000**, Med. Chem., 294.
24. End, D. W. *Investigational New Drugs* **1999**, *17*, 241..
25. End, D.; Skrzat, S. G.; Devine, A.; Angibaud, P.; Venet, M.; Sanz, G.; Bowden, C. *Proc. Am. Assoc. Cancer Res.* **1998**, *39*, 269.
26. Njoroge, F. G.; Taveras, A. G.; Kelly, J.; Remiszewski, S. W.; Mallams, A. K.; Wolin, R.; Afonso, A.; Cooper, A. B.; Rane, D.; Liu, Y.-T.; Wong, J.; Vibulbhan, B.; Pinto, P.; Deskus, J.; Alvarez, C.; Del Rosario, J.; Connolly, M.; Wang, J.; Desai, J. A.; Rossman, R. R.; Bishop, W. R.; Patton, R.; Wang, L.; Kirschmeier, P.; Bryant, M. S.; Nomeir, A. A.; Lin, C.-C.; Liu,

M.; McPhail, A. T.; Doll, R. J.; Girijavallabhan, V.; Ganguly, A. K. *J. Med. Chem.* **1998**, *41*, 4890-4902.
27. Liu, M.; Bryant, M. S.; Chen, J.; Lee, S.; Yaremko, B.; Lipari, P.; Malkowski, M.; Ferrari, E.; Nielsen, L.; Prioli, N.; Dell, J.; Sinha, D.; Syed, J.; Korfmacher, W.; Nomeir, A. A.; Lin C.-C.; Wang, L.; Taveras, A. G.; Doll, R. J.; Njoroge, F. G.; Mallams, A. K.; Remiszewski, S.; Catino, J. J.; Girijavallabhan, V. M.; Kirschmeier, P.; Bishop, W. R. *Cancer Research* **1998**, *58*, 4947-4956.
28. Wong, J. K.; Piwinski, J. J.; Green, M. J.; Ganguly, A. K.; Anthes, J. C.; Billah, M. M. *Bioorganic & Medicinal Chemistry Letters* **1993**, *3*, 1073-1078.
29. Njoroge, F. G; Vibulbhan, B., Rane, D. F.; Bishop, W. R .; Petrin, J.; Patton, R.; Bryant, M. S.; Chen, K. -J.: Nomeir, A. A.; Lin, C. -C.; Liu, M.; King, I.; Chen, J.; Lee, S.; Yaremko, B.; Dell, J.; Lipari, P.; Malkowski, M.; Li, Z.; Catino, J.; Doll, R. J.; Girijavallabhan, V.; and Ganguly, A. K. *J. Med. Chem.* **1997**, *40*, 4290-4301.
30. Mallams, A. K.; Rossman, R. R.; Doll, R. J.; Girijavallabhan, V. M.; Ganguly, A. K.; Petrin, J.; Wang, L.; Patton, R.; Bishop, W. R.; Carr, D. M.; Kirschmeier, P.; Catino, J. J.; Bryant, M. S.; Chen K-J.; Korfmacher, W. A.; Nardo, C.; Wang, S.; Nomeir, A. A.; Lin, C.-C.; Li, Z.; Chen, J.; Lee, S.; Dell, J.; Lipari, P.; Malkowski, M.; Yaremko, B.; King, I.; Liu, M. *J. Med. Chem.* **1998**, *41*, 877-893.
31. Njoroge, F. G.; Doll, R. J.; Vibulbhan, B.; Alvarez, C.; Bishop, W. R.; Petrin, J.; Kirschmeier, P.; Carruthers, N. I.; Wong, J. K.; Albanese, M. M.; Piwinski, J. J.; Catino, J.; Girijavallabhan, V.; Ganguly, A. K. *Bioorg. & Med. Chem.* **1997**, *5*, 101-114.
32. Bishop, W. R.; Bond, R.; Petrin, J.; Wang, L.; Patton, R.; Doll, R.; Njoroge, N.; Windsor, W.; Syto, R.; Schwartz, J.; Carr, D.; James, L.; Kirschmeier, P. *J. Biol. Chem.* **1995**, *270*, 30611-30618.
33. Strickland, C. L.; Weber, P. C.; Windsor, W. T.; Wu, Z.; Le, H. V.; Albanese, M. M.; Alvarez, C. S.; Cesarz, D.; del Rosario, J.; Deskus, J.; Mallams, A. K.; Njoroge, F. G.; Piwinski, J. J.; Remiszewski, S.; Rossman, R. R.; Taveras, A. G.; Vibulbhan, B.; Doll, R. J.; Girijavallbhan, V. M.; Ganguly, A. K. *J. Med. Chem.* **1999**, *42*, 2125-2135.
34. The x-ray crystal structures shown represent FPT-FPP complexes with 3, 8, 10-trihalo FTI Sch-66336 (B) and with a 3, 7, 8-trihalo analog of Sch-66336 (A). The amide substituents from the FTIs shown in Figure 6 (A) and (B) have been removed for clarity.
35. Taveras, A. G.; Vaccaro, C.; Chao, J.; Doll, R. J.; Girijavallabhan, V.; Ganguly, A. K.; Strickland, C.; Weber, P.; Hollinger, F.; Snow, M.; Bishop, W. R.; Patton, R.; Kirschmeier, P.; James, L.; Liu, M.; Nomeir, A.; Heimark, L.; Puar, M. (*manuscript to be submitted to J. Med. Chem.* **2000**).
36. Taveras, A. G.; Deskus, J.; Chao, J.; Vaccaro, C. J.; Njoroge, F. G.; Vibulbhan, B.; Pinto, P.; Remiszewski, S.; del Rosario, J.; Doll, R. J.; Alvarez, C.; Lalwani, T.; Mallams, A. K.; Rossman, R. R.; Afonso, A.; Girijavallabhan, V. M.; Ganguly, A. K.; Pramanik, B.; Heimark, L.; Bishop, W. R.; Wang, L.; Kirschmeier, P.; James, L.; Carr, D.; Patton, R.; Bryant, M. S.; Nomeir, A. A.; Liu, M. *J. Med. Chem.* **1999**, *42*, 2651-2661.

37. Ashar, H. R.; James, L.; Gray, K.; Carr, D.; McGuirk, M.; Maxwell, E.; Armstrong, L.; Doll, R. J.; Taveras, A. G.; Bishop, W. R. and Kirschmeier, P. *Experimental Cell Research* (in press,**2000**).
38. Adjei, A. A.; Erlichman, C.; Davis, J. N.; Cutler, D. L.; Sloan, J. A.; Marks, R. S.; Hanson, L. J.; Svingen, P. A.; Atherton, P.; Bishop, W. R.; Kirschmeier, P.; Kaufmann, S. H. *Cancer Res.* **2000**, *60*, 1871-1877.
    Nielsen, L. L.; Shi, B.; Hajian, G.; Yaremko, B.; Lipari, P.; Ferrari, E.; Gurnani, M.; Malkowski, M.; Chen, J.; Bishop, W. R.; Liu, M. *Cancer Res.* **1999**, *59*, 5896-5901.
39. Nielsen, L. L.; Shi, B.; Hajian, G.; Yaremko, B.; Lipari, P.; Ferrari, E.; Gurnani, M.; Malkowski, M.; Chen, J.; Bishop, W. R.; Liu, M. *Cancer Res.* **1999**, *59*, 5896-5901.

Chapter 14

Pyrrolo[2,3-*d*]pyrimidine and Pyrazolo[3,4-*d*]pyrimidine Derivatives as Selective Inhibitors of the EGF Receptor Tyrosine Kinase

G. Caravatti[1], J. Brüggen[1], E. Buchdunger[1], R. Cozens[1], P. Furet[1], N. Lydon[2], T. O'Reilly[1], and P. Traxler[1]

[1]Novartis Pharma AG, TA Oncology, K–136–4–26, CH–4002 Basel, Switzerland
[2]Kinetix Pharmaceuticals Inc., Medford, MA

The EFG receptor tyrosine kinase (EGFR) is an attractive target for the development of agents directed against tumors which either overexpress the EGFR or which have a mutated or amplified gene encoding the EGFR. Several ATP-competitive inhibitors of this kinase have shown promising *in vitro* and *in vivo* efficacy and are currently in different stages of clinical development. One of them is PKI166, a pyrrolo-[2,3-*d*]pyrimidine, which has been selected from a large series of pyrrolo[2,3-*d*]pyrimidines and structurally related pyrazolo-[3,4-*d*]pyrimidines. The discovery and preclinical data of PKI166 are summarized.

The phosphorylation and dephosphorylation of intracellular proteins is a general, very frequently observed reaction during signal transduction processes regulating fundamental cellular functions, such as cell growth, differentiation, and cell death. The interplay of these two events functions as a biological switch, i.e. signaling pathways are turned on or shut down depending on the phosphorylation state of a given protein. Protein kinases catalyze the transfer of the γ-phosphate of ATP to hydroxyl groups of substrate proteins. They are divided into serine/threonine and tyrosine kinases depending on the residue to

© 2001 American Chemical Society

which the phosphate group is attached. In this chapter we concentrate exclusively on protein tyrosine kinases and more specifically on inhibitors of the epidermal growth factor receptor (EGFR) tyrosine kinase.

Rationale for EGFR Tyrosine Kinase Inhibitors in Cancer

The EGF receptor, also known as ErbB1, is a transmembrane glycoprotein that mediates the mitogenic response to epidermal growth factor[1], transforming growth factor α (TGF-α)[2], epinephrin and amphiregulin[3;4]. The receptor consists of an extracellular ligand-binding domain, a short hydrophobic transmembrane domain, and an intracellular region which contains an activatable protein tyrosine kinase domain. The EGFR and its ErbB family members ErbB2, ErbB3, and ErbB4, are widely expressed in epithelial, mesenchymal and neuronal tissues and play fundamental roles during development. Interest was dramatically raised when it was found during the last two decades that c-*erb*B1, the gene encoding the EGFR, undergos gene amplification or overexpression in many human cancers of epithelial[5-10] or neuroepithelial[5;11] origin. In addition to amplification, rearranged or truncated forms of the EGFR gene have been identified. These truncated receptors all have alterations in the extracellular ligand binding domain of the EGFR[12-16], which are reminiscent of alterations found in the v-*erb*B oncogene. In an early study of a limited number of tumors, the more common type III deletion mutant was found in 16% of NSCLC, 17-52% of glioblastomas and 73% of breast tumors[16]. Like the v-*erb*B encoded protein, the type III truncated EGFR is unable to bind EGF but has constitutive enzymatic activity[12]. These findings suggest that deregulation of the EGFR signaling pathway is involved in many epithelial malignancies.

One of the first steps following the activation of the ErbB receptor family by their respective extracellular ligands, involves their homo- and heterodimerization. ErbB2 appears to be the preferred heterodimerisation partner for all other ErbB family members[17;18]. Dimerization of the ErbB proteins is followed by the trans-phosphorylation of their intracellular carboxy-terminal tails. The tyrosine phosphorylated peptide stretches formed by this process function then as high-affinity binding sites for SH2 domain-containing proteins[19;20]. This leads to protein-protein complex formation and the potent stimulation of major mitogenic signaling cascades, such as mitogen activated protein kinase (MAP kinase) through the ras pathway or, the PI3 kinase[21]. Inhibition of the tyrosine kinase of the ErbB receptors will suppress the formation of these signaling complexes and block downstream mitogenic signals. Due to its central role and frequent deregulation in cancers, the EGF receptor has become a widely studied pharmacological target.

ATP-competitive Inhibitors of the EGFR Tyrosine Kinase

Conceptually there are many options to inhibit the enzymatic activity of a receptor tyrosine kinase. First, one could think of inhibitors which prevent their activation by blocking the binding of the external ligand to the receptor. This is a very attractive approach since the inhibitor would act outside the cell. To date, no low molecular weight compounds have been described which act by this mechanism, however, specific monoclonal antibodies targeting the EGF receptor or ErbB2 have been developed successfully (e.g., C225, Herceptin™)[22-26]. Intracellular points of attack would be substrate mimics, and mimics of the phosphorylated tail of the receptor which would prevent the binding of SH2 domain containing proteins to the receptor[27;28]. The vast majority of work, however, has been invested in the development of ATP competitive inhibitors of the kinase domain of the EGFR. That approach has gained additional momentum with the increasing amount of SAR data and with the availability of structural information obtained from X-ray structure analyses[29].

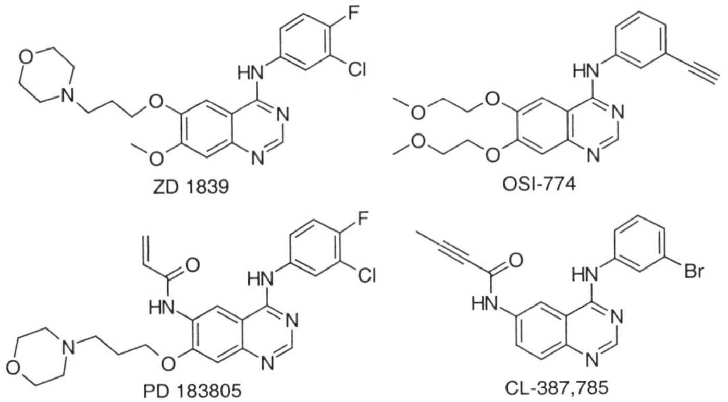

Figure 1: 4-Phenylamino-quinazolines: ATP competitive inhibitors of the EGFR tyrosine kinase.

A number of compound classes have been described in the literature which inhibit the EGFR tyrosine kinase[30-34]. Potent compounds from the phenylamino-quinazoline[35-37], phenylamino-pyrido-pyrimidine[38], and phenyl-amino-pyrimido-pyrimidine[39] classes have been reported which possess the desired *in vitro* and *in vivo* biological profile of an EGFR kinase inhibitor (Figure 1). Compounds from these classes are in various stages of preclinical or clinical development[40-42]: ZD 1839 (Iressa™, AstraZeneca) has successfully passed Phase IA trials in volunteers and is currently in Phase III trials in patients [40;41;43]. OSI-774 (formerly CP-358,774, OSI Pharmaceuticals) is in Phase II trials[42]. An irreversible inhibitor from the same 4-phenylamino-quinazoline class, PD 183805 (CI1033)[44;45], is being developed by Pfizer/Warner-Lambert and has entered Phase I trials. The acrylamide moiety of PD 183805 is reported to irreversibly react with the thiol group of cysteine 773 which is unique for the EGF receptor and ErbB2[44;46]. Similar compounds, e.g. CL-387,785[47], with a butynamide as the trapping group for the thiol were also reported. A critical point during the optimization of this entire compound class was the improvement of their physicochemical properties, which has been solved by the attachment of solubilizing groups to the molecules[44].

SAR of Pyrrolo[2,3-*d*]pyrimidines and Pyrazolo[3,4-*d*]pyrimidines: PKI166

Pyrrolo[2,3-*d*]pyrimidines and pyrazolo[3,4-*d*]pyrimidines were identified in a random screening and further optimized using a pharmacophore model of the ATP binding site of the EGFR tyrosine kinase[48;49]. A summary of a very extensive structure activity determination[48] is given in Tables I to IV. Due to a dramatic increase in potency observed very early on by the attachment of a chlorine atom to the *meta*-position of the phenylamino moiety (**1** vs. **3**) this modification was maintained during the optimization work. As indicated in Table I, even large substituents are tolerated in the 5 and 6 positions of the pyrrolo-pyrimidine system (**3–6**). The relatively good activity of the 6-phenyl substituted derivative **6** triggered the synthesis of a whole range of analogs (Table II). With the exception of the biphenyl derivative **13**, most compounds retained the potency. Interestingly, substituents capable of donating a hydrogen bond seemed to have a beneficial effect (**10–12** vs. **8, 9**). The IC_{50} of the ethyl ester of carboxylic acid derivative **14** was in the low μ-molar range (data not shown), indicating as seen with **13** that there may be steric restrictions in this part of the molecule.

Finally, and with the aim to replace the 4-anilino moiety in these molecules, a series of benzyl substituted analogs was investigated. Representative examples are shown in Table III. It was found that a benzylamino group is able to replace the *m*-chlorophenylamino with no change in activity (**12** vs. **15**). Due to the frequently encountered metabolic instability of benzylic methylenes the study was extended to 1-phenethylamines (**16–19**). Interestingly, the (*R*)-configurated derivatives were consistently more potent than the corresponding (*S*)-isomers. This led to the discovery of PKI166 (**18**), a very potent inhibitor of the EGFR tyrosine kinase.

Table I: Effect of 5 and 6 substituents of the pyrrolo-pyrimidine

Compound no.	R1	R2	R3	EGFR IC$_{50}$ [μM]
1	CH$_3$	CH$_3$	H	1.9
2	H	H	Cl	0.01
3	CH$_3$	CH$_3$	Cl	0.045
4	-(CH$_2$)$_4$-		Cl	0.033
5	Ph	Ph	Cl	0.096
6	H	Ph	Cl	0.013

Table II: Effects of substituents at the 6-phenyl group of the pyrrolo-pyrimidine

Compound no.	R	EGFR IC$_{50}$ [μM]
7	H	0.013
8	4-OCH3	0.015
9	2-OCH3	0.019
10	4-OH	0.002
11	3-NH2	0.004
12	4-NH2	0.003
13	4-Ph	> 100
14	4-COOH	0.001

Table III: Replacements of the anilino group

Compound no.	R1	R2	EGFR IC$_{50}$ [μM]
15	4-NH2	Benzyl	0.003
16	4-NH2	(R)-1-phenethyl	0.002
17	4-NH2	(S)-1-phenethyl	0.046
18 (PKI166)	4-OH	(R)-1-phenethyl	0.0017
19	4-OH	(S)-1-phenethyl	0.043

In the pyrazolo[3,4-d]pyrimidine series the SAR parallels that of the corresponding pyrrolo[2,3-d]pyrimidines[50]. A 40-fold increased activity is obtained with the m-chlorophenylamino versus the phenylamino group at the 4 position of the core structure (**20** vs. **22**). Substitutions at the meta position are widely tolerated (compounds **23-29**). In contrast to the pyrrolo-pyrimidine series, the 4-phenylamino group cannot be replaced by a benzylamino or 1-phenethylamino group. In both cases there is a significant ($\geq$ 10-fold) loss in activity and, interestingly, the two enantiomers in the 1-phenethylamino series have similar activity (data not shown).

Table IV: Pyrazolo[3,4-d]pyrimidines

Compound no.	R1	R2	EGFR IC$_{50}$ [μM]
20	H	Ph	1.1
21	NH-Ph	Ph	0.058
22	H	3-Cl-Ph	0.025
23	NH$_2$	3-Cl-Ph	0.021
24	NH-n-Bu	3-Cl-Ph	0.002
25	NH-Bn	3-Cl-Ph	0.007
26	NH-CH$_2$-c-C$_6$H$_{11}$	3-Cl-Ph	0.11
27	morpholino	3-Cl-Ph	0.058
28	piperazinyl	3-Cl-Ph	0.4
29	Ph	3-Cl-Ph	0.019

Binding Model of PKI166

The three dimensional structure of the kinase domain of EGFR has not yet been determined. However, more than sixty X-ray crystal structures of other protein kinases, in unligated or ligated forms, have been solved to date[51]. Based on this information, homology models of kinases for which the experimental structure is lacking can be constructed. A few years ago, we derived a model of the EGFR kinase using the X-ray structure of the cyclic-AMP dependent protein kinase, the only kinase structure available at that time[49]. This model was instrumental for the optimization of the pyrrolopyrimidine class of inhibitors to which PKI166 belongs[48]. We discuss here the binding mode of PKI166 in the context of our model, updated using a recent crystal structure of the more closely related fibroblast growth factor receptor kinase[52].

It is now firmly established that key interactions in the binding of ATP by protein kinases are a pair of bidentate hydrogen bonds involving the adenine moiety of the cofactor and the backbone of the protein in the so-called hinge region. The hinge region is the stretch of six or seven amino acids which connects the N- and C-terminal domains of the kinase at the interface of which the binding site of ATP is located. In the case of the EGFR kinase, one can assume that the adenine N1 atom accepts a hydrogen bond from the backbone NH of residue Met 769 while the 6-amino group donates one to the backbone carbonyl of Gln 767. A representation of ATP docked in our model of the EGFR kinase in the corresponding orientation is shown in Figure 2.

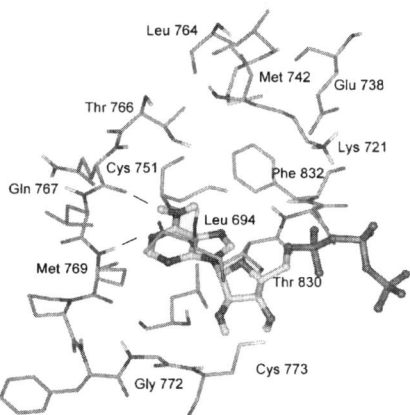

Figure 2: Model of the ATP binding site of the EGFR kinase

Mimicking the cofactor donor-acceptor hydrogen bonding system with heterocyclic scaffolds other than adenine was the basis for the discovery of the pyrrolopyrimidine class of EGFR kinase inhibitors. Figure 3 (1) shows PKI166 docked in the ATP binding site of the EGFR kinase model according to this binding mode hypothesis. The pyrimidine N1 atom of the inhibitor forms the hydrogen bond with Met 769 while its pyrrole NH group plays the role of the 6-amino group of the adenine ring of ATP in donating a hydrogen bond to Gln 767. These interactions position the phenol moiety of PKI166 in a hydrophobic pocket extending in a direction corresponding to that of the lone pair of the adenine N7 atom of ATP. This hydrophobic pocket, not exploited by ATP, is formed by the side chains of residues Thr 766, Cys 751, Leu 764, Met 742, Thr 830, Phe 832 and the hydrocarbon part of the side chain of Lys 721. In addition, the hydroxy phenol group can donate a hydrogen bond to Glu 738, a residue proximal to the pocket. On the other side of the inhibitor, the benzylic moiety occupies the region of the active site normally involved in binding the ribose group of ATP. Noteworthy, is the possibility of a hydrophobic contact between the phenyl ring and Cys 773. Cysteine at this position is a distinctive feature of the members of the ErbB family. The corresponding amino acid in the other protein kinases usually has a polar side chain. Thus, the binding model, through this putative Cys 773 hydrophobic interaction, provides an explanation for the high selectivity displayed by PKI166 towards the EGFR and ErbB2 kinases.

X-ray crystallography has shown that some kinase inhibitors presenting a donor-acceptor hydrogen bonding system in their structure bind to the hinge region somewhat differently than the adenine ring of ATP[52;53]. Although, like ATP, they accept a hydrogen bond from the backbone NH of the residue corresponding to Met 769, they make a donor interaction with the carbonyl of the latter residue and not with that of the residue corresponding to Gln 767.

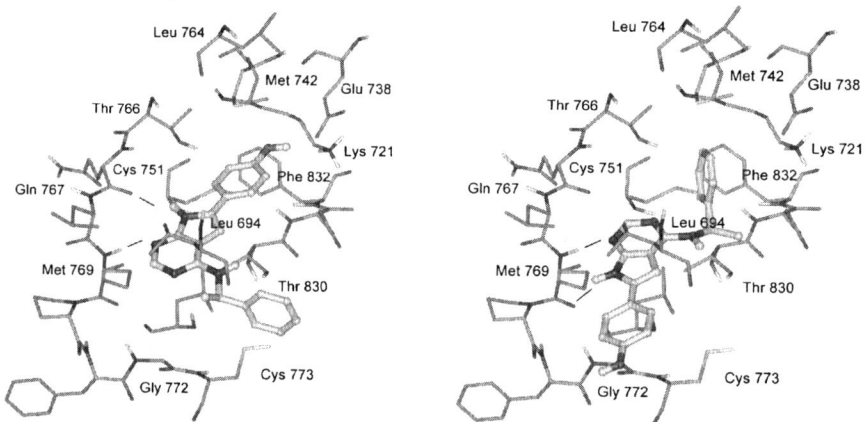

Figure 3: Proposed (l) and alternative (r) binding mode of PKI166 in the ATP binding site of the EGFR kinase

Docking studies show that an alternative binding mode based on this hydrogen bonding pattern is possible for PKI166 as represented in Figure 3 (r). In this binding mode, the benzylic moiety of the inhibitor adopts a different conformation and is now located in the hydrophobic pocket adjacent to the adenine N7 atom of ATP. The new orientation of the molecule places the phenol ring in a hydrophobic channel, opened to solvent, formed by the side chain of Leu 694 and the α carbon of Gly 772.

Both proposed binding modes provide a plausible structural basis to the high potency of PKI166 in inhibiting the EGFR kinase. In both cases, high complementarity with the ATP binding site in terms of hydrophobic and hydrogen bond interactions is observed. However, the first binding mode, by invoking the unique character of Cys 773 in the erbB family of tyrosine kinases, may better explain the high selectivity of the inhibitor.

Synthetic Access to Pyrrolo[2,3-*d*]pyrimidines

A general, very convenient route to the pyrrolo[2,3-*d*]pyrimidines is exemplified by the synthesis of PKI166 (**18**) (Figure 4). The intermediate 2-amino-3-carboethoxypyrrole **30**, obtained following the procedure of E. Toja *et al.*[54], was converted in high yield to the 4-hydroxy-pyrrolo[2,3-*d*]pyrimidine **31** by heating with a mixture of formamide, formic acid and DMF. Heating **31** with phosphoryl chloride gave the corresponding halogenated derivative **32** which reacted readily with (*R*)-1-phenethylamine in *n*-butanol to the protected intermediate **33**. Final cleavage of the methyl ether was achieved using boron tribromide in dichloromethane. The 5,6-disubstituted analogs of table I were prepared by published procedures[48].

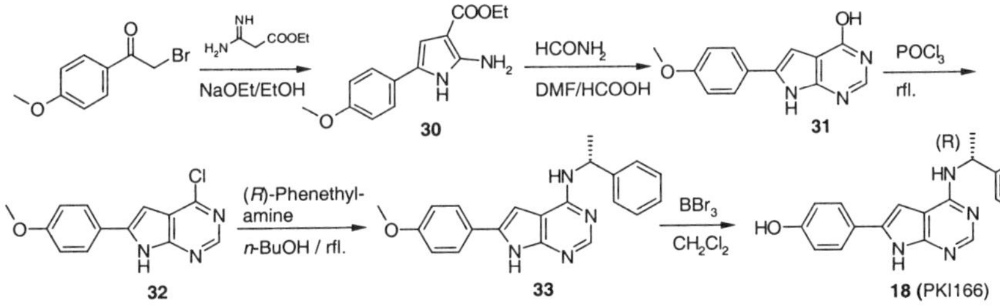

Figure 4: Synthesis of PKI166

The pyrazolo[3,4-*d*]pyrimidines shown in table IV were synthesized following published procedures[50].

Preclinical Studies with PKI166[55]

PKI166 (**18**) is a highly potent inhibitor of both the intracellular domain (ICD) of the EGFR and the ErbB2 tyrosine kinases with IC_{50} values of 1.7 and 14 nM, respectively (Table V). Enzyme kinetic studies (data not shown) are in accordance with ATP-competitive inhibition of the EGFR tyrosine kinase.

PKI166 showed a high degree of selectivity with respect to inhibition of the serine/threonine kinases CDK-1, PKC-α and PKA. On the other hand, the compound has moderate selectivity against some tyrosine kinases, inhibiting a number of them at submicromolar concentrations. However, a selectivity ratio of ≥ 100 is still achieved for all the tyrosine kinases tested.

Table V: Enzymatic profile of PKI166

Enzyme	PKI166 (**18**) IC$_{50}$ [μM] ± SEM	Kinase type
EGFR ICD	0.0017 ± 0.0005	Tyr
ErbB2	0.007 ± 0.0009	Tyr
KDR	0.39 ± 0.14	Tyr
c-Met	> 10	Tyr
c-Abl	0.26 ± 0.077	Tyr
c-Src	0.18 ± 0.020	Tyr
c-Kit	0.88 ± 0.23	Tyr
Cdk1/Cyc.B	> 10	Ser/Thr
PKC-α	> 10	Ser/Thr
PKA	> 10	Ser/Thr

All IC$_{50}$ values are averages of at least three determinations.

As expected from the *in vitro* enzyme data, PKI166 inhibited signaling through the ligand-activated EGFR in cells. In a cell-based ELISA assay[56], EGF-induced autophosphorylation of the EGFR was inhibited with an IC$_{50}$ value of 51 nM. Western blot analysis using anti-phosphotyrosine antibodies gave similar results (IC$_{50}$ value between 10–30 nM). Immunoblotting with anti-EGFR antibodies showed that the compound did not affect the overall level of EGFR. PKI166 also inhibited ErbB2 autophosphorylation in two human breast carcinoma cell lines (SK-BR-3 and BT-474) and in c-*erb*B2 transfected NIH3T3 cells with IC$_{50}$ values between 0.1-1 μM. In contrast, PKI166 did not affect PDGF-stimulated autophosphorylation of the PDGFR in cells (IC$_{50}$ >10 μM). PKI166 effectively inhibited EGF-induced c-*fos* induction with an IC$_{50}$ value between 0.1-1 μM, whereas c-*fos* mRNA induction in response to FGF or PMA was much less sensitive to inhibition.

In cell proliferation assays, PKI166 selectively inhibited the growth of EGF-dependent BALB mouse keratinocytes, EGFR-overexpressing A431 and NCI-H596 cell lines and the ErbB2-overexpressing SK-BR-3 and BT-474 cell lines (Table VI). In contrast, inhibition of EGF-independent cell lines such as T24 bladder carcinoma cells occurred at somewhat higher concentrations.

Table VI: Antiproliferative activity of PKI166 (IC$_{50}$ in μM)

	BALB/MK	NCI-H596	A-431	SK-BR-3	BT474	T24
PKI166	0.45	0.94	0.26	3.05	1.29	7.1

All IC$_{50}$ values are averages of at least three determinations.

When administered orally, PKI166 produced significant and dose-dependent antitumor effects against A-431 tumors growing in nude mice (Figure 5). At 30 mg/kg/day PKI166 produced a final T/C of 29%. Moreover, the higher

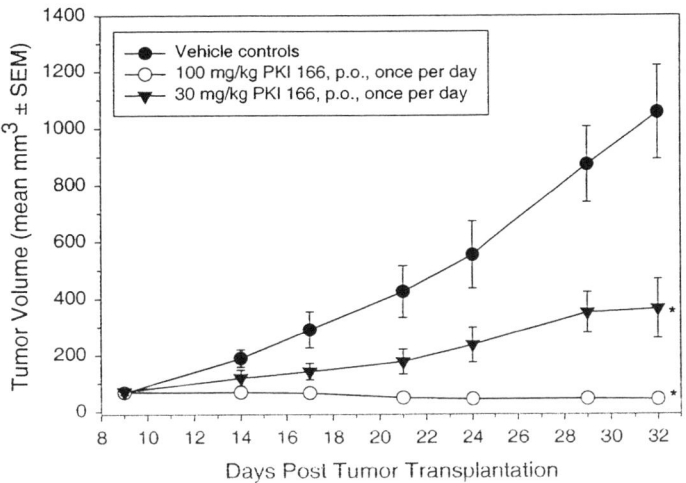

Treatments were started on day 9 after tumor transplantation (n = 7 per group). PKI166 was administered p.o. once per day at 30 or 100 mg/kg until 32 days post tumor transplantation. *p < 0.05 *versus* vehicle controls.

Figure 5: Antitumor effects of PKI166 against s.c. transplanted human A-431 epidermoid carcinoma in female BALB/c nude mice

dose of 100 mg/kg/day induced durable regressions (- 38%). Both doses were well tolerated, and there was no mortality or body weight loss.

Peak plasma and tumor concentrations of PKI166 following administration of a single 100 mg/kg oral dose to A-431 tumor-bearing athymic nude mice were 9.6 μM and 21.9 μM, respectively. The time to C_{max} was delayed in tumors (2 h) as compared to plasma (1h). 24 hrs after dosing, substantial concentrations of the drug could still be detected in tumors (C_{24h}: 1.5 μM), exceeding by a factor of 150 the IC50 for inhibition of autophosphorylation in A431 cells (measured by Western blotting) . Elimination of PKI166 from the tumor was slower than from plasma, consequently the $AUC_{0.25-24h}$ was greater in tumors (~ 153 h•nmol/g) than in plasma (~ 29 h•μmol/L). In all plasma and tumor samples, the glucuronide of PKI166 was detected as the main metabolite.

Using two different methods, immunoblotting and immunohistochemistry, PKI166 was shown to inhibit the EGF-stimulated autophosphorylation also *in vivo* (A-431 tumors transplanted to nude mice). A single oral dose of 100 mg/kg, PKI166 produced a complete and long-lasting inhibition of EGF-stimulated EGFR autophosphorylation. Even 8 hrs post-treatment, EGFR autophosphorylation was still completely blocked, and 24 hrs post treatment, the inhibition was still > 50%.

These data strongly support that PKI166 acts predominantly through inhibition of the EGFR-mediated signal transduction pathway and provide the basis for the selection of PKI166 for clinical trials.

References

1. Todaro, G. J.; Rose, T. M.; Spooner, C. E.; Shoyab, M.; Plowman, G. D. *Semin. Cancer Biol.* **1990**, *1*, 257-263.
2. Carpenter, G. *Annu. Rev. Biochem.* **1987**, *56*, 881-914.
3. Shoyab, M.; Plowman, G. D.; McDonald, V. L.; Bradley, J. G.; Todaro, G. J. *Science* **1989**, *243*, 1074-1076.
4. Johnson, G. R.; Kannan, B.; Shoyab, M.; Stromberg, K. *J. Biol. Chem.* **1993**, *268*, 2924-2931.
5. Gullick, W. J. *Br. Med. Bull.* **1991**, *47*, 87-98.
6. Yamamoto, T.; Kamata, N.; Kawano, H.; Shimizu, S.; Kuroki, T.; Toyoshima, K.; Rikimaru, K.; Nomura, N.; Ishizaki, R. *Cancer Res.* **1986**, *46*, 414-416.
7. Ishitoya, J.; Toriyama, M.; Oguchi, N.; Kitamura, K.; Ohshima, M.; Asano, K.; Yamamoto, T. *Br. J. Cancer* **1989**, *59*, 559-562.
8. Lacroix, H.; Iglehart, J. D.; Skinner, M. A.; Kraus, M. H. *Oncogene* **1989**, *4*, 145-151.
9. Gullick, W. J.; Marsden, J. J.; Whittle, N.; Ward, B.; Bobrow, L.; Waterfield, M. D. *Cancer Res.* **1986**, *46*, 285-292.
10. Ullrich, A.; Coussens, L.; Hayflick, J. S.; Dull, T. J.; Gray, A.; Tam, A. W.; Lee, J.; Yarden, Y.; Libermann, T. A. *Nature* **1984**, *309*, 418-425.
11. Libermann, T. A.; Nusbaum, H. R.; Razon, N.; Kris, R.; Lax, I.; Soreq, H.; Whittle, N.; Waterfield, M. D.; Ullrich, A.; Schlessinger, J. *Nature* **1985**, *313*, 144-147.
12. Ekstrand, A. J.; Longo, N.; Hamid, M. L.; Olson, J. J.; Liu, L.; Collins, V. P.; James, C. D. *Oncogene* **1994**, *9*, 2313-2320.
13. Wong, A. J.; Ruppert, j. M.; Bigner, S. H.; Grzeschik, C. H.; Humphrey, P. A.; Bigner, D. S.; Vogelstein, B. *Proc. Natl. Acad. Sci. USA* **1992**, *89*, 2965-2969.
14. Garcia de Palazzo, I.; Adams, G. P.; Sundareshan, P.; Wong, A. J.; Testa, J. R.; Bigner, D. D.; Weiner, L. M. *Cancer Res.* **1993**, *53*, 3217-3220.
15. Sugawa, N.; Ekstrand, A. J.; James, C. D.; Collins, V. P. *Proc. Natl. Acad. Sci. USA* **1990**, *87*, 8602-8606.
16. Moscatello, D. K.; Holgado, M.; Godwin, A. K.; Ramirez, G.; Gunn, G.; Zoltick, P. W.; Biegel, J. A.; Hayes, R. L.; Wong, A. J. *Cancer Res.* **1995**, *55*, 5536-5539.
17. Tzahar, E.; Waterman, H.; Chen, X.; Levkowitz, G.; Karunagaran, D.; Lavi, S.; Ratzkin, B. J.; Yarden, Y. *Mol. Cell. Biol.* **1996**, *16*, 5276-5287.

18. Olayioye, M. A.; Neve, R. M.; Lane, H. A.; Hynes, N. E. *EMBO J.* **2000**, *19*, 3159-3167.
19. Sudol, M. *Oncogene* **1998**, *17*, 1469-1474.
20. Shoelson, S. E. *Curr. Opin. Chem. Biol.* **1997**, *1*, 227-234.
21. Daly, R. J. *Growth Factors* **1999**, *16*, 255-263.
22. Mendelsohn, J. *Clin. Cancer Res.* **1997**, *3*, 2703-2707.
23. Goldstein, N. I.; Prewett, M.; Zuklys, K.; Rockwell, P.; Mendelsohn, J. *Clin.Cancer Res.* **1995**, *1*, 1311-1318.
24. Anderson, V.; Cooper M.; O'Chei, E.; Gilly, J.; Falcey, J.; Waksal, H. W. *Proc. Am. Assoc. Cancer Res.* **1998**, *39*,3561.
25. Mendelsohn, J.; Shin, D. M.; Donato, N.; Khuri, F.; Radinsky, R; Glisson, B. S.; Shin, H. I.; Metz, E.; Pfister, D.; Perez-Soler, R.; Lawhorn, K.; Matsumoto, T.; Gunnett, K.; Falcey, J.; Waksal, H.; Hong, W. K. A. *Proc. Am. Soc. Clin. Oncol.* **1999**, *18*, 1502.
26. Gunnett, K.; Motzer, R.; Amato, R.; Todd, M.; Poo, W.-J.; Cohen, R.; Baselga, J.; Cooper M.; Robert, F.; Falcey, J.; Waksal, H. *Proc. Am. Soc. Clin. Oncol.* **1999**, *18*, 1309.
27. Schlessinger, J.; Ullrich, A. *Neuron* **1992**, *9*, 383-391.
28. Pawson, T.; Schlessinger, J. *Curr. Biol.* **1993**, *3*, 434-442.
29. Sowadski, J. M.; Epstein, L. F.; Lankiewicz, L.; Karlsson, R. *Pharmacol. Ther.* **1999**, *82*, 157-164.
30. Levitzki, A.; Gazit, A. *Science* **1995**, *267*, 1782-1788.
31. Traxler, P.; Lydon, N. *Drugs Future* **1995**, *20*, 1261-1274.
32. Traxler, P. M. *Expert Opin. Ther. Pat.* **1997**, *7*, 571-588.
33. Bridges, A. J. in: *Emerging Drugs: The Prospect for Improved Medicines.* Bowman, W. C.; Fitzgerald, J. B.; Taylor, J. B. (eds): London, UK, Ashley Publications Ltd.; **1998**:
34. Traxler, P. *Expert Opin. Ther.Pat.* **1998**, *8*, 1599-1625.
35. Barker, A. J.; Davies, D. H. UK 92-305703(520722), 21. 1992. EP. 1992.
36. Ward, W. H. J.; Cook, P. N.; Slater, A. M.; Davies, D. H.; Holdgate, G. A.; Green, L. R. *Biochem. Pharmacol.* **1994**, *48*, 659-666.
37. Fry, D. W.; Kraker, A. J.; McMichael, A.; Ambroso, L. A.; Nelson, J. M.; Leopold, W. R.; Connors, R. W.; Bridges, A. J. *Science* **1994**, *265*, 1093-1095.
38. Thompson, A. M.; Bridges, A. J.; Fry, D. W.; Kraker, A. J.; Denny, W. A. *J. Med. Chem.* **1995**, *38*, 3780-3788.
39. Rewcastle, G. W.; Bridges, A. J.; Fry, D. W.; Rubin, J. R.; Denny, W. A. *J. Med. Chem.* **1997**, *40*, 1820-1826.
40. Woodburn, J. R.; Barker, A. J.; Gibson, K. H.; Ashton, S. E.; Wakeling, A. E. *Proc.Amer.Assoc.Cancer Res.* **1997**, *38*, 6333-6333.
41. Kelly, A. C.; Laight, A.; Morris, C. O.; Woodburn, J. R.; Richmond, G. H. P. EORTC Meeting **1998**.

42. Moyer, J. D.; Barbacci, E. G.; Iwata, K. K.; Arnold, L.; Boman, B.; Cunningham, A.; Diorio, C.; Doty, J.; Morin, M. J.; Moyer, M. P.; Neveu, M.; Pollack, V. A.; Pustilnik, L. R.; Reynolds, M. M.; Sloan, D.; Theleman, A.; Miller, P. *Cancer Res.* **1997**, *57*, 4838-4848.
43. Ciardiello, F.; Caputo, R.; Bianco, R.; Damiano, V.; Pomatico, G.; De Placido, S.; Bianco, A. R.; Tortora, G. *Clin. Cancer Res.* **2000**, *6*, 2053-2063.
44. Smaill, J. B.; Rewcastle, G. W.; Loo, J. A.; Greis, K. D.; Chan, O. H.; Reyner, E. L.; Lipka, E.; Showalter, H. D. H.; Vincent, P. W.; Elliott, W. L.; Denny, W. A. *J. Med. Chem.* **2000**, *43*, 1380-1397.
45. Fry, D. W. *Anti-Cancer Drug Des.* **2000**, *15*, 3-16.
46. Fry, D. W.; Bridges, A. J.; Denny, W. A.; Doherty, A.; Greis, K. D.; Hicks, J. L.; Hook, K. E.; Keller, P. R.; Leopold, W. R.; Loo, J. A.; McNamara, D. J.; Nelson, J. M.; Sherwood, V.; Smaill, J. B.; Trumpp-Kallmeyer, S.; Dobrusin, E. M. *Proc. Natl. Acad. Sci. USA* **1998**, *95*, 12022-12027.
47. Discafani, C. M.; Carroll, M. L.; Floyd, M. B., Jr.; Hollander, I. J.; Husain, Z.; Johnson, B. D.; Kitchen, D.; May, M. K.; Malo, M. S.; Minnick, A. A., Jr.; Nilakantan, R.; Shen, R.; Wang, Y. F.; Wissner, A.; Greenberger, L. M. *Biochem. Pharmacol.* **1999**, *57*, 917-925.
48. Traxler, P. M.; Furet, P.; Mett, H.; Buchdunger, E.; Meyer, T.; Lydon, N. *J. Med. Chem.* **1996**, *39*, 2285-2292.
49. Furet, P.; Caravatti, G.; Lydon, N.; Priestle, J. P.; Sowadski, J. M.; Trinks, U.; Traxler, P. *J. Comput. -Aided Mol. Des.* **1995**, *9*, 465-472.
50. Traxler, P.; Bold, G.; Frei, J.; Lang, M.; Lydon, N.; Mett, H.; Buchdunger, E.; Meyer, T.; Mueller, M.; Furet, P. *J. Med. Chem.* **1997**, *40*, 3601-3616.
51. Toledo, L. M.; Lydon, N. B.; Elbaum, D. *Curr. Med. Chem.* **1999**, *6*, 775-805.
52. Mohammadi, M.; Froum, S.; Hamby, J. M.; Schroeder, M. C.; Panek, R. L.; Lu, G. H.; Eliseenkova, A. V.; Green, D.; Schlessinger, J.; Hubbard, S. R. *EMBO J.* **1998**, *17*, 5896-5904.
53. Schulze-Gahmen, U.; Brandsen, J.; Jones, H. D.; Morgan, D. O.; Meijer, L. *Proteins: Struct., Funct., Genet.* **1995**, *22*, 378-391.
54. Toja, E.; DePaoli, A.; Tuan, G.; Kettenring, J. *Synthesis* **1987**, 272-274.
55. Buchdunger, E.; Cozens, R. M.; Di Giovanna, M.; Furet, P.; Lydon, N.; Mett, H.; O'Reilly, T.; Woods-Cook, K.; Traxler, P. *in preparation* **2000**.
56. Trinks, U.; Buchdunger, E.; Furet, P.; Kump, W.; Mett, H.; Meyer, T.; Mueller, M.; Regenass, U.; Rihs, G. *J. Med. Chem.* **1994**, *37*, 1015-1027.

Chapter 15

STI571: A New Treatment Modality for CML?

Jürg Zimmermann, Pascal Furet, and Elisabeth Buchdunger

Pharma Research, Novartis, S–507.7.51 Postfach, CH–4002 Basel, Switzerland

STI571 is a protein-tyrosine kinase inhibitor which potently inhibits the Abl tyrosine kinase *in vitro* and *in vivo*. The compound specifically inhibits proliferation of *v-abl* and *bcr-abl* expressing cells, suggesting that it is not a general antimitotic agent. In addition, STI571 is a potent inhibitor of the platelet-derived growth factor receptor kinase (PDGF-R) and of the receptor kinase for stem cell factor (SCF), c-Kit, and inhibits PDGF- and SCF-mediated biochemical events. In contrast, it does not affect signal transduction mediated by other stimuli including epidermal growth factor (EGF), insulin and phorbol esters. Pharmacokinetic studies in various animal species demonstrate that pharmacologically relevant concentrations are achieved in the plasma following oral administration of the drug. STI571 shows anti-tumor activity as a single agent in animal models at well tolerated doses. Promising data from phase I clinical trails in CML patients support the notion that STI571 represents a new treatment modality for CML.

Introduction

The unifying concept of growth regulation and its disorders in cancer cells has a genetic background. Many of the genes that are mutated in cancer cells, including both oncogenes and tumor suppressors, encode proteins that are crucial regulators in various signal transduction pathways. Protein tyrosine kinases play a fundamental role in intracellular signaling and deregulation of these enzymes has been observed in cancer and benign proliferative disorders. One of the classical examples of protein tyrosine kinases is the non-receptor kinase Abl. There is considerable evidence that this kinase plays an important role in the pathogenesis of several human leukaemias. Chronic myeloid leukaemia (CML) is a haematological stem cell disorder characterized by excessive myeloid proliferation *(1)*. The hallmark of CML is the Philadelphia

© 2001 American Chemical Society

(Ph) chromosome which is detected in virtually all cases of CML and in 20% of cases of adult acute lymphoblastic leukaemia (ALL) *(2)*. It is formed by a reciprocal translocation between chromosome 9 and 22 [t(9;22) (q34;q11)] *(3-8)*, the molecular consequence of which is the replacement of the first exon of c-*abl* with sequences from the BCR gene resulting in a Bcr-Abl fusion protein with enhanced tyrosine kinase activity *(9-11)*. Bcr-Abl expression can induce a disease resembling CML in mice *(12-15)*, providing strong evidence that the Bcr-Abl protein is a major factor in the pathophysiology of CML. Due to its relatively clear etiology, CML presents an attractive target for therapy using an inhibitor of the Abl tyrosine kinase. This approach could yield a selective anti-cancer agent that would be useful for *ex vivo* bone marrow purging in preparation for autologous bone marrow transplantation of Bcr-Abl positive leukemias and for the *in vivo* therapy of these leukemias. Support for this approach also comes from a study, which shows that inactivation of the bcr-abl oncogene by an antisense approach selectively kills leukemia cells harboring this gene *(16)*.

Low molecular weight inhibitors of the Abl kinase have been reported: arylvinylamides, 2-oxinodole *(17)*, homophtalimides, benzoisothiazo-line-2,2-dioxides, catechols [for a review see *(18)*], tyrphostines *(19, 20)*, benzopyranones *(21)* and pyridol[2,3-d]pyrimidines *(22)*. In this review we report our findings on a series of novel phenylamino-pyrimidine (PAP)-derivatives capable of inhibiting phosphorylation of the v-Abl kinase at nanomolar concentration. Eventually, STI571 was identified as the most promising candidate for clinical development.

Chemistry and SAR

Extensive screening led us to the identification of a new class of active protein kinase inhibitors: the phenylamino-pyrimidines (PAPs). The compounds were prepared via the classical route *(23)* for the synthesis of the pyrimidine ring system. The acetyl derivatives of different heterocycles were converted into the enaminones using *N,N*-dimethylformamide-diethylacetal (a) which were then allowed to react with phenylguanidines (b) to give the phenylamino-pyrimidines. Functional group transformation of the side chain eventually gave the desired products (*Schemes 1 and 2, Table 1*).

In general the compounds are poorly soluble in water. The presence of basic groups in the side chain, however, dramatically increase aqueous solubility. The weak basicity of the anilinopyrimidine template does not allow extensive salt formation at pH7. Compound 1 is a polyvalent base ($pK_a,1$: 8.07, pK_a, 2: 3.73, $pK_a,3$: 2.56, pK_a, 4: 1.52) and the corresponding methane sulfonate salt shows high solubility in water (> 100 g/l) yielding an acidic solution (pH 4.2). At pH 7.4 however the aqueous solubility is much lower (~ 50 mg/l).

Scheme 1: General synthesis of the PAPs

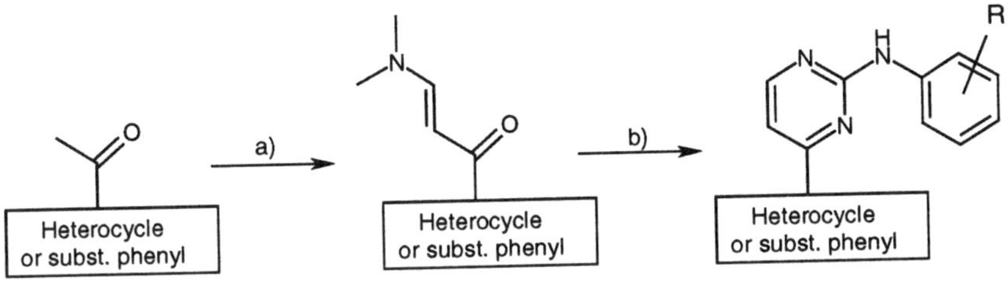

Scheme 2: Structural formulas of the compounds prepared

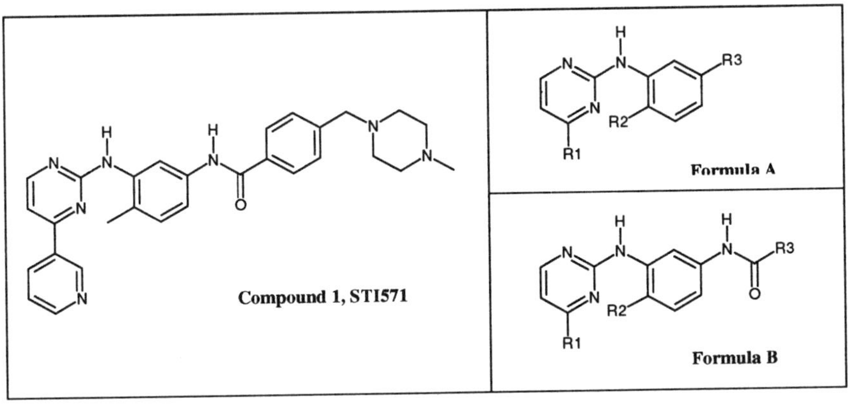

Compound 1, STI571

Formula A

Formula B

Table I: Compounds prepared

No.		R1	R2	R3	mp[°C]
1	1	Scheme 2	"	"	207-212
2	B	3-pyridyl	H	identical as in compound 1	198-201
3	B	3-pyridyl	H	3-pyridyl	217-220
4	B	3-pyridyl	CH$_3$	4-methylphenyl	102-220
5	B	3-pyridyl	H	pentyl	180-184
6	B	3-pyridyl	CH$_3$	2-naphtyl	97-101
7	B	3-pyridyl	H	4-fluorophenyl	215-216
8	B	3-pyridyl	H	2-thiophenyl	139-141
9	A	3-indolyl	H	tetrafluoroethoxy	140-142
10	B	3-pyridyl	H	phenyl	207-209
11	B	3-pyridyl	H	phenyl	179-180
12	A	3-pyridyl	H	3-aminopropylaminocarbonyl	142-146
13	B	3-pyridyl	H	cyclohexyl	205-206
14	B	3-pyridyl	H	4-pyridyl	224-226
15	A	4-pyridyl	H	amino	200-202
16	A	3-pyridyl	H	amino	58-68
17	B	3-pyridyl	CH$_3$	2-methoxyphenyl	88-92
18	B	3-pyridyl	CH$_3$	4-chlorophenyl	216-219
19	A	3-pyridyl	H	2-(1-imidazoly)ethoxy	amorph.
20	B	3-pyridyl	H	4-methylphenyl	214-216
21	A	3-pyridyl	H	nitro	213-219
22	B	3-pyridyl	H	4-canophenyl	220-219
23	A	3-pyridyl	CH$_3$	amino	143-144
24	A	3-pyridyl	H	chloro	146-147
25	A	3-pyridyl	H	hydrogen	143-144
26	B	3-pyridyl	H	2-methoxyphenyl	147-150
27	B	3-pyridyl	H	2-carboxyphenyl	206-209
28	A	3-pyridyl	H	1-imidazolyl	132-134
29	B	3-pyridyl	H	2-pyridyl	187-190
30	A	3-pyridyl	H	methoxycarbonyl	192-195
31	B	3-pyridyl	CH$_3$	methyl	220-222
32	A	3-pyridyl	H	carboxy	292-294
33	A	3-pyridyl	H	nitro	212-213
34	A	4-Cl-phenyl	H	tetrafluoroethoxy	133-135
35	A	4-pyridyl	CH$_3$	methyl	282-284
36	A	4-pyridyl	H	tetrafluoroethoxy	193-194

This class of compounds proved to be an excellent lead for the development of selective protein kinase inhibitors with high potency. The ability to produce a large number of compounds applying simple chemistry and the 'drug-likeness' of the core structure allowed the identification of selective inhibitors of protein kinase C (*24*) and PDGF-R kinase autophosphorylation (*25*) through exploitation of the diversity of various substituents at 4-position of the pyrimidine-ring and at the phenyl group. All attempts to replace the phenylamino-pyrimidine core structure yielded inactive compounds when assayed against various protein kinases. The slightly acidic proton on the amino group is also essential for activity, alkylation abolished completely the inhibitory activity (*24*). Various aromatic and heteroaromatic rings are tolerated at the 4-position of the pyrimidine. This "generic SAR" is valid for many protein kinases, including various subtypes of PKCs, PDGF-R, EGF-R and v-Abl.

The compounds were initially assayed for the inhibition of four tyrosine kinases (v-Abl, PDGF-R-kinase, EGFR-kinase and c-Src) and three serine/threonine kinases (PKCα, PKCδ and cAMP-dependent protein kinase, PKA), see Table II. The objective was aimed at optimizing potency for the inhibition of the phosphorylation by the v-Abl kinase, whereas the other kinases tested served as selectivity criteria. Compounds **1-18** inhibited the v-Abl-kinase *in vitro* with an IC_{50} value of less than 1 µM, the most active compound being derivative **1** (IC_{50} = 38 nM). The most potent inhibitors of the v-Abl kinase (IC_{50} < 300 nM) are members of the structural class B, which contain an amide function on the phenyl ring. The nature of the acyl-substitutents is of minor importance: substituted phenyl (**1, 4, 6** and **7**), heterocycles (**3** and **8**) and alkyl (**5**) are tolerated.

The compounds were further screened for selectivity. None of the PAPs showed inhibition (IC_{50} < 1 µM) of the PK-A and the EGFR kinase. The c-Src catalyzed phosphorylation was inhibited by only a few derivatives. PDGF-R-kinase activity, however, was inhibited with similar potency as the v-Abl kinase. But also here it was possible to gain selectivity: for example the potent v-Abl kinase inhibitor **3** was devoid of PDGF-R activity. The selectivity against PKCα and -δ was not very difficult to achieve within this series. Compound **1** did not inhibit PKC at all and the v-Abl inactive derivative **36** showed a selective PKCα inhibition. The methyl group in the 6-position of the phenyl ring helps to improve on selectivity: compounds **1, 4** and **6** bearing this "flag-methyl" do not inhibit PKCα and -δ to a significant extent, whereas compounds lacking this group are dual inhibitors. (e.g. **3**). Compound **1** (**STI571**) showed the most

Table II: Inhibition of protein kinases, IC$_{50}$ [μM][1]

No.	v-Abl [2]	EGFR [3]	c-Src [4]	PDGFR [3]	PKA [4]	PKCα [4]	PKCδ [4]
1	0.038	>100	>100	0.05	>500	>100	>100
2	0.1	>100	7.8	0.2	475	n.d.	n.d.
3	0.15	>100	9	50	>500	1.7	17
4	0.2	>100	>100	0.01	>100	>100	>500
5	0.2	>100	3.7	0.6	n.d.	2.0	19
6	0.2	>100	>100	0.05	>100	>100	>500
7	0.2	>50	>100	3	>500	n.d.	n.d.
8	0.3	>50	3	20	>500	n.d.	n.d.
9	0.37	1.8	2.2	>10	73	0.29	1.4
10	0.4	>50	15.7	5	>500	1.2	23
11	0.4	65	>100	0.1	>500	72	>500
12	0.4	45	9.8	>10	n.d.	n.d.	n.d.
13	0.45	>100	46	0.8	>500	n.d.	n.d.
14	0.45	>100	78	>100	>500	1.3	24
15	0.6	n.d.	16	>10	n.d.	n.d.	n.d.
16	0.7	50	14	>100	6.0	210	>500
17	0.8	>100	>100	0.3	>500	80	>100
18	0.8	>100	>100	0.01	>100	>100	>100
19	1.0	7.1	4.9	>10	340	1.5	34
20	1.0	>100	>100	0.8	>500	n.d.	n.d.
21	1.2	>100	>100	>100	>500	2.5	>500
22	1.3	>100	>100	1.5	>500	0.35	>500
23	1.5	>100	>100	>10	>500	240	>500
24	1.5	>100	>100	>10	>500	1.6	23
25	1.8	100	>100	>10	>500	10.5	39
26	1.	100	>100	1.0	>500	0.32	>500
27	2.0	n.d.	44	>100	>500	n.d.	n.d.
28	3.3	>100	1.6	18	>500	1.0	21
29	3.8	n.d.	90	>100	>500	n.d.	n.d.
30	5.1	n.d.	90	>10	>500	2.5	39
31	5.8	>100	>100	50	>500	500	>500
32	8.0	n.d.	126	>10	n.d.	45	>100
33	34	100	93	>100	>500	0.79	>500
34	>50	>100	>100	>10	>500	0.85	>100
35	61	>100	>100	>100	n.d.	2.6	>500
36	>100	>100	>100	>10	n.d.	0.3	>100

1) The concentration of the compound resulting in 50% reduction in kinase activity (IC50) are given. 2) Utilizing a bacterial construct of v-Abl and [Val5]-angiotensin as substrate *(42)*. 3) Inhibition of EGF- and PDGF-stimulated total cellular tyrosine phosphorylation in A431 cells and BALB/c 3T3, respectively was measured using a microtiter ELISA *(43)*. 4) Inhibition of substrate phosphorylation by purified enzymes was assayed as described in the literature *(44)*.

promising enzymatic profile and was characterized in a number of preclinical models.

Structural Basis of the Inhibition of Abl kinase by STI571

To rationalize the inhibitory activity of STI571, we undertook docking studies based on a model of the ATP binding site of the Abl kinase. This was constructed by homology to an available X-ray crystal structure of the activated form of the fibroblast growth factor receptor (FGF-R) kinase in complex with an inhibitor (PD 173074) also containing an aminopyrimidine moiety in its structure (26). Our studies resulted in a binding model consistent with all structure-activity relationships (Plate 1). Its main features are summarized below.

- The N1 atom and 2-amino substituent of the pyrimidine moiety are engaged in bidentate hydrogen bond interactions with the peptidic backbone of residue Met 318 of the hinge loop of the enzyme. These interactions are postulated by analogy to the bidentate hydrogen bonds formed by the aminopyrimidine moiety of PD 173074 with Ala 564 in the ligated X-ray structure of the FGF-R kinase. They can explain why alkylation of the 2-amino substituent abolishes the inhibitory activity of this class of compound (24).
- The pyrimidine ring makes a major hydrophobic contact with the C-terminal domain residue Leu 370 while the pyridyl group is in hydrophobic interaction with Val 256 of the glycine-rich loop. The pyridyl nitrogen atom is located at a distance from the side chain of Thr 315 allowing the formation of a water mediated hydrogen bond.
- Leu 248, another residue of the glycine rich loop, contacts the two phenyl rings of the inhibitor, supporting a contribution of these rings to its inhibitory potency.
- A van der Waals contact is made between the methyl substituent in the *ortho* position of the anilino phenyl ring and the side chain of Leu 370. The fact that the equivalent residue in PKC is a bulkier methionine that would sterically interfere with the methyl group provides an explanation for the lack of activity of STI571 in inhibiting this enzyme.
- The piperazinyl-solubilizing moiety makes a salt bridge interaction, through its terminal amine function, with the side chain carboxylate group of Glu 258, a residue also belonging to the glycine-rich loop. This salt bridge is a key interaction in the proposed model. It is consistent with the observation that parent compounds of STI571 lacking a piperazinyl group (such as compounds 4 and 18 of Tables I and II) are one order of magnitude less active in inhibiting the Abl kinase. In addition, the fact that Glu 258 is conserved in the c-kit and PDGF-R kinases, both also potently inhibited by

STI571, and not in the EGFR and c-Src kinases which are insensitive to the compound, makes this salt bridge interaction a plausible determinant of selectivity.

Very recently, Schindler et al. have reported the crystal structure of the Abl kinase in complex with an analogue of STI571 (27). The co-crystallized analogue lacks the piperazinyl group and has a 3-pyridyl ring replacing the phenyl acyl substituent. The main finding of this work is that the STI571 analogue binds to an inactive conformation of the kinase in which the activation loop, non-phosphorylated at Tyr 383, is folded into the ATP binding site. In the observed conformation, the shape of the binding site differs from that presented by the activated form of the kinase as can be inferred from homology models based upon X-ray structures of activated kinases. Therefore, it is conceivable that the binding mode of the STI571 analogue seen in this structure is different from that we propose above for STI571 using the structure of an activated kinase as a template. Plate 2 shows the relative orientations of the two compounds in these binding modes. In the structure of Schindler et al., the inhibitor also hydrogen bonds to the backbone of hinge residue Met 318 but using the nitrogen atom of the 3-pyridyl moiety and not the pyrimidine N1 atom as acceptor. More striking is the difference in the positions of the anilino and acyl moieties of the inhibitors in the two binding modes. In the crystal structure of the inactive form of the kinase, they sit deep in the cavity in the vicinity of residues Lys 271, Met 290, Glu 286, Ile 313, Thr 315, Ala 380 and Phe 382 while in the activated kinase model they are located in the region of the binding site opened to solvent, mainly interacting with Leu 248 and Glu 258. The binding mode observed by Schindler et al. is clearly incompatible with the structure of the ATP binding site in the activated kinase model. Docking STI571 into the model, in the corresponding orientation, leads to severe steric clashes with some of the residues of the back of the cavity such as Phe 382 and Glu 286 that have different positions compared to the inactive conformation. In contrast, it is possible to dock STI571 in the inactive kinase structure according to the binding mode shown in Plate 1 without creating steric conflicts.

The question naturally arises as to which conformation of the Abl kinase, inactive or active, is the target of STI571 and which binding mode is the most relevant to account for the high potency and selectivity of the inhibitor towards this enzyme.

Schindler et al. provide arguments in favor of a mechanism of inhibition involving the inactive conformation presented by Abl in their crystal structure. The authors found that phosphorylation of Tyr 393, which should stabilize the activation loop in the conformation encountered in active protein kinases, makes Abl less sensitive to STI571. Thus, STI571 seems to preferentially bind to an

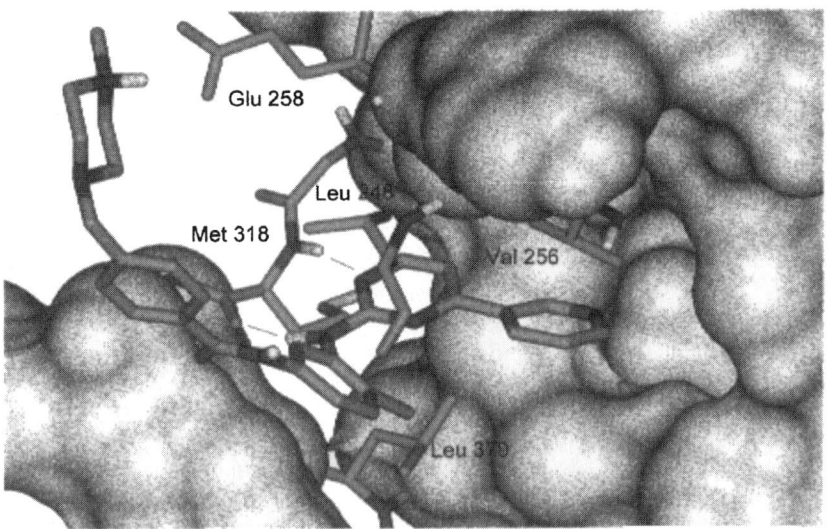

Plate 1: Proposed binding mode of STI571 in a model of the activated form of the Abl kinase.
(Color figure can be found in color insert.)

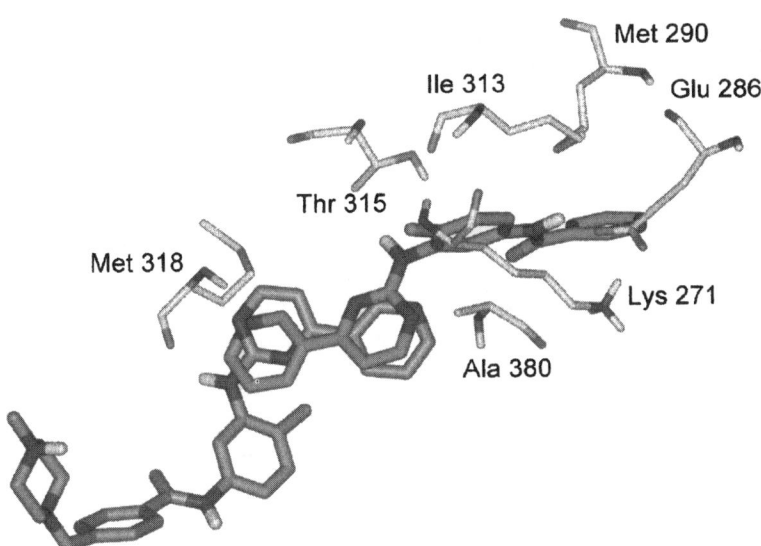

Plate 2: Relative orientations of STI571 (brown) and its analogue (green) in the activated kinase model and in the crystal structure of Schindler *et al.* *(27)*. The protein residues represented have the positions seen in the crystal structure.
(Color figure can be found in color insert.)

inactive form of the kinase. In addition, their crystal structure provides a reasonable explanation for the inability of STI571 to inhibit c-Src, a kinase presenting a very high homology to Abl in the region of the ATP binding site. The available X-ray structure of inactive c-Src shows that the activation loop adopts a different conformation compared to that observed in the Abl structure. This conformation would prevent the binding of STI571 to c-Src in the orientation reported by the authors.

However, there is evidence that the binding mode determined by Schindler *et al.* is not the only one possible for inhibitors of the phenylamino-pyrimidine class. Some structure-activity relationships clearly demonstrate it. For instance, compound 9 of Tables I and II, in which the 3-pyridyl group of STI571 is replaced by a 3-indolyl moiety, is not able to form the interactions seen in the X-ray structure. Still, this phenylamino-pyrimidine derivative inhibits Abl in the submicromolar range with selectivity (one order of magnitude) over c-Src. In this case, the data is consistent with the binding orientation we propose in the model but not with the orientation adopted by the STI571 analogue in the crystal structure.

In summary, the crystal structure of Schindler *et al.* suggests that STI571 inhibits the Abl kinase by binding with high affinity and high specificity to an inactive form of the kinase. Nevertheless, the contribution of an alternative binding mode, compatible with the structures of both the activated and inactivated forms of the kinase and more general within the phenylamino-pyrimidine class of inhibitors, cannot be excluded. Additional structure-activity relationships in the STI571 series in conjunction with mutational studies or X-ray structures of the protein should shed more light on this question.

Pharmacological Profile of STI571

STI571 was further assayed *in vitro* for inhibition of a panel of protein kinases *(28, 29, 30)*. The compound potently inhibited the v-Abl protein-tyrosine kinase (IC_{50} = 0.038 µM) but was inactive against serine/threonine kinases such as PKC and Cdc2/cyclin B. STI571 did not inhibit the EGF receptor intracellular domain and showed no or weak inhibition of kinase activity of the receptors for vascular endothelial growth factor (Kdr, Flt-1), Tek, c-Met and of protein kinases of the Src family (c-Src, c-Fgr, c-Lyn, Lck).

STI571 was assayed in cellular systems to evaluate specificity and mechanism of action. The compound inhibited autophosphorylation of v-Abl with an IC_{50} value of 0.25 µM *(28)* as well as the kinase activity of activated Abl fusion proteins such as p210bcr-abl, p185bcr-abl and TEL-Abl *(28, 32, 33)*. When tested against other tyrosine kinases in cell-based assays, STI571 preferentially inhibited signal output from the ligand-activated PDGF receptor

(29, 30). Inhibition of the constitutively active TEL-PDGF-R fusion protein, was observed at an IC$_{50}$ of 0.15 µM (32). The compound also potently inhibited stem cell factor (SCF)-mediated events such as c-Kit autophosphorylation and MAP kinase activation in MO7e cells (30, 31) as well as in small cell lung cancer cells (34). In contrast, signal transduction mediated by EGF, insulin, insulin-like growth factor I (IGF-I), fibroblast growth factor (FGF) and phorbol ester was insensitive to STI571 (29). Furthermore, STI571 did not affect Flt-3 or the receptor for colony stimulating factor 1 (CSF-1), c-Fms, or the non-receptor tyrosine kinases Src and Jak-2 (30), a tyrosine kinase which mediates signaling from a number of cytokine receptors, including receptors for IL-3, G-CSF and erythropoietin.

STI57I was tested for antiproliferative activity on factor-independent p210bcr-abl-expressing 32D (32p210) and MO7e cells (MO7p210) and their growth factor-dependent parental cells. STI571 showed potent inhibition of p210bcr-abl-expressing cells. In contrast, incubation with up to 10 µM STI571 did not inhibit growth of parental or v-src transformed cells (28). STI571 also induced erythroid differentiation and death of K562 cells, a human p210bcr-abl –positive cell line (28) and showed growth inhibition of p185bcr-abl and TEL-Abl expressing cells (32). Importantly, STI571 selectively blocked proliferation of Bcr-Abl-positive CML and ALL cell lines, inducing apoptosis (33, 35, 36, 37). Cell killing was also observed in fresh leukemic cells from Ph positive CML and ALL patients (33, 35). In contrast, STI571 had no effect on Bcr-Abl-negative ALL and AML primary blast cells (33).

Selective inhibition of Bcr-Abl-positive progenitor cells from CML patients was observed in colony forming assays. At 1µM, STI571 selectively inhibited colony formation from blood and bone marrow of Bcr-Abl-positive CML patients with a 92-98% decrease in Bcr-Abl-positive colonies but little effect on normal hematopoiesis (28, 36). The findings were confirmed by assessing the effects of STI571 on proliferation of progenitors using long-term culture (LTC) (40). Short-term incubation with 10 µM STI571 over one week under LTC resulted in sustained 80-90% inhibition of Bcr-Abl-positive progenitors. Continuous exposure for 2-8 weeks led to 90% inhibition of primitive Bcr-Abl-positive progenitors with marginal toxicity to normal cells.

The efficacy and specificity of STI571 has also been demonstrated in animal model systems. 32D cells were used to test STI571 for antitumor activity in syngenic mice (28). Once daily intraperitoneal treatment with doses of 2.5-50 mg/kg starting one week after cell injection caused dose-dependent inhibition of tumor growth. In contrast, STI571 50 mg/kg p.o. showed no antitumor activity against tumors derived from v-src transformed 32D cells (28), a finding compatible with the lack of inhibition of Src kinase activity by STI571. The *in vivo* antitumor activity has also been confirmed in nude mice injected with the Bcr-Abl-positive KU812 cell line derived from a CML patient in blast crisis

(38). Early pharmacokinetic studies in various animal species had demonstrated that STI571 is rapidly absorbed following oral administration and pharmacologically relevant concentrations are achieved in the plasma (e.g. peak plasma level of 13.4 µM 10 min after administration of 50 mg/kg to mice; Novartis, unpublished data). Optimizing the treatment schedule and assuring continuous blockade of Bcr-Abl tyrosine phosphorylation in the tumor, resulted in tumor-free survival of mice that had been injected with KU812 cells *(38)*. Furthermore, 160 mg/kg p.o. every eight hours for 11 days inhibited tumor growth despite the presence of advanced disease. Tumor nodules began to regress 48 h after beginning of treatment and by day 8 no treated animal had a measurable tumor. The antitumor effect of STI571 was specific for p210bcr-abl expressing cells as no growth inhibition occurred in mice given injections of U937, a Bcr-Abl-negative myeloid cell line.

Clinical proof of concept

Based on the attractive preclinical profile, a dose escalating phase I trial of STI571 was initiated in CML patients in mid 1998. Treatment with STI571 was given as daily oral therapy to patients who had failed interferon therapy. The compound has been generally well tolerated and no maximally tolerated dose was identified in this study. At doses of 300 mg or greater, all patients had complete hematological responses, and a significant number of patients also achieved major cytogenetic responses *(41)*.

In summary, STI571 is a hopeful example of how targeting the molecular abnormality known to drive the deregulated growth of cancer cells can lead to an effective, yet well tolerated treatment.

References

1. Deisseroth A.B., Andreeff M., Champlin R., Keating, M.J., Kantarjian, H., Khouri, I.F.; Talpaz, M. Chronic Leukemias. In: V.T. DeVita (ed.), Cancer: *Principles and practice of oncology*, pp. 1965-1979, Philadelphia: JB Lippincott Company. 1994.
2. Nowell P.C., Hungerford D.A. *Science* **1960**, *132*, 1497.
3. Groffen, J.; Stephenson, J.R.; Heisterkamp, N.; De-Klein, A.; Bartram, C.R.; Grosfeld, G. *Cell* **1984**, *36*, 93-99.

4. Shtivelman, E.; Lifshitz, B.; Gale, R.P.; Canaani, E. *Nature* **1985**, *315*, 550-554.
5. Clark, S.S.; McLaughlin, J.; Timmons, M.; Pendergast, A.M.; Ben Neriah, Y.; Dow, L.W.; Crist, W.; Rover, G.; Smith, S.D.; Witte, O.N. *Science* **1988**, *239*, 775-777.
6. Clark, S.S., McLaughlin, J.; Crist, W.M.; Champlin, R.; Witte, O.N. *Science* **1987**, *235*, 85-88.
7. Fainstein, E.; Marcelle, C.; Rosner, A.; Canaani, E.; Gale, R.P.; Dreazen, O.; Smith, S.D.; Croce, C.M. *Nature* **1987**, *330*, 386-388.
8. Heisterkamp, N.; Stam, K.; Groffen, J.; De-Klein, A.; Grosfeld, G. *Nature* **1985**, *315*, 758-761.
9. Kurzrock, R.; Gutterman, J.U.; Talpaz, M. *N. Engl. J. Med.* **1988**, *319*, 990-998.
10. Lugo, T.G.; Pendergast, A.M.; Muller, A.J.; Witte, O.N. *Science* **1990**, *247*, 1079-1082.
11. Konopka, J.B.; Watanabe, S.M.; Witte, O.N. *Cell* **1984**, *37*, 1035-1042.
12. Daley, G.Q.; Van Etten, R.A.; Baltimore, D. *Science* **1990**, *247*, 824-830.
13. Kelliher, M.A.; McLaughlin, J.; Witte, O.N.; Rosenberg, N. *Proc. Natl. Acad. Sci.* **1990**, *87*, 6649-6653.
14. Daley, G.Q. *Leuk. Lymphoma* **1993**, *1*(11 Suppl): 57-60.
15. Heisterkamp, N.; Jenster, G.; ten Hoeve, J.; Zovich, D.; Pattengale, P.K.; Groffen J. *Nature* **1990**, *344*, 251-253.
16. Szczylik, C.; Skorski, T.; Nicolaides, N.; Manzella, L.; Malaguarnera, L.; Venturelli, D.; Gewirtz, A.; Calabretta, B. *Science* **1991**, *253*, 562-565.
17. Buzetti, F.; Brasca, M.G.; Crugnola, A.; Fustinoni, S.; Longo, A.; Penco, S. *Il Farmaco* **1993**, *48*, 615-636.
18. Spada, A.P.; Myers, M.R. Exp. *Opin. Ther. Patents*, **1995**, 5, 805-817.
19. Levitzki, A. *FASEB J.* **1992**, *6*, 3275-3282.
20. Anafi, M.; Garzit, A.; Zehavi, A.; Ben-Neriah, A.; Levitzki, A. *Blood*, **1993**, *82*, 3524-3529.
21. Geissler, J.F.; Roesel, J.L.; Meyer, Th.; Trinks, U.P.; Traxler, P.; Lydon, N.B. *Cancer Research* **1992**, *52*, 4492-4498.
22. Dorsey, J.F.; Jove, R.; Kraker, A.J.; Wu, J. *Cancer Res.*, **2000**, *60*, 3127-3131.
23. Bredereck, H.; Effenberger, F.; Bosch, H. *Ber. Dtsch. Chem. Ges.* **1964**, *97*, 3397-3406.
24. Zimmermann, J.; Caravatti, G.; Mett, H.; Meyer, Th.; Müller, M.; Lydon, N. B.; Fabbro, D. *Arch. Pharm. Pharm. Med. Chem.* **1996**, *7*, 371-376.
25. Zimmermann, J.; Buchdunger, E.; Mett, H.; Meyer, Th.; Lydon, N.; Traxler, P. *Bioorganic & Medicinal Chemistry Letters* **1996**, *6/11*, 1221-1226.

26. Mohammadi, M.; Froum, S.; Hamby, J. M.; Schroeder, M. C.; Panek, R. L.; Lu, G. H.; Eliseenkova, A.V.; Green, D.; Schlessinger, J.; Hubbard, S.R. *EMBO J.* **1998**, *17*, 5896-5904.
27. Schindler, T.; Bornmann, W.; Pellicena, P.; Todd Miller, W.; Clarkson, B.; Kuriyan, J. *Science* **2000**, *289*, 1938-1942.
28. Druker, B.J.; Tamura, S.; Buchdunger, E.; Ohno, S.; Segal, G.M.; Fanning, S.; Zimmermann, J.; Lydon. N.B. *Nature Medicin* **1996**, *2*, 561-566.
29. Buchdunger, E.; Zimmermann, J.; Mett, H.; Meyer, Th.; Müller, M.; Druker, B.; Lydon, N.B. *Cancer Res.* **1996**, *56(1)*, 100-104.
30. Buchdunger E.; Cioffi C.L.; Law N.; Stover D.; Ohno-Jones S.; Druker B.J.; Lydon N.B. *J. Pharmacol. Exp. Ther.*, **2000**, *295*, 139-145.
31. Heinrich, M.C.; Griffith, D.J.; Druker, B.J.; Wait, C.L.; Ott, K.A.; Zigler, A.J. *Blood* **2000**, *96*, 925-932.
32. Carroll, M.; Ohno-Jones, S.; Tamura, S.; Buchdunger, E.; Zimmermann, J.; Lydon, N.B.; Gilliland, D.G.; Druker, B.J. *Blood* **1997**, *90*, 4947-4952.
33. Beran, M.; Cao, X.; Estrov, Z.; Jehan, M.; Jin, G.; O'Brien, S.; Talpaz, M.; Arlinghaus, R.B.; Lydon, N.B.; Kantarjiian, H. *Clin. Cancer Res.* **1998**, *4*, 1661-1672.
34. Krystal, G.W.; Honsawek, S.; Litz, J.; Buchdunger, E. *Clin Cancer Res.* **2000**, *6*, 3319-3326.
35. Gambacorti-Passerini C.; Le Courte, P.; Mologni, L.; Fanelli, M.; Belotti, D.; Pogliani, E.; Lydon, N.B. *Blood Cell, Molecules and Diseases*, **1997**, *23*, 380-394.
36. Deininger, M.W.N.; Goldman, J.M.; Lydon, N.B.; Melo, J. *Blood*, **1997**, *9*, 3691-3698.
37. Dan, S.; Naito, M.; Tsuruo, T. *Cell Death and Differentiation*, **1998**, *5*, 710-715.
38. Le Coutre, P.; Mologni, L.; Cleris, L.; Marchesi, E.; Giardini, R.; Formelli, F.; Gambacorti-Passerini, C. *J. Natl. Cancer Inst.*, **1999**, *91*, 163-168.
39. Uhrbom,, L.; Hesselager, G.; Oestman, A.; Nistér, M.; Westermark, B. *Int. J. Cancer*, **2000**, *85*, 398-406.
40. Kasper, B.; Fruehauf, S.; Schiedlmeiter, B.;Buchdunger, E.; Ho, A.D.; Zeller, W.J. *Cancer Chemother. Pharmacol.* **1999**, *44*, 433-438.
41. Druker, B.J; Talpaz, M.; Resta, D.; Peng, B.; Buchdunger, E.; Fortd, J.; Sawyers, C.L. *Blood*, **1999**, *94*, 368a.
42. Lydon, N.B.; Adams, B.; Poschet, J.F.; Gutzwiller, A.; Matter, A. *Oncogene Research*, **1990**, *5*, 161-173.
43. Trinks, U.; Buchdunger, E.; Furet, P.; Kump, W.; Mett, H.; Meyer, Th.; Müller, M.; Regenass, U.; Rihs, G.; Lydon N.B.; Traxler, P. *J. Med. Chem.* **1994**, *37*, 1015-1027.

44. Buchdunger, E.; Trinks, U.; Mett, H.; Regenass, U.; Müller, M.; Pinna, L.; McGlynn, E.; Traxler, P.; Lydon, N.B. *Proc. Natl. Acad. Sci. USA,* **1994**, *91*, 2334-2338.

Acknowledgement

We kindly acknowledge Nick Lydon, Brian Druker, Helmut Mett, Thomas Meyer, Robert Cozens, Terence O'Reilly and the STI571 project team.

Chapter 16

The Discovery and Development of Second-Generation Matrix Metalloproteinase Inhibitors for the Treatment of Cancer

Andy Baxter and John Montana

Celltech Chiroscience Ltd., Granta Park, Great Abington, Cambridge CB1 6GS, Great Britain

A significant number of the matrix metalloproteinase (MMP) gene family of zinc-dependent enzymes have been implicated in a variety of diseases. This has fuelled considerable interest from the phamaceutical industry in the identification and development of inhibitors of this class of enzyme. A number of such compounds have entered development over the last 10 years and several have progressed to late-stage clinical studies, particularly in the field of oncology. These so-called first generation inhibitors were derived from substrate recognition peptide sequences and have a very broad inhibitory profile within the MMP family. As the field of metalloproteinase research has expanded and our knowledge has increased, new gene sub-families such as the ADAM and ADAM-TS metalloproteinases have been identified, and the first generation inhibitors have also been shown to inhibit a number of these new enzymes. Thus, it is hypothesized that the broad inhibitory profile of these first generation compounds, could explain some of the side effects observed in man. Therefore, more recent work in this field has centered on the identification of inhibitors with different selectivity profiles and the effect of such inhibitors in models of a variety of different diseases. Another drawback of the early development candidates was their modest pharmacokinetic profile which, in part, was due to their peptidomimetic nature. Hence, another area of focus for researchers in this field has been to improve DMPK parameters and to move away from peptidomimetics to truly non-substrate-based inhibitors. This review focuses on recent advances in these two key areas in MMP oncology research.

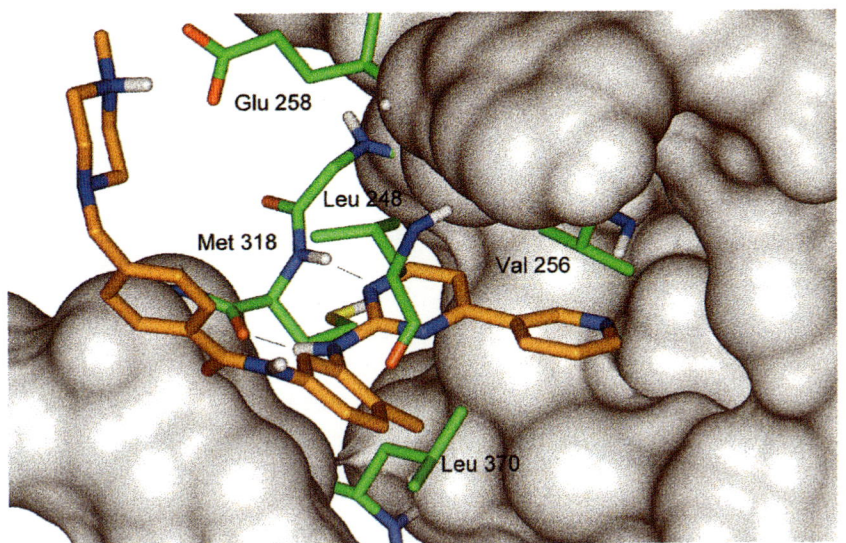

Plate 1: Proposed binding mode of STI571 in a model of the activated form of the Abl kinase.

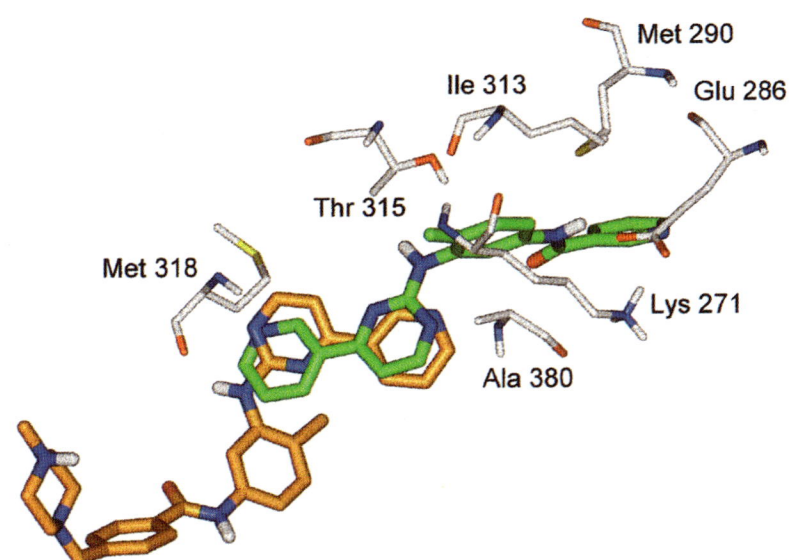

Plate 2: Relative orientations of STI571 (brown) and its analogue (green) in the activated kinase model and in the crystal structure of Schindler *et al.* (27). The protein residues represented have the positions seen in the crystal structure.

Plate 1. Representation of ATP docked in its binding site in VEGFR-2. (For reasons of clarity, residues Ala864 and Val846 are not depicted since these lie directly above the adenine and ribose rings respectively).

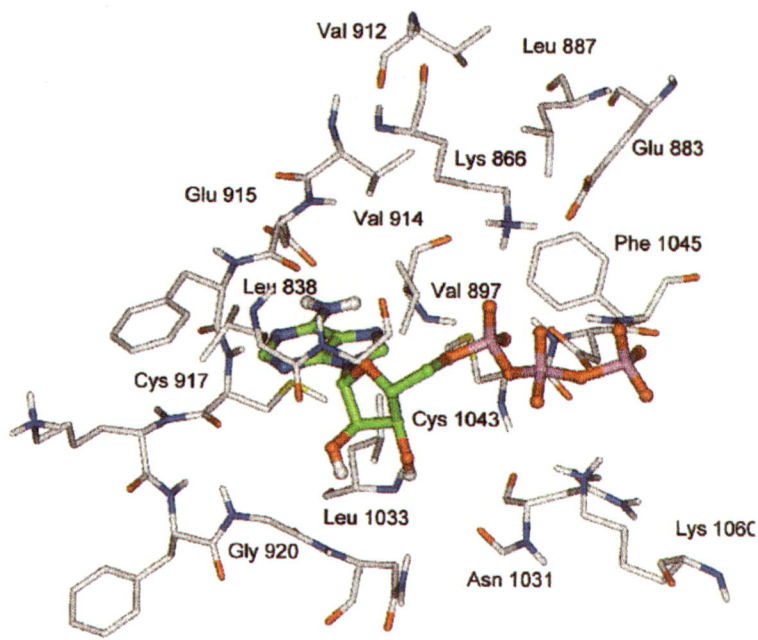

Plate 2. Portrait of ZD4190 bound to the VEGFR-2 ATP-binding site.

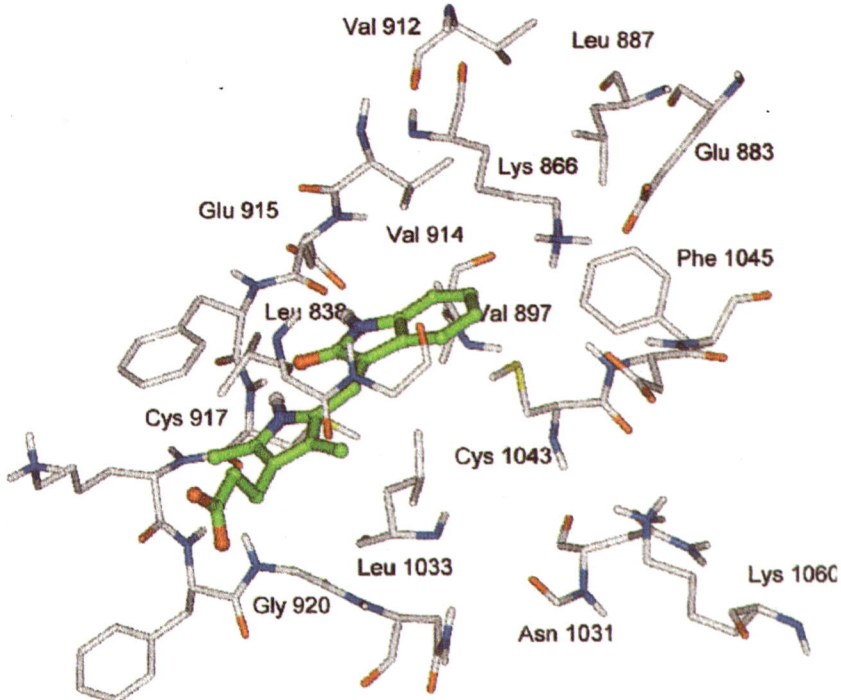

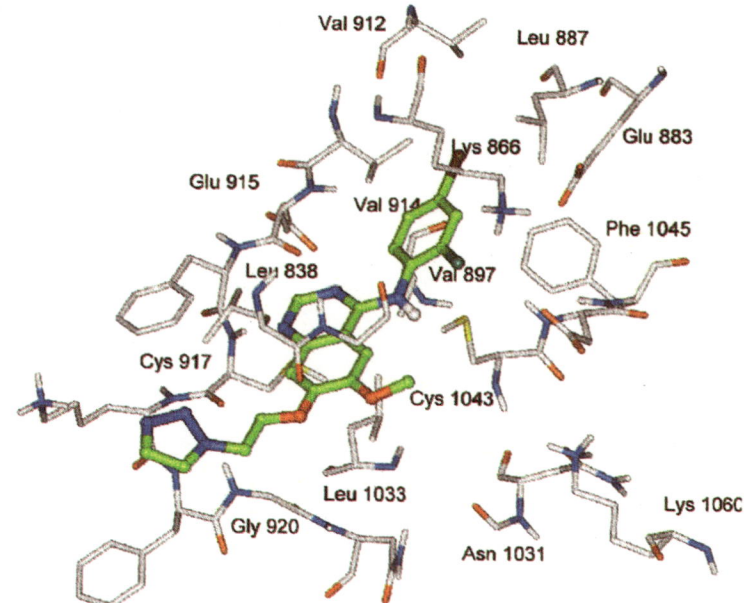

Plate 3. Portrait of SU6668 docked in the VEGFR-2 ATP-binding site.

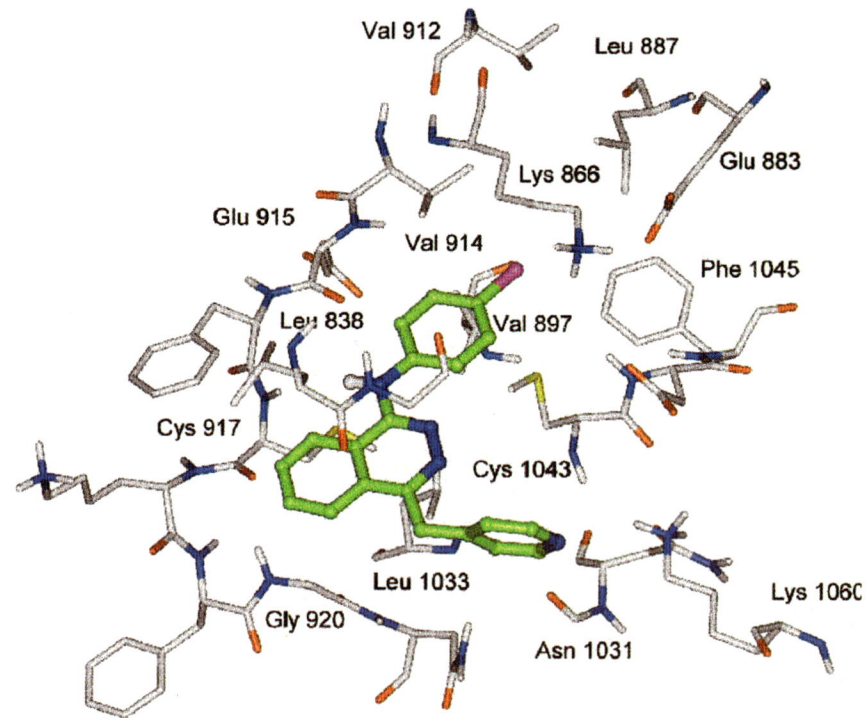

Plate 4. Portrait of PTK787 docked in the VEGFR-2 ATP-binding site.

Introduction:

Matrix metalloproteinases (MMPs) are extremely destructive, zinc-dependent enzymes involved in the turnover and re-modelling of extracellular matrix (ECM) [1-2]. The degradation and turnover of the ECM is a key element of many physiological processes, but under normal conditions, the activity of these catabolic enzymes is tightly controlled. Zymogens of MMPs are often processed by a series of proteolytic events to reveal the active form of the enzyme [3]. Furthermore, the activity of MMPs is regulated by specialized inhibitory proteins called tissue inhibitors of metalloproteinases (TIMPs) [4-5]. Accelerated breakdown of ECM occurs in various pathological processes, including inflammation, chronic degenerative diseases and tumor invasion. In many patho-physiological conditions, modulation of the ECM is required to allow cellular proliferation, migration and invasion to take place. These fundamental cellular processes coupled with un-checked matrix degradation are characteristic of many disease states such as rheumatoid arthritis, osteoarthritis, osteoporosis, periodontal disease, inflammatory bowel disease, psoriasis, multiple sclerosis, congestive heart failure, atherosclerosis, asthma, chronic obstructive pulmonary disease (COPD), age-related macular degeneration (ARMD) and cancer. Thus MMPs are implicated in the pathology of a wide range of diseases and inhibition of these enzymes represents an attractive target for therapeutic intervention [6].

The MMPs belong to a large and expanding gene family. In general they are divided into four broad groups, the collagenases (MMP-1, 8 and 13), gelatinases (MMP-2 and MMP-9), stromelysins (MMP-3, 10 and 11) and the MT-MMPs (MMP-14 through 17 and MMP-25). A number of these enzymes are implicated in the processes of tumor growth, invasion, and metastasis [7-9]. This is highlighted by the ability of TIMPs to inhibit these processes in experimental models that have been associated with their MMP inhibitory activity [10]. In addition MMPs are also implicated in the regulation of neovascularization highlighting the further potential for inhibitors in the treatment of angiogenesis driven diseases [11]. Taken together these observations highlight the potential of MMP inhibition in the treatment of cancer [12-14]. Although individual or combinations of MMPs have been implicated in association with specific cancers [15], it has proved a significant challenge for medicinal chemists to provide 'selective' inhibitors of these enzymes, targeted at a particular disease. Most often, inhibitors are broad spectrum in their action, in that they inhibit all or at least several members of the MMP gene family. Efforts in the design and optimization of MMP inhibitors have been extensively reviewed elsewhere [16-21].

A Drug Discovery Paradigm:

Further complexity is provided by the more recent characterization of related metalloproteinase gene families. It was noted that a number of first generation MMP inhibitors, as typified by Marimastat (**1**) were capable of inhibiting the processing of pro-TNFα to the pro-inflammatory cytokine (**Figure 1**) [22-23]. The identity of this TNFα convertase (TACE) activity has since been confirmed as a member of the ADAM gene family (ADAM-17) [24-25]. The ADAM family is related to the snake venom metalloproteinases, containing both a disintegrin and metalloproteinase domain. They form a large group of cell surface proteins that are thought to be involved in proteolysis, cell adhesion, signalling and fusion [26]. It is thought that a combined MMP/TACE inhibitor could offer additional benefit for the treatment of inflammatory/autoimmune diseases such as rheumatoid arthritis and multiple sclerosis, where both MMPs and TNFα are pathogenic [27]. However, this additional activity is probably not desirable in an anti-cancer therapy since TNFα itself has anti-tumor properties.

Figure 1 – Enzyme Inhibitory Profile of Marimastat

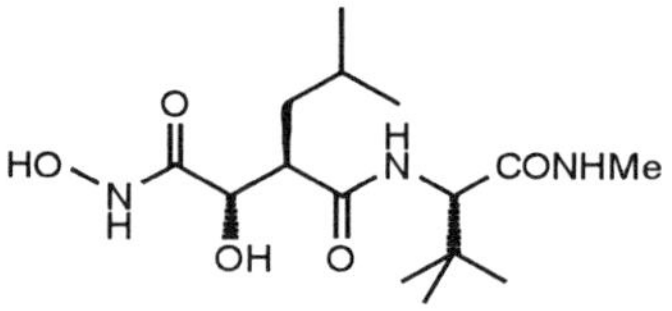

Marimastat (**1**)

IC_{50} (nM) [28].							
MMP1	MMP2	MMP3	MMP7	MMP8	MMP9	MMP13	MMP14
3	9	36	2	4	7	2	14

IC_{50} (μM) [28].				
TNFα release	TNF RII shedding	L-selectin shedding	IL1 RII shedding	IL6 receptor shedding
4	6	1	3.4	3.6

In addition to TNFα, a number of other extracellular proteins are known to be cleaved by metalloproteinase enzymes [29]. These so called 'sheddase' enzymes, are also thought to be members of the ADAM family and a number of these shedding events are known to be inhibited by first generation MMP inhibitors [30-31]. Whilst in some cases inhibition of these 'sheddases' may be of benefit in certain disease areas it is also possible that such non-selective action may be undesirable and lead to the side-effects that have become associated with such first generation compounds in clinical trials. Most apparent is the propensity of such compounds to cause tendinitis in animal models that manifests itself as joint pain in the clinic [32].

The primary purpose of this review is to discuss the medicinal chemistry issues associated with MMP inhibitors that have led to the identification of several clinical candidates in the oncology area. The review will discuss these clinical candidates individually in some detail and outline the design of more selective and more bioavailable MMP inhibitors, highlighting how drug discovery paradigms have changed over the last five years.

Design of MMP inhibitors:

Substrate-based inhibitors of the MMP enzymes have been designed around the cleavage site of the natural substrate. The most potent inhibitors have traditionally been centered on a hydroxamic acid zinc-binding group with peptide recognition being provided by occupation of the S1' pocket (**Figure 2**) [17]. Further interactions are then provided by hydrogen bonding interactions between the amide characteristics present in the inhibitor and residues present in the backbone of the enzyme active site. The hydroxamic acid group provides a strong bidentate interaction with the active site zinc atom with an additional hydrogen bond being provided by the NH. While they confer good potency, first generation hydroxamic acids became associated with very broad-spectrum activity and, in general, inadequate pharmacokinetic parameters. It is well recognized that drugs based on peptides and peptidomimetics often display poor absorption characteristics due to their highly polar nature and their structural similarilty to the natural substrates of peptidases present in the gastrointestinal tract [33]. The challenge for medicinal chemistry has therefore been to either provide compounds with an alternative zinc binding group that display these same potency enhancing characteristics whilst improving DMPK (Drug Metabolism and Pharmacokinetics) parameters, or design non-substrate based MMP inhibitors where most of the amide bond characteristics are removed. In practice most groups have employed both of these strategies in parallel.

Figure 2 - The Design of Substrate Based MMPIs around the Substrate Cleavage Site.

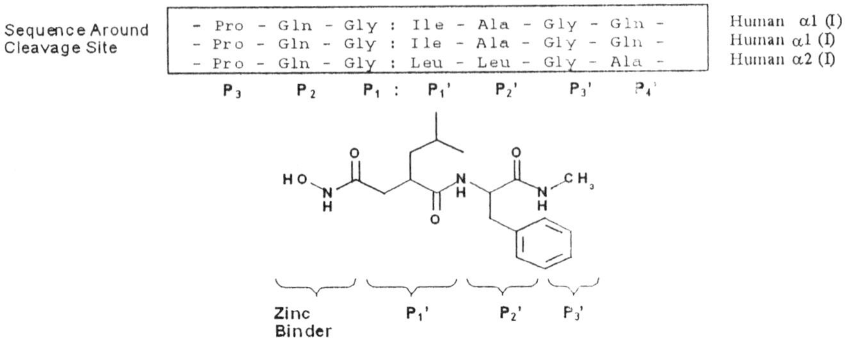

Sequence Around Cleavage Site	- Pro - Gln - Gly : Ile - Ala - Gly - Gln - - Pro - Gln - Gly : Ile - Ala - Gly - Gln - - Pro - Gln - Gly : Leu - Leu - Gly - Ala -	Human α1 (I) Human α1 (I) Human α2 (I)
	P_3 P_2 P_1 : P_1' P_2' P_3' P_4'	

Zinc Binder P_1' P_2' P_3'

Early MMP Inhibitors:

It was an early observation from X-ray crystal structures that the P2' substituent formed no significant interaction with the enzyme and workers at Roche and British Biotech were able to exploit this in their prototypical inhibitors, Ro 31-9790 (**2**) and Batimastat (**3**) [17].

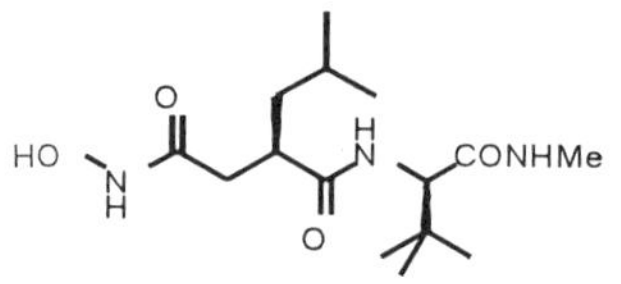

(**2**) 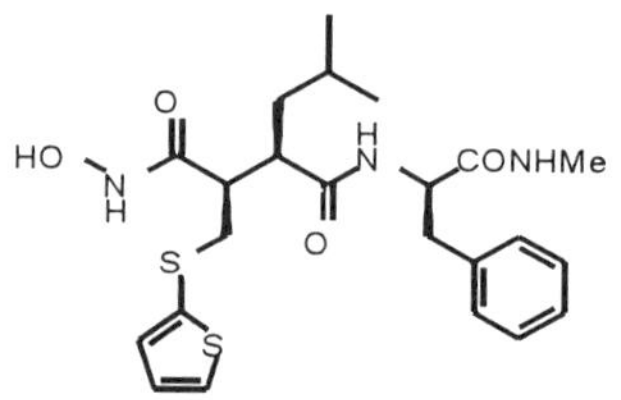   (**3**)

Whilst Roche selected Ro 31-9790 (**2**) for development for arthritis, Batimastat (**3**) was selected for development as an anti-cancer agent on the basis of promising results in models of metastasis [34], angiogenesis [35] and tumor progression [36-38]. In particular intraperitoneal administration of Batimastat (**3**) was very effective at resolving the malignant ascites that formed in the peritoneal cavity in a murine xenograft model of human ovarian carcinoma. In these experiments an increase in survival was noted [36]. This work led to Batimastat (**3**) being pursued as far as phase II clinical trials utilizing intraperitoneal administration to patients with malignant ascites. An earlier Phase I study of Batimastat (**3**) in breast cancer patients resulted in relatively low blood levels of the compound being reached following oral administration.

The combination of a P2' *tert*-butyl substituent (a sterically demanding group that protects vulnerable amide bonds from hydrolysis) and a group α- to the hydroxamic acid (that protects the hydroxamic acid from first pass metabolism) was employed in the more bioavailable inhibitor, Marimastat (**1**). In a phase I, single dose rising study (25 – 800 mg) a terminal elimination half-life of approximately 9.5 hours was determined [39]. In a second study in healthy volunteers, the compound was given twice daily for six days at doses of 50, 100 and 250 mg. During this relatively short dosing regimen the compound was well tolerated, and the lower dose was able to maintain a plasma level in excess of the IC_{90} for fibroblast collagenase (MMP-1) throughout the six-day period [40]. This compound leads the field in that it is currently in phase III clinical trials targeted at pancreatic, ovarian, colorectal, gastric, non-small cell lung and brain cancers. However, Marimastat (**1**) is also a 'sheddase' inhibitor and during the phase I/II clinical trials was demonstrated to show dose-limiting joint pain side-effects in the clinic after three months [32]. Whilst being effective in animal models and in Phase II studies, reducing the rate in rise of cancer antigens and pro-longing survival in cancer patients, even a relatively low dose of Marimastat results in patients being required to take dosing holidays following chronic treatment [41-44]. These effects were considered manageable at the dose range selected for future studies. The condition generally resolved rapidly on discontinuation of drug and several patients restarted treatment after a dosing holiday of 2-4 weeks. Marimastat (**1**) was licensed to Schering-Plough in September 1999. Most recently, however, British Biotech revealed that in four of the pivotal phase III clinical trials, in patients with very advanced disease, the compound was not efficacious at the doses employed. It is thought that patients with less advanced disease are more likely to respond to treatment with Marimastat (**1**). The patient population in two more phase III clinical trials in non-small cell lung cancer therefore represents the most suitable setting in which the compound is currently being evaluated. Results from these trials are expected late in 2000 [45].

BB-3644 (Structure not published) is a second generation compound and is also subject of the collaboration between British Biotech and Schering-Plough. This compound has previously been into development for the treatment of multiple sclerosis, and in pre-clinical models, BB-3644 is reported to show potent anti-cancer properties without the joint-pain seen with Marimastat (**1**) [45]. BB-3644 has completed a phase Ia study in healthy volunteers that showed it to be well tolerated and orally absorbed. A phase Ib maximum tolerated dose trial is underway to determine whether the benefit shown in pre-clinical models is observed in cancer patients. If this study is positive, BB-3644 will likely move into phase II trials during 2001.

Second Generation MMP Inhibitors:

In the first generation compounds such as Marimastat (**1**) the *tert*-butyl residue in P2' is utilised to protect the P2'/P3' amide bond from metabolism.

Figure 3 – Schematic Representation of key Interactions of a First Generation MMP Inhibitor and MMP8
(Reproduced from reference 46. Copyright 1995 American Chemical Society.)

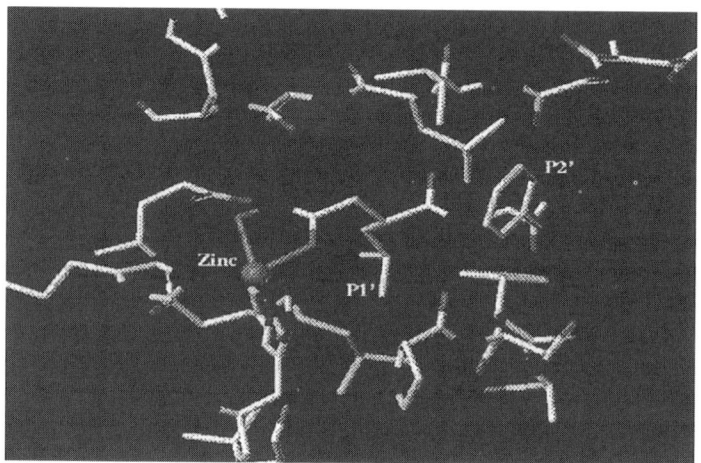

Later studies revealed however that it is the P1'/P2' amide bond that is critical for providing potency and that the P2' group and P2'/P3' amide group are not pre-requisites for activity (**Figure 3**) [46].

Some companies have utilized these observations and invested effort in removing the peptide character from the first generation substrate-based inhibitors by truncation. To this end, the Roche compound, Trocade (**4**), represents the most advanced compound, until recently in phase III clinical trials for rheumatoid arthritis [47]. Truncating the substrate-based template in the form of a P2' piperidine compromises potency against some of the MMP enzymes. However, utilising the hydrogen bonding acceptor capacity of the α-hydantoin restores potency, modifies MMP selectivity (**Figure 4**) and improves the metabolic stability of the hydroxamic acid group compared to first generation inhibitors such as Marimastat (**1**) [48]. This compound demonstrates good pharmacokinetics in rats (26% oral bioavailability and a half-life of 3h). Whilst the compound had no effect in the adjuvant arthritic rat model it has been reported to inhibit cartilage degradation *in vitro* and cartilage and bone changes in the STR/OTR murine model of osteoarthritis [49]. In addition this compound does not demonstrate activity against TNFα release or the 'sheddases' and in rheumatoid arthritis patients, Trocade was well tolerated with no evidence of tendinitis side-effects [50-51].

Figure 4 - Enzyme Inhibitory Profile of Trocade

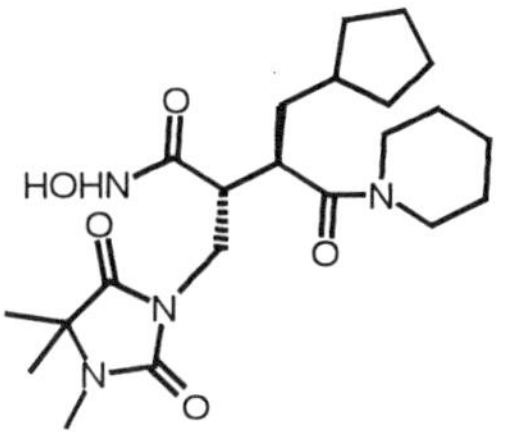

Trocade (**4**)

IC$_{50}$ (nM) [52]					
MMP 1	MMP 2	MMP 3	MMP 8	MMP 9	MMP 13
3	154	527	4.4	59.1	3.4

IC$_{50}$ (µM) [28] (* IA = inactive at 50µM)				
TNFα release	TNF RII shedding	L-selectin shedding	IL1 RII shedding	IL6 receptor shedding
IA*	IA*	IA*	IA*	IA*

Other companies have investigated replacing the hydroxamic acid with alternative zinc-binding groups. BMS-275291 (**5**), discovered at Chiroscience and currently in phase II clinical trials with Bristol-Meyers Squibb for cancer replaces the hydroxamic acid with a mercaptoamide group [53]. This compound and its derivatives represent potent broad-spectrum MMP inhibitors with no 'sheddase' activity (**Figure 5**). Selected compounds from this series, such as BMS-275291, are free of tendinitic side-effects in both of the recognized animals models, the rat and marmoset. BMS-275291 given as 5 doses of 10-90 mg/kg over 72 hours after tumor implantation produced dose-dependent reductions in the number of lung metastases, and once-daily treatment with the drug was associated with dose-dependent inhibition of angiogenesis in a murine model. Plasma levels of unchanged compound in mice given therapeutic doses were above the IC$_{50}$ values for gelatinases for extended periods [54].

Figure 5 - Enzyme Inhibitory Profile of BMS-275291

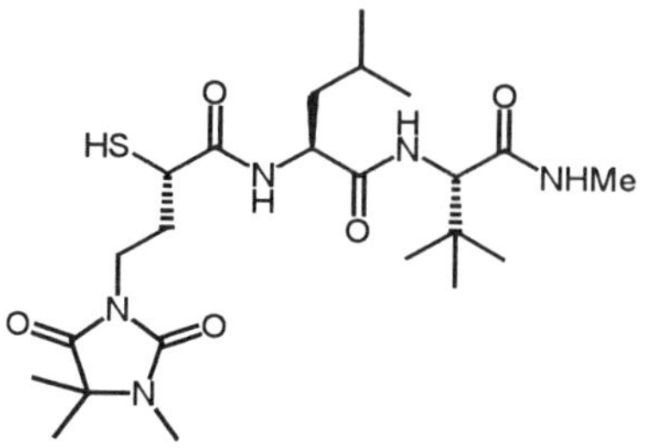

BMS-275291 (**5**)

IC$_{50}$ (nM) [53]				
MMP1	MMP2	MMP3	MMP8	MMP9
26	41	157	10	25

IC$_{50}$ (μM) [53]				
TNFα release	TNF RII shedding	L-selectin shedding	IL1 RII shedding	IL6 receptor shedding
IA*	IA*	IA*	IA*	IA*

* IA = inactive at 50μM

Non-substrate Based MMP Inhibitors:

Since the publication of Ciba-Geigy's patent in 1994 [55], much effort in the MMP inhibitor area has been devoted to the design of non-substrate based inhibitors. Such inhibitors that are designed to present less peptidic characteristics have been optimised to improve on physicochemical properties and the pharmacokinetics of the first generation inhibitors. CGS 27023A (**6**) represents the prototype compound from this class [56] and many companies have followed up with programmes of their own based on this lead. In particular, one of the most advanced compounds in this area is the Agouron cyclic analogue, AG3340, Prinomastat (**7**). This compound is currently in phase III clinical trials for both cancer and ARMD (**Figure 6**) [57-58].

Figure 6 - Enzyme Inhibitory Profile of CGS27023A and Prinomastat

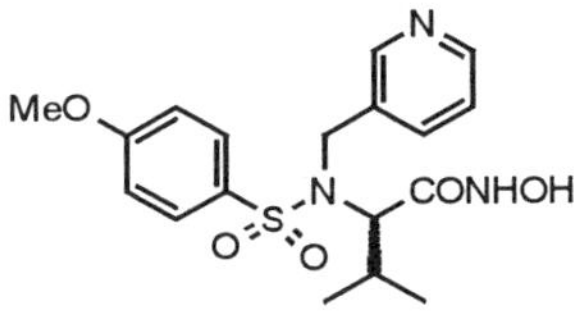

CGS27023A (6) 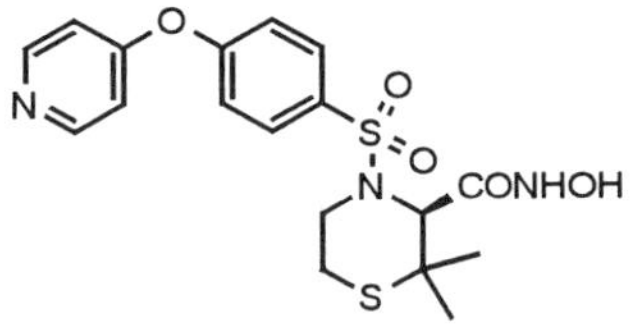   Prinomastat (7)

	Ki (nM) [56-58]				
Compound	MMP1	MMP2	MMP3	MMP9	MMP13
CGS27023A	33	20	43	8	4.3*
Prinomastat	8.3	0.05	0.3	0.26	0.03

* = IC_{50} value [59]

	IC_{50} (µM) [60]				
Compound	TNFα release	TNF RII shedding	L-selectin shedding	IL1 RII shedding	IL6 receptor shedding
CGS27023A	8	7	11	5	6.5
Prinomastat	1.6	14	0.8	1.8	12

Whilst these compounds represent potent MMP inhibitors with improved pharmacokinetic properties over early substrate based compounds, both are relatively broad spectrum in their action to the extent that they also inhibit 'sheddases' (**Figure 6**). The side-effect liability of these compounds is therefore considered to be high.

Although originally described as a clinical candidate for arthritis, CGS27023A (**6**) has also entered clinical trials in oncology. Preclinical data on CGS27023A (**6**) suggests that the compound is anti-angiogenic and inhibits the growth of a broad range of human tumor xenografts grown subcutaneously in nude mice. In addition CGS27023A (**6**) when dosed in combination with a range of standard chemotherapy agents has been shown to demonstrate a synergistic response [61]. In a phase Ib study in 36 patients with advanced solid tumors, the compound, dosed at regimens between 150 mg, bid and 600 mg tid, resulted in plasma levels more than an order of magnitude greater than the IC_{50} values for the target enzymes. A small proportion of patients (19%) had stable disease and remained on treatment for twelve weeks with no evidence of tumor regression. Minor side-effects included rash and nausea but by far the most significant side-effect reported was tendinitis in two-thirds of the treated groups [62]. Similarly, Prinomastat (**7**) has demonstrated excellent activity in *in vitro* and *in vivo* models of cancer, but it is clear that this compound is also limited in its clinical use by the joint pain side-effects even at relatively low doses [58]. Once again, Pfizer who recently acquired Prinomastat (**7**) by their take over of

Warner Lambert has indicated that the compound, currently in phase III clinical trials in advanced hormone refractory prostate cancer and advanced (Stage IV) non-small cell lung cancer, did not meet primary efficacy objectives. Neither detrimental nor convincing beneficial effect of the combination of Prinomastat (**7**) with standard chemotherapy was observed. Consequently, Pfizer has halted these two phase III trials. Based on input from the Data Safety Monitoring Board (DSMB), patients having earlier stage (Stage IIIB) disease recruited into a second on-going non-small cell lung cancer trial will continue to be studied. Pfizer has reported that the company intends to continue exploration of Prinomastat (**7**) in other tumor types and most importantly, in earlier stage disease, where oncologists believe inhibition of angiogenesis may have greater utility. Four phase II trials are currently underway and two additional phase II trials will begin shortly [63].

The Design of More Selective MMPIs:

The results obtained so far on the most advanced clinical candidates does not provide good confidence that an MMPI will eventually find utility in the treatment of cancer. However, more selective compounds, such as BMS-275291 (**5**), may well lead to drugs that, without presenting the dose limiting joint pain side-effects, allow dosing at a level suitable to obtain efficacy in advanced cancer patients.

When considering selectivity, a significant problem, until recently, has been the lack of published structure-based drug design data on non-substrate based inhibitors. It was originally thought that the pyridylmethyl substituent of CGS27023A occupied the P1' pocket and that the methoxyaryl unit was a P2' substituent. Subsequent studies have, however, demonstrated that it is the methoxyaryl substituent that occupies P1' [64]. This has led to a revolution in the design of non-substrate based inhibitors with a range of selectivities [16-21]. One of the greatest challenges with this class of inhibitor has been to design more selective compounds that display acceptable physicochemical, drug metabolism and pharmacokinetic profiles. One of the medicinal chemistry issues associated with the hydroxamic acid moiety is that they are metabolically labile to both hydrolysis and reduction. A successful strategy employed to address this has been to introduce sterically demanding substituents α- to the labile hydroxamic acid in order to protect this group from metabolism.

The pyran group in Ro113-0830 (**8**) serves such a purpose [65-66]. Ro113-0830 (**8**) has good oral bioavailability and pharmacokinetic properties in pre-clinical studies (approx. 40% oral bioavailability in rats and monkeys with a half-life of 3-4 hours). This compound is a potent inhibitor of MMPs 2, 3, 9 and 13 but is not reported to inhibit TACE, sheddases, matrilysin and collagenase-1 (**Figure 7**) [66]. Ro113-0830 has not been directed towards the cancer endpoint but the compound protects against cartilage breakdown in the rat *in vivo* cartilage plug model using bovine nasal cartilage and articular cartilage (ED_{50} of 6.5 and <4.5 mg/kg/day, respectively). Efficacy was also demonstrated in a rheumatoid arthritis model, the murine collagen induced arthritis (MCIA) model (ED_{50} 5 - 50 mg/kg qid po). Its pharmacokinetic profile in mice and rats allows for double the effective dose to be administered every other day instead of a single dose every day with the same efficacy (rat *in vivo* cartilage plug assay and MCIA assays). One of the studies involved examination of Ro113-0830 in a 3 month rabbit model of osteoarthritis in which the compound showed efficacy at 3 and 10 mg/kg/day qd po. Examination of the microscopic histopathology of the medial tibial plateau showed a significant reduction in the overall incidence and severity of fissuring and erosion lesions relative to the control groups. In a similar fashion, examination of the gross pathology of the femoral condoyle revealed a significant reduction in the incidence and severity of ulcers and pitting lesions. The similar chondroprotective efficacy of Ro113-0830 (**8**) in murine, rat and rabbit studies of cartilage degradation suggests that the drug could be effective in treating osteoarthritis in humans. This compound is in phase II clinical trials for osteoarthritis [66].

More recently, a significant amount of structure based drug design has been carried out around this series of molecules and it is recognised that the 4-chlorophenyl substituent resides in the P1' pocket and can be modified to address selectivity [67]. One feature of this is that the compounds from this series, in general, are not capable of potent inhibition of MMP-1 and MMP-7. These enzymes are recognised to present a shallow P1' pocket and the 4-chlorophenyl residue is thought to be too large to fit into the P1' pocket of these enzymes. These data compare favorably with that generated on the early substrate based inhibitors employing large bulky P1' substituents.

Figure 7 - Enzyme Inhibitory Profile of Ro113-0830

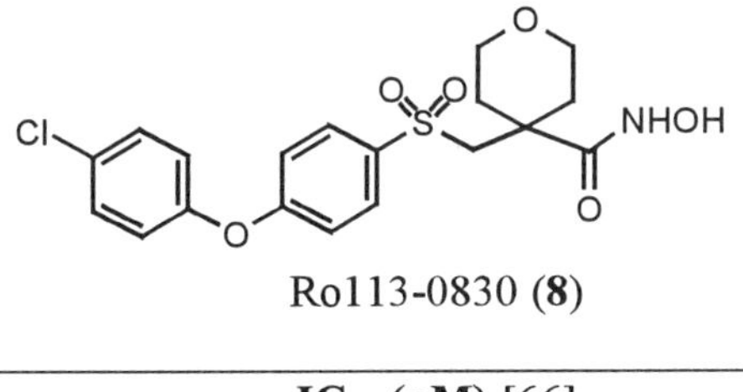

Ro113-0830 (**8**)

IC_{50} (nM) [66]				
MMP1	MMP2	MMP3	MMP9	MMP13
590	0.22	9.3	0.58	0.52

In common with the substrate based inhibitors it is possible to replace the hydroxamic acid group completely. There are many examples of replacing the hydroxamic acid with a carboxylic acid [16-21]. In most cases there is a significant loss in potency due to the fact that on converting a hydroxamic acid to a carboxylic acid two hydrogen bonding interactions with the enzyme are lost [68]. One benefit of some of the carboxylic acid based compounds is that it is often possible to achieve selectivity both over the so called 'sheddases' and within the MMP gene family. Indeed early work from the Merck Group in substrate-based inhibitors revealed that carboxylic acids could also be utilized to provide potent inhibitors of stromelysin-1 (MMP-3) and gelatinase-A (MMP-2) [69]. At the time there was a strong feeling in the MMP community that MMP-3 was a crucial enzyme in the progression of arthritis. The lack of efficacy of Merck's inhibitors in animal models of disease was therefore disappointing. However, follow up studies on a MMP-3 'knock-out' mouse revealed that these animals were in fact more susceptible to collagen induced arthritis than the wild type animals [43]. This excellent work, however, highlights the danger associated with the design of more selective MMP inhibitors due to the apparent redundancy present in the system. In the case of the Bayer compound, BAY12-9566 (**9**) the carboxylic acid also confers a considerable refinement in pharmacokinetics demonstrating excellent oral bioavailability, high trough plasma concentrations and a long half-life [71]. *In vitro* and *in vivo* studies with this compound show anti-invasive and antiangiogenic activity, and efficacy in a range of murine xenograft cancer models, including melanoma, lung, ovary, colon, pancreas and breast. In animals the compound is highly protein bound and in a phase I pharmacokinetic study in man plasma levels are less than dose proportional suggesting saturation of absorption. Dose levels for further trials were split in order to improve drug delivery [71]. This compound was, until recently, in phase III clinical trials for cancer, osteoporosis and osteoarthritis. BAY12-9655 (**9**) is a modestly selective inhibitor of gelatinase-A (MMP-2) (**Figure 8**) and while this compound does display dose-limiting thrombocytopenia, in patients there are no reports of tendinitis following chronic dosing [71].

There still remains a significant amount of effort in the MMP inhibitor field aimed at replacing hydroxamic acids with alternative zinc binding groups. As highlighted above, most of the activity to date has resulted in the identification of potent and bioavailable carboxylic acids and thiols.

Reverse hydroxamates have been known to provide MMP inhibitors from the substrate-based class for many years. The Group at Abbott recognized the potential of this zinc binding group in providing more selective MMP inhibitors aimed at producing clinically effective agents lacking the dose-limiting side effects associated with the first generation, broad-spectrum inhibitors. In the discovery of ABT-770 (**10**), enhancing gelatinase-A (MMP-2)

Figure 8 - Enzyme Inhibitory Profile of BAY12-9655

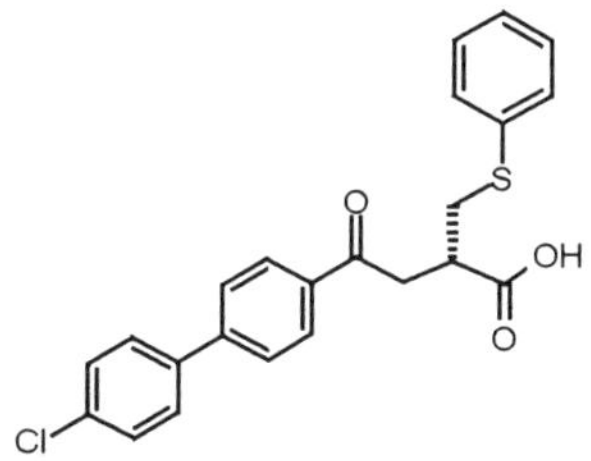

BAY12-9655 (9)

Ki (nM) [71]				
MMP1	MMP2	MMP3	MMP9	MMP13
>5000	11	134	301	1470

Figure 9 - Enzyme Inhibitory Profile of ABT-770

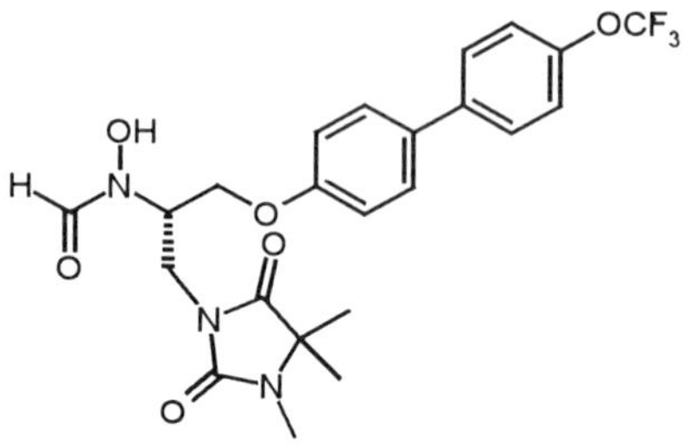

ABT-770 (**10**)

IC$_{50}$ (nM) [72]							
MMP 1	MMP 2	MMP 3	MMP 7	MMP 8	MMP 9	MMP 13	MMP 14
4600	3.2	42	10000	7.8	170	45	580

potency while eliminating 'sheddase' activity was their primary concern in the design of an effective anti-cancer agent with low side effect liability. Consequently, ABT-770 (**10**) is a potent inhibitor of MMP-2 and MMP-8, moderately potent against MMP-3 and MMP-13, weakly active against MMP-9 and MMP-14 whilst having little activity at all against MMP-1 and MMP-7 (**Figure 9**). This compound is reported not to inhibit 'sheddases' at concentrations of >50 µM and is demonstrated to be orally effective in animal models of metastasis [72].

ABT-770 (**10**) represents the first example of a traditional non-substrate based template with an alternative zinc-binding group capable of conferring nM potency against the MMP enzymes.

To explore this issue further some companies have either undertaken the synthesis of metalloproteinase focussed combinatorial libraries or a broader screening effort against a defined metalloproteinase target. Pharmacia and Upjohn employed the latter strategy in the identification of novel stromelysin inhibitors with surprising results [73]. The initial screening campaign provided several weak, competitive stromelysin (MMP-3) inhibitors represented by a class of thiadiazole ureas (**Figure 10**). Elaboration of this interesting template employing traditional peptidomimetic SAR improved the potency of this series significantly.

Figure 10 - SAR of Pharmacia Upjohn Thiadiazoles

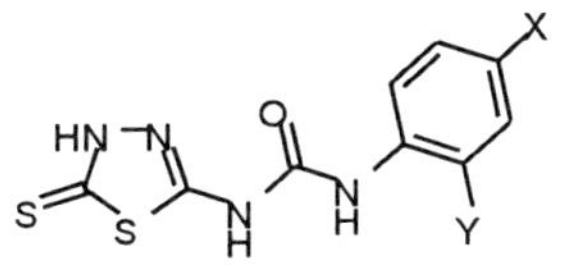

X = Cl; Y = Me (**11**)
X = Y = H (**12**)

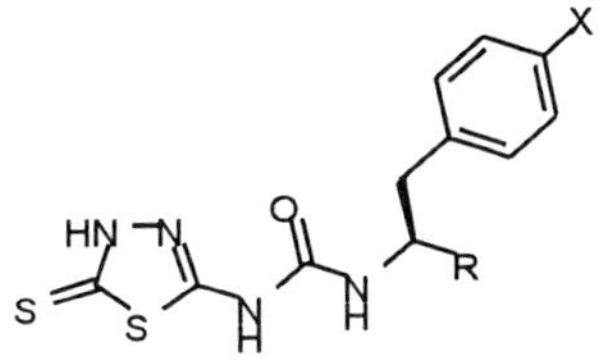

X = R = H (**13**)
X = H; R = CONHMe (**14**)
X = F; R = CONHMe (**15**)

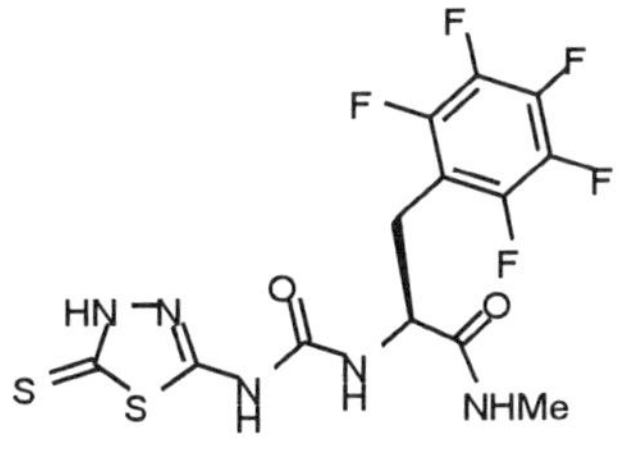

(**16**)

For instance, the screening hit (**11**) [IC$_{50}$ vs MMP-3 = 125 µM] could be simplified to the unsubstituted aryl amide (**12**) with retention of potency [IC$_{50}$ vs MMP-3 = 99 µM].

Chain extension to the phenethylamine derivative (**13**) improved potency significantly [K$_i$ vs MMP-3 = 7 µM]. Introduction of an amide substituent (**14**) provided a further 10-fold increase in potency and this was initially thought to be due to realizing the P2'/P3' amide interaction present in many substrate-based inhibitors. In this series it is possible to improve potency significantly by introducing a 4-fluoro substituent into the aromatic ring (**15**) [IC$_{50}$ vs MMP-3 = 100 nM; MMP-2 = 8.9 µM] and this is contrary to SAR in traditional phenylalanine P2' replacements, suggesting that their initial hypothesis was incorrect. Further fluorine substitution provides a highly potent and selective MMP-3 inhibitor (**16**) [IC$_{50}$ vs MMP-3 = 18 nM; MMP-2 = 3 µM].

When this work was almost complete, an X-ray crystal structure of (**16**) complexed with MMP-3 revealed that the compound was bound completely on the unprimed side of the enzyme substrate binding cleft. In this unusual mode of binding, the thiocarbonyl interacts with the catalytic zinc and the NH on the thiadiazole participates in a bifurcated hydrogen bond to the carboxylate of Glu 202. Other hydrogen bonds are formed with the unprimed backbone in a unique orientation that explains some anomalies in the SAR compared to traditional peptidomimetics. The aryl group is thought to form a π-π stacking arrangement with Tyr 155 that explains the improved potency observed with phenylalanine analogues compared to other traditional replacements employed in the field of MMP inhibitors. Unlike the prime sites, significant enzyme differences are found for the MMPs in the unprimed region. For instance in the collagenases, the Tyr 155 residue is mutated to a serine, effectively eliminating the π-π stacking arrangement observed in MMP-3 with (15). Hence, this class of compound is the first to provide highly potent MMP inhibitory activity via binding to the prime sites of the enzyme.

Whilst the pharmacokinetics and efficacy of this series of compounds have not been reported, these thiadiazoles represent the first small molecule template capable of providing unprimed site binding interactions. Whilst there is no TACE activity associated with this series, information provided by the binding of this template may well allow opportunities to identify further selective MMP inhibitors as well as selective inhibitors of ADAM and ADAM-TS gene family members.

Future Directions for MMPIs in Cancer:

The biological effects of MMPIs are interesting and suggest that this class of compounds should have therapeutic utility in the treatment of human cancers. Several MMP inhibitors with varied selectivity profiles are now in pre-clinical or in clinical trials for the treatment of cancer and chronic degenerative diseases. It remains to be seen whether these compounds will be efficacious in humans and whether the therapeutic index will be improved over the first generation clinical candidates. Initial results from the phase III studies on Marimastat (1), Prinomastat (7) and BAY 12-9566 (9) are not encouraging for MMPIs in the treatment of advanced malignancies. This is most likely due to the fact that these compounds do not have a sufficient therapeutic window. However, MMPIs also have great potential for treating early stage disease when tumors are smaller and may be more dependent upon angiogenic factors. The physical association of MMP-2 with the $\alpha_v\beta_3$ integrin receptor [74] and the ability of VEGF to upregulate MMP-2 gene expression both *in vitro* and *in vivo* suggests a prominent role for the MMPs in neovascularization [75]. Hence, there is good confidence that MMPIs may find utility in diseases where neovascularisation is the driving process.

In addition, a major unmet medical need in the treatment of cancer is the process of tumor metastasis to bone. Osteolysis during bone metastasis is characterised by an increase in osteoclast activity in the vicinity of the secondary tumor at the surface of the bone. MMPIs have been shown to reduce osteoclastic bone resorption *in vitro* and *in vivo*. TIMPs have been utilized in conjunction with bisphosphonates to synergistically reduce the process of bone resorption in models of osteotropic tumor metastasis [76]. Thus, there may be some utility for MMPIs in the treatment of bone metastasis and in the treatment of bone pain in osteotropic cancer patients

Meanwhile, the area continues to present opportunities and challenges to medicinal chemists. The continuing expansion of the MMP gene family and the more recent discovery of the ADAM and ADAM-TS gene families mean that the need for more selective inhibitors will be ever greater. In addition, further characterization of the therapeutic relevance of these newer enzymes will result in a further increase in interest in the area of metalloproteinase inhibitors. For instance there is increasing evidence that members of the ADAM and ADAM-TS gene family may be involved in angiogenesis [77-78] and in the process of bone resorption [79-81]. Companies that have a diverse range of structural types in their metalloproteinase inhibitor compound portfolio should be well placed to capitalize on developments in these new and exciting areas of research.

Summary:

Many groups have now recognized that the next generation MMP inhibitors will need to be more selective. There are a number of relatively broad-spectrum inhibitors either in the clinic or in advanced pre-clinical studies and these should allow proof of concept within their target disease areas. There are also X-ray crystal structures available now for most of the interesting MMPs, including MMP-1, MMP-2, MMP-3, MMP-7, MMP-8, MMP-13 and most recently MMP-14, thus providing greater opportunities for structural/computational studies to be undertaken to further differentiate activity within the MMP gene family [67-68, 82-83].

Similarly, the recent publication of the TACE crystal structure illustrates that structural data is already emerging from the ADAM gene family [83]. It is possible that future generations of metalloproteinase inhibitors may be either mixed ADAM/ADAM-TS/MMP inhibitors targeted at specific diseases. However, if medicinal chemists are to seek truly selective inhibitors of these new enzymes, information about active site interactions between ligands and target proteins will be required. This information should ultimately allow the design of selective ADAM and ADAM-TS inhibitors that not only differentiate between these gene family members and the MMPs but that are also capable of differentiating between members of these respective gene families.

Literature Cited:

1. Birkedal-Hansen, H. *Curr. Opin. Cell. Biology* **1995**, *7*, 728-735.
2. Woessner, J. F. *FASEB J.* **1991** *5*, 2145-2154.
3. Nagase, H. *Biol. Chem.* **1997**, *378*, 151-160.
4. Murphy, G. *Acta Orthop. Scand.* **1995**, *66*, 55-60.
5. Toi, M.; Ishigaki, S.; Tominaga, T. *Breast Cancer Research and Treatment* **1998**, *52*, 113-124.
6. Levy, D. E.; Ezrin, A. M. *Emerging Drugs* **1997**, *2*, 205-230.
7. Kahari, V.; Saarialho-Kere, U. *The Finnish Medical Society Duodecim, Ann. Med.* **1999**, *31*, 34-45.
8. Chambers, A. F.; Matrisian, L. *Journal of the National Cancer Institute* **1997**, *89*, 1260-1270.
9. Coussens, L. M.; Werb, Z. *Current Biology* **1996**, *3*, 895-904.
10. Gomez, D. E.; Alonso, D. F.; Yoshiji, H.; Thorgeirsson, A. *Eur. J. Cell Biol.* **1997**, *74*, 111-122.
11. Moses, M. A. *Stem Cells* **1997**, *15*, 180-189.
12. Yu, A. E.; Hewitt, R. E.; Connor, E. W.; Stetler-Stevenson, W. G. *Drugs and Aging* **1997**, *11*, 229-244.
13. Brown, P. D. *Medical Oncology* **1997**, *14*, 1-10.
14. Brown, P. D. *Breast Cancer Research and Treatment* **1998**, *52*, 125-136.
15. Giambernardi, T. A.; *et al. Matrix Biology* **1998**, *16*, 483-496.
16. Morphy, J. R.; Millican, T. A.; Porter, J. R. *Current Medicinal Chemistry* **1995**, *2*, 743-762.
17. Beckett, P. R.; Davidson, A. H.; Drummond, A. H. *Drug Discovery Today* **1996**, *1*, 16-26.
18. White, A. D.; Thomas, M. A.; Bocan, P. A. *Current Pharmaceutical Design* **1997**, *3*, 45-58.
19. Beckett, P. R.; Whittaker, M. *Exp. Opin. Ther. Patents* **1998**, *8*, 259-282.
20. Michaelides, M. R.; Curtin, M. L. *Current Pharmaceutical Design* **1999**, *5*, 787-819.
21. Skotnicki, J. S.; Levin, J. I.; Zask, A.; Killar, L. M. *Metalloproteinases as targets for anti-inflammatory drugs.* Edited by Bottomley, K. M. K.; Bradshaw, D.; Nixon, J. S. **1999**, 17-57, Berkhauser Verlag Basel Switzerland.
22. McGeehan, G. M.; Becherer, J. D.; Bast, R. C.; *et al. Nature* **1994**, *370*, 558-561.
23. Gearing, A. J.; Beckett, P.; Christodoulou, M.; *et al. Nature* **1994**, *370*, 555-557.
24. Black, R. A.; Rauch, C. T.; Kozlosky, C. J.; *et al. Nature* **1997**, *385*, 729-733.
25. Moss, M. L.; Jin, S. L.; Milla, M. E.; *et al. Nature* **1997**, *385*, 733-736.
26. Black, R. A.; White, J. M. *Curr. Opin. Cell Biol.* **1998**, *10*, 564-659.
27. Nelson, F. C.; Zask, A. *Exp. Opin. Invest. Drugs* **1999**, *8*, 383-392.

28. Bird, J. *What is the Appropriate Selectivity for Development of MMP Inhibitors for Therapeutic use?* Presented at *Gordon Research Conference - Matrix Metalloproteinases, USA*, **1999**.
29. Hooper, N. M.; Karran, E. H.; Turner, A. J. *Biochem. J.* **1997**, *321*, 265-279.
30. Gallea-Robache, S.; Morand, V.; Millet, S.; *et al. Cytokine* **1997**, *9*, 340-346.
31. Lombard, M. A.; Wallace, T. L.; Kubicek, M. F.; *et al. Cancer Research* **1998**, *58*, 4001-4007.
32. Praga-Wojtowicz, S.; Torri, J.; Johnson, M.; *et al. J. Clin. Oncology* **1998**, *16*, 2150-2156.
33. Chan, O. H.; Stewart, B. H. *Drug Discovery Today* **1996**, *1*, 461-473.
34. Chirivi, R. G. S.; Garofalo, A.; Crimmin, M. J.; Bawden, L. J.; Brown, P. D.; Giavazzi, R. G. *Int. J. Cancer* **1994**, *58*, 460-464.
35. Taraboletti, G.; Garofalo, A.; Belotti, D.; Drudis, T.; Borsotti, P.; Scanziani, E.; Brown, P.; Giavazzi, R. *J. Natl. Cancer. Inst.* **1995**, *87*, 293-298.
36. Eccles, S. A.; Box, G. M.; Court, W. J.; Bone, E. A.; Thomas, W.; Brown, P. *Cancer Res.* **1996**, *56*, 2815-2822.
37. Sledge, G. W.; Qulali, M.; Goulet, R.; Bone, E. A.; Fife, R. *J. Natl. Cancer Res.* **1995**, *87*, 1546-1550.
38. Wang, X.; Fu, X.; Brown, P. D.; Crimmin, M. J.; Hoffman, R. M.; *Cancer Res.* **1994**, *54*, 4726-4728.
39. Millar, A. W.; Brown, P. D.; Moore, J.; Galloway, W. A.; Cornish, A. G.; Lenehan, T. J.; Lynch, K. P. *Br. J. Clin. Pharm.* **1998**, *45*, 21-26.
40. Drummond, A. H.; *et al. Proc. Am. Assoc. Cancer Res.* **1995**, *36*, 595 (Abstr.).
41. Primrose, J.; Bleiberg, H.; Daniel, F.; Johnson, P.; Mansi, J.; Neoptolemos, J.; Seymour, M.; Van Belle, S. *Ann. Oncol.* **1996**, *7*, 35 (Abstr.).
42. Poole, C.; Adams, M.; Barley, V.; Graham, J.; Kerr, D.; Louviaux, I.; Perren, T.; Piccart, M.; Thomas, H. *Ann. Oncol.* **1996**, *7*, 68 (Abstr.).
43. Millar, A.; Brown, P. D. *Ann. Oncol.* **1996**, *7*, 123 (Abstr.).
44. Parsons, S. L.; Watson, S. A.; Griffin, N. R.; Steele, R. J. C. *Ann. Oncol.* **1996**, *7*, 47 (Abstr.).
45. British BioTech PLC: Preliminary results for the year ended 30th April 2000. Presentation and press announcement. 5th July **2000**.
46. Grams, F.; Crimmin, M.; Bode, W.; *et al, Biochemistry* **1995**, *34*, 14012-14020.
47. Lorenz, H-M. *Current Opinion in Anti-inflammatory and Immunomodulatory Investigational Drugs* **2000**, *2*, 47-52.
48. Broadhurst, M. J.; Brown, P. A.; Lawton, G.; *et al. Bioorg. Med. Chem. Lett.* **1997**, *7*, 2299-2302.
49. Brewster, M.; Lewis, E. J.; Bottomley, K. M. K.; *et al, Arthritis Rheumatism.* **1998**, *41*, 1639-1644.

50. Wood, N. D.; Aiken, M.; Durston, S.; *et al. Agents Actions Suppl.* **1997**, *49*, 49-55.
51. *EULAR*. Vienna, Austria, November 1997, Abstract 48.
52. Lewis, E. J.; Bishop, J.; Bottomley, K. M.; *et al. Br. J. Pharmacol.* **1997**, *121*, 540 -546.
53. Baxter, A. D.; Bird, J.; Bannister, R.; *et al. Cancer Drug Discovery and Development: Matrix Metalloproteinase Inhibitors in Cancer Therapy,* Edited by Clendeninn, N. J.; Appelt, K. Humana Press Inc., Totowa NJ. **2000**, Chapter 8, 193-221.
54. Naglich, J. G.; *et al. Proc. Amer. Assoc. Cancer Res.* [91st Annu. Meet. Amer. Assoc. Cancer Res. (April 1-5, San Francisco)] **2000**, *41*, 3122 (Abstr.).
55. MacPherson, L. J.; Parker, D. T. Patent Application, EP606046A1 (Ciba Geigy). Filed 13th July **1994**.
56. MacPherson, L. J.; Bayburt, E. K.; Capparelli, M. P.; *et al. J. Med. Chem.* **1997**, *40*, 2525-2532.
57. Shalinsky, D. R.; Varki, N. M.; Brekken, J.; *et al. Annals of the New York Academy of Sciences* **1999**, *878*, 236-270.
58. Sorbera, L. A.; Castaner, J. *Drugs of the Future* **2000**, *25*, 150-158.
59. Pikul, S.; McDow, K. L.; Dunham, N. G.; *et al. J. Med. Chem.* **1999**, *42*, 87-94.
60. Bird, J.; Bhogal, R.; Baxter, A.; *et al. Activity of matrix metalloproteinase inhibitor (MMPI) in models of inflammation.* Presented at Inflammation '99, Paris, France (**1999**).
61. Wood, J. *CGS27023A, a matrix metalloproteinase inhibitor with anti-tumour activity.* Presented at Protease Inhibitors, New Therapeutics and Approaches (1996), Baltimore, Maryland, USA.
62. Levitt, N. C.; Eskens, F.; Propper, D. J.; Harris, A. L.; Denis, L.; Ganesan, T. S.; Mather, R. A.; McKinley, L.; Planting, A.; Talbot, D. C. *Proc. Am. Soc. Clin. Oncol.* **1998**, *17*, 213a (Abstr.).
63. Pfizer Press Announcement, 4th August 2000.
64. Gonnella, N. C.; Li, Y-C.; Zhang, X.; *et al. Biochemistry* **1998**, *40*, 14048-14056.
65. Bender, S. L.; Broka, C. A.; Campbell, J. A.; *et al. Eur. Patent Appl.* 780386 (**1997**).
66. Lollini, L.; Haller, J.; DeLustro, B.; *et al.* Inhibition Of Matrix Metalloproteinases: Therapeutic Applications **1998** October 21-24, Tampa, Florida. Abs: 20.
67. Lovejoy, B.; Welch, A. R.; Carr, S.; *et al. Nature Structural Biology* **1999**, *6*, 217-221.
68. Babine, R. E.; Bender, S. L. *Chem. Rev.* **1997**, *97*, 1359-1472.
69. Esser, C. K.; Bugianesi, R. L.; Caldwell, C. G.; *et al. J. Med. Chem.* **1997**, *40*, 1026-1040.

70. Muidgett, J. S.; Hutchinson, N. I.; Chartrain, N. A.; *et al. Arthritis Rheumatism* **1998**, *41,* 110-121.
71. Sorbera, L. A.; Graul, A.; Silvestre, J.; *et al. Drugs of the Future* **1999**, *24,* 16-21.
72. Morgan, D. W.; Albert, D. H.; Magoc, T.; *et al*: Poster at Amer. Assoc. Cancer Res. [91st Annu. Meet. Amer. Assoc. Cancer Res. (April 1-5, San Francisco)] **(2000)**.
73. Jacobsen, E. J.; Mitchell, M. A.; Hendges, S. K.; *et al. J. Med. Chem.* **1999**, *42,* 1525-1536.
74. Brooks, P. C.; Stromblad, S.; Sanders, L. C.; von Schalsch, T. L.; Aimes, R. T.; Stetler-Stevenson, W. G.; Quigley, J. P.; Cheresh, D. A. *Cell* **1996**, *85,* 683-693.
75. Zucker, S.; Mirza, H.; Conner, C. E.; Lorenz, A. F.; Drews, M. H.; Bahou, W. F.; Jesty, J. *Int. J. Cancer* **1998**, *75,* 780-786.
76. Yoneda, T.; Sasaki, A.; Dunstan, C.; Williams, P. J.; Bauss, F.; DeClerck, Y. A.; Mundy, G. R. *J. Clin. Invest.* **1997**, *99,* 2509-2517.
77. Trochon,V.; Li, H.; Vasse, M.; Frankenne, F.; Thomaidis, A.; Soria, J.; Lu, H.; Gardner, C., Soria, C. *Angiogenesis* **1998**, *2,* 277-285.
78. Vasquez, F.; Hastings, G.; Ortega, M-A.; Lane, T. F.; Oikemus, S.; Lombardo, M.; Iruela-Arispe, M. L. *J. Biol. Chem.* **1999**, *274,* 23349-23357.
79. Inoue, D.; Reid, M.; Lum, L.; Kratzschmar, J.; Weskamp, G.; Myung, Y. M.; Baron, R.; Blobel, C. P. *J. Biol. Chem.* **1998**, *273,* 4180-4187.
80. Dallas, D. J.; Genever, A.; Patton, A. J.; Millichip, M. I.; Mckie, N.; Skerry, T. M. *Bone* **1999**, *25,* 9-15.
81. Lum, L.; Wong, B. R.; Josien, R.; Becherer, J. D.; Erdjument-Bromage, H.; Schlondorff, J.; Tempst, P.; Choi, Y.; Blobel, C. P. *J. Biol. Chem.* **1999**, *274,* 13613-13618.
82. Browner, M. F. *Perspectives in Drug Discovery and Design* **1994**, *2,* 343-351.
83. Borkakoti, N. *Metalloproteinases as targets for anti-inflammatory drugs.* Edited by Bottomley, K. M. K.; Bradshaw, D.; Nixon, J. S. **1999**, 1-16, Berkhauser Verlag Basel Switzerland.

Chapter 17

Prospects for Antiangiogenic Therapies Based upon VEGF Inhibition

Pascal Furet and Paul W. Manley*

Preclinical Research, Novartis Pharma Ltd., CH-4002 Basel, Switzerland

Anti-cancer agents currently in clinical development, targeted at blocking the angiogenic activity of VEGF, include VEGFR-2 kinase inhibitors, such as SU5416 and PTK787 / ZK222584. The rationale behind these agents, their mechanism of action and their biological profiles are reviewed. Based upon current knowledge, the prospects for the design of second generation agents are addressed, together with their potential in cancer therapy.

The Angiogenesis Process in Cancer

Solid tumors comprise of a heterogeneous assembly of proliferating tumor cells, host tissue stromal fibroblasts and immune cells, cooperating together and served by a vasculature. The role of the vasculature within such a neoplasm is to provide a blood supply to the tumor cells, thereby supplying oxygen and other nutrients, as well as removing metabolic waste products, such as lactic acid. Tumors are subject to hypoxia and acidosis due to factors such as inadequate vascularization and flow abnormalities within tumor vessels, resulting from their chaotic architecture, and leakiness (1). In response to hypoxia/acidosis both neoplastic and stromal cells stimulate the sprouting, growth and maturation of new capillaries from existing blood vessels, via a process termed angiogenesis (2). Angiogenesis can be summarized as involving the production of cytokines from effector cells, leading to vascular endothelial cell activation and division, degradation of the surrounding extracellular matrix by proteases, matrix invasion by endothelial cells of the sprouting vessel, recruitment of accessory cells and the laying down of a basement membrane in the maturing vessel.

Whereas angiogenesis is essential for embryogenesis and embryonic development, it is a tightly controlled process which only regularly occurs in adult tissues during the female reproductive cycle and tissue repair. Consequently, whereas most current cancer therapies must be applied intermittently due to their inherent toxicities, anti-angiogenic therapy promises to be safe enough for chronic administration to patients (3). Such treatment should inhibit the growth of primary tumors and prevent the development of metastases beyond the 20-24 cell diameters (< 1 mm) governed by the diffusion of oxygen through cells (4). Since most cancer deaths result from metastases, and patients with highly vascularized primary tumors have a significantly higher frequency of metastatic disease than those with low angiogenic cancers, anti-angiogenic therapy is emerging as an extremely promising, broad spectrum cancer treatment (5).

Important pro-angiogenic cytokines include vascular endothelial growth factors (VEGFs), angiopoietins, ephrins, basic fibroblast growth factor (bFGF), platelet-derived growth factor (PDGF), transforming growth factor-β (TGF-β), insulin-like growth factor-1 (IGF-1), tumour necrosis factor-α (TNF-α) and the interleukins IL-1 and IL-8. VEGF is amongst the most important of these and is the focus of this article (6).

VEGF and Its Role in Tumour Angiogenesis

The VEGF family of proteins includes VEGF-A (first termed vascular permeability factor, but now simply referred to as VEGF), VEGF-B, -C, -D, -E and placenta growth factor, PlGF (7 - 9). VEGF increases microvessel permeability and is a mitogenic and chemotactic factor, selective for endothelial cells derived from arteries, veins and lymph vessels. VEGF levels are elevated in the majority of human tumors, with expression being highest in hypoxic tumor cells adjacent to necrotic areas. VEGF mRNA expression is upregulated by a number of mechanisms associated with angiogenesis during embryonic development, growth, tissue repair and reproduction. In tumor angiogenesis, oncogenes and deletion of the p53 tumor suppressor gene can upregulate VEGF expression (10). In response to tumor hypoxia, VEGF expression is increased in both neoplastic and host stromal cells through hypoxia-responsive elements in the promoter region of the VEGF gene, which interact with hypoxia-inducible factor, HIF-1 (11, 12). Hypoxia can also increase VEGF expression by countering the intrinsic instability of VEGF mRNA through neutralizing destabilizing elements within its structure (7, 13).

The human VEGF gene gives rise to splice variants encoding proteins containing 115, 121, 145, 165, 189 and 206 amino acids. The predominant and most active of these mitogens is VEGF$_{165}$, which in its native form is a heparin-binding, disulphide-linked antiparallel homodimeric 45 kDA glycoprotein (7, 8, 14). Although VEGF$_{165}$ is excreted, significant amounts of protein remain bound within the extracellular matrix and to the cell surface which, upon proteolysis, can be released in a biologically active form.

Two high affinity VEGF receptors are expressed in vascular endothelial cells: VEGFR-1 (originally termed Flt-1: Fms-like tyrosine kinase) and VEGFR-2 (or human KDR: kinase insert domain-containing receptor, or murine Flk-1: fetal liver kinase-1). VEGFR-1 binds VEGF, VEGF-B and PlGF, whereas VEGFR-2 binds VEGF, VEGF-C, -D and –E. A third receptor, VEGFR-3 (Flt-4), expressed on lymphatic endothelial cells, upon binding either VEGF-C or –D, mediates lymphangiogenesis (9). Although VEGFR-1 and –2 are predominantly expressed on the vascular endothelium, VEGFR-1 is also expressed on trophoblast cells, renal cells and monocytes and VEGFR-2 on retinal cells, stem cells and megakaryocytes. Expression levels of VEGFR-1 and VEGFR-2 are normally low, but are greatly increased in situations involving enhanced vascular permeability and angiogenesis, with both receptors being upregulated by VEGF and VEGFR-2 being upregulated by bFGF (15). In angiogenesis, hypoxia plays an important role in receptor expression, with HIF-1 increasing VEGFR-1 and HIF-2 increasing VEGFR-2 gene transcription (11, 16).

The VEGF receptors are members of the receptor tyrosine kinase family and comprise of an extracellular ligand binding domain (containing seven immunoglobulin-like domains), a transmembrane domain and a cytoplasmic, split tyrosine kinase domain (7, 9). Although VEGFR-1 and VEGFR-2 bind differently, both involve one homodimeric VEGF molecule bridging two receptors, via predominantly hydrophobic interactions extending across two rather flat interfaces (≈ 820 Å^2) between the receptors and the "poles" of the ligand (17).

Upon binding, the tyrosine kinases of the dimerized VEGF receptors become activated, whereby they bind adenosine 5'-triphosphate (ATP; Figure 1) and catalyze the transfer of the γ-phosphate to the hydroxy-group of a tyrosine residue leading to receptor autophosphorylation. These cytoplasmic phosphorylated sites can then serve as binding sites for other substrates, which in turn may also be phosphorylated, leading to their activation. VEGF can thereby induce the phosphorylation of a number of proteins, with VEGFR-1 and VEGFR-2 inducing different signaling cascades. VEGFR-1 has been implicated in cell migration, however transgenic mice lacking the kinase domain of VEGFR-1 develop a normal vasculature and appear healthy, whereas embryos lacking VEGFR-2 fail to develop a vasculature (18, 19). The key role of VEGFR-2 in endothelial cell proliferation and angiogenesis, is further supported by the anti-angiogenic activities seen with antibodies directed against VEGFR-2 and the VEGF-VEGFR-2 complex, as well as with specific inhibitors of VEGFR-2 kinase (6). The prospects for anti angiogenic therapies based upon such agents is discussed below.

Structure of VEGFR-2 Kinase and Its Interaction with ATP

The VEGFR receptors are structurally and evolutionarily related to the platelet-derived growth factor receptor family: PDGFR-α, PDGFR-β, stem cell growth factor (c-Kit) and colony stimulating factor-1 (CSF-1R), which are characterized by the presence of a kinase insert domain, which is not involved in intrinsic catalytic activity but plays a role in signal transduction (*20*). In VEGFR-2, which has a total of 1354 residues, this hydrophilic and highly charged insert (Asn931 – Leu998) incorporates two tyrosine residues postulated to be phosphorylation sites (*21*). Two other autophosphorylation sites (Tyr1052 and Tyr1057) lie in a flexible activation loop (Asp1044 – Glu1073), the conformation of which probably regulates catalytic activity by opening access to the ATP-binding site following VEGF-binding and receptor dimerization. A crystal structure of the unligated catalytic domain, derived from a construct containing residues 804 – 937 and 988 – 1169, but lacking 50 residues from the insert domain, has been published (*22*). [The numbering of the VEGFR-2 sequence in this review is that of the sequence deposited in the SWISS-PROT database (entry P35968) and differs by -2 units from that of McTigue *et al.* (*22*)]. Like other protein kinases the protein folds into two lobes with the ATP-binding site and the catalytic loop, containing the invariable kinase aspartic acid residue (Asp1026), located in a cleft between the lobes. The ATP-binding site primarily consists of residues Glu915 – Asn921 and the glycine-rich loop of Leu838 – Ile851. The crystal structure of the binding site closely resembles that of the catalytic domain of FGFR-1 (*23*), with which it has 55% homology and since x-ray crystal structures of FGFR-1 ligated with ATP-competitive inhibitors are available, one of these (*24*) was used to construct models of the ATP-binding site of VEGFR-2 in complex with ATP and the inhibitors discussed in this review.

In the model (Plate 1), ATP is anchored to VEGFR-2 in the same way as in other kinases, by a highly conserved bidentate donor-acceptor H-bonding motif, with the N-1 of the adenine ring bonding to the backbone-NH of Cys917 and the NH_2 interacting with the carbonyl of Glu915. In addition, the adenine ring is sandwiched between two hydrophobic residues Ala864 and Leu1033, belonging respectively to the N- and C-terminal lobes; a third hydrophobic interaction is made up between the ribose ring and Val846, a residue of the glycine-rich loop.

VEGFR-2 Kinase Inhibitors

The nature of the interactions between VEGF and its receptors, as well as those between the phosporylated kinases and their substrates, make them unattractive targets for the design of small molecular weight antagonists. In

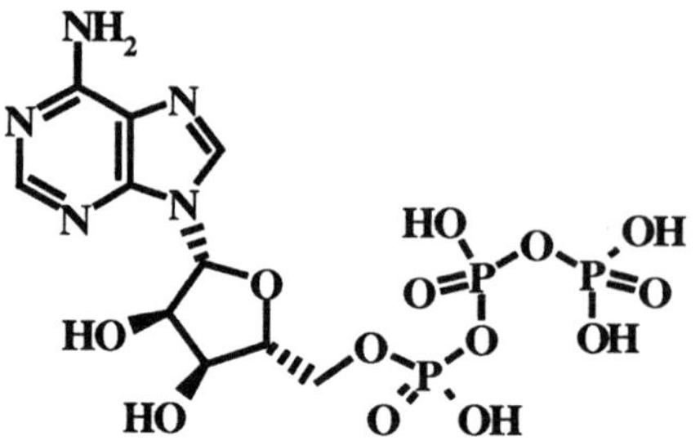

Adenosine 5'-triphosphate (ATP)

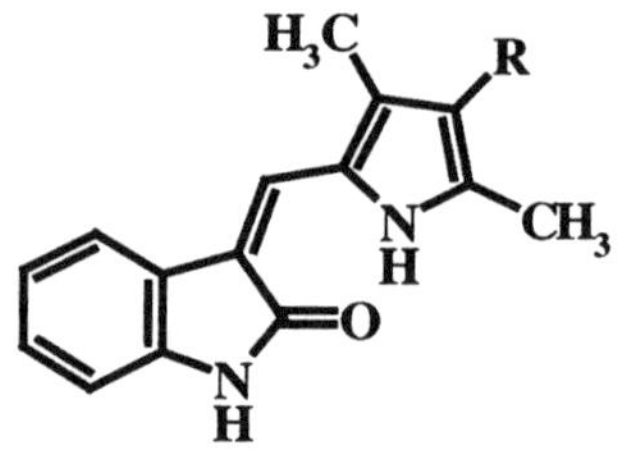

SU5416, R = H

SU6668, R = CH_2CH_2COOH

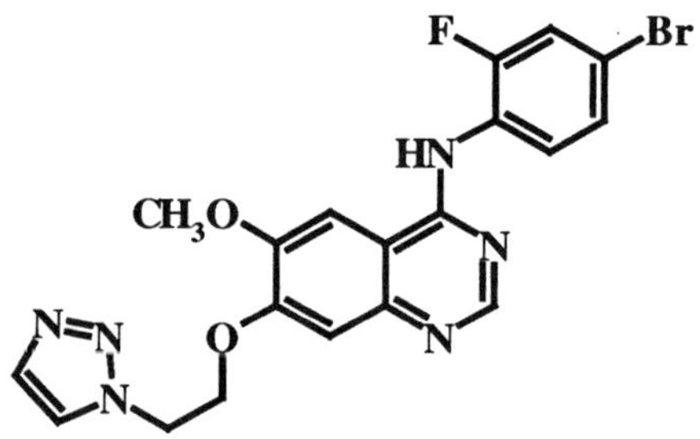

ZD4190

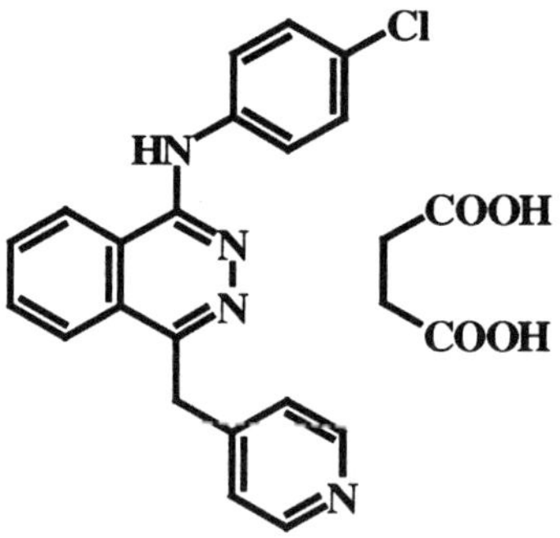

PTK787 / ZK222584

Figure I. Structures of ATP and some inhibitors of VEGFR-2 kinase.

contrast, the prospects for ATP-competitive kinase inhibitors are much more attractive. However, since kinase-catalyzed protein phosphorylation plays a key role in intracellular signaling and, with the human genome potentially encoding over 1000 such kinases, the design of selective, therapeutically useful inhibitors is a major challenge, which relies to a large extent on an understanding of the structural biology of the target enzymes. This is particularly important since the amino-acid residues involved in ATP-recognition (adenine binding, ribose and triphosphate recognition) are highly conserved throughout the kinases, such that proximal amino-acids and the topology of the protein surface in the vicinity of the ATP-binding site become critical for drug selectivity.

The potency and selectivity profiles of putative kinase inhibitors is typically evaluated by incubating compounds with enzyme in the presence of

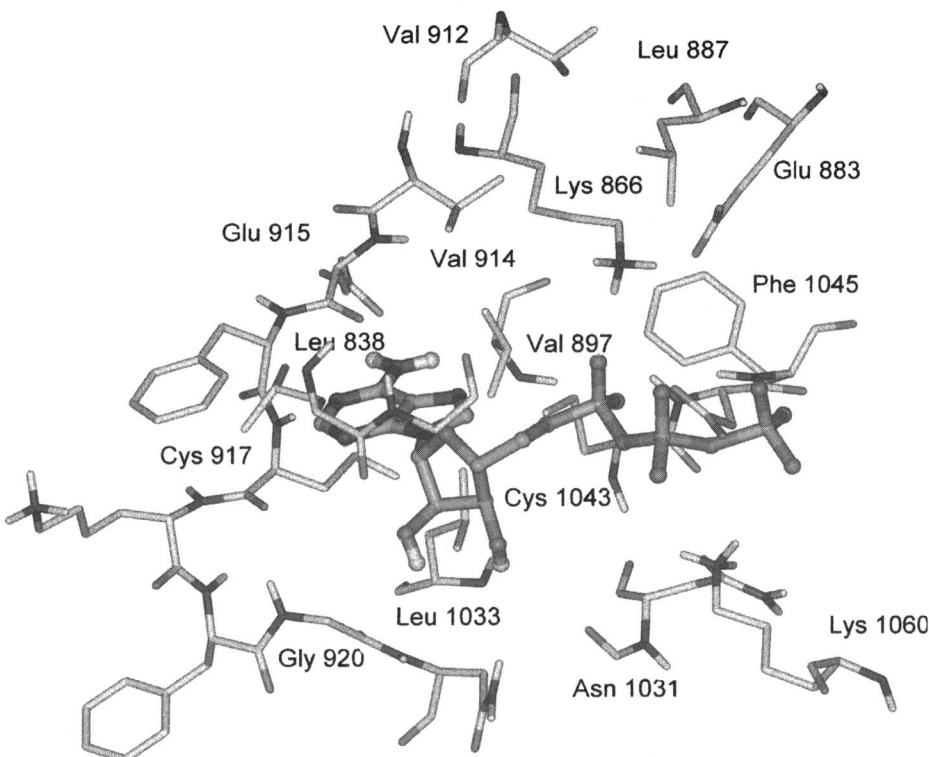

Plate 1. Representation of ATP docked in its binding site in VEGFR-2. (For reasons of clarity, residues Ala864 and Val846 are not depicted since these lie directly above the adenine and ribose rings respectively).
(Color figure can be found in color insert.)

[^{33}P]-ATP and an artificial substrate, using optimized buffer and salt conditions. Phosphorylated tyrosine on the substrate is then detected by means of a β-scintillation counter. Differences in assay conditions, make it difficult to compare enzyme assays between laboratories and consequently Table I compares data generated in the same assay in parallel for the four compounds discussed below; some published data is also included for comparison.

Table I. Inhibitory profiles of VEGFR-kinase inhibitors. [a]

KINASE [b]	ZD4190	SU5416	SU6668	PTK787
VEGFR-1 (Flt-1)	430 ± 230 (708 ± 63) [c]	43 ± 11	15 ± 3	110 ± 27
VEGFR-2 (KDR, Flk-1)	200 ± 19 (29 ± 4) [c]	220 ± 34 (700) [d]	200 ± 15 (2430) [d]	42 ± 3
PDGFR-ß	1900 ± 850	2220 ± 1500 (10500) [d]	39 ± 1 (60) [d]	490 ± 52
c-Kit	810 ± 290	660 ± 165	750 ± 120	620 ± 56
CSF-1R (c-Fms)	3500 ± 870	84 ± 4	45 ± 2	1240 ± 180
EGFR (HER-1, erbB)	69 ± 58	> 10000	> 10000	> 10000
FGFR-1	5800 ± 2100 (5300 ± 1700) [c]	> 10000 (7080) [d]	> 10000 (3040) [d]	> 10000
CDK-1	> 10000	> 10000	> 10000	> 10000
Tie-2 (Tek)	1300 ± 490	> 10000	> 10000	> 10000
c-Met	7600 ± 300	> 10000	1700 ± 520	> 10000
IGF-1R	> 10'000	> 10000	4000 ± 420	> 10000
c-Src	380 ± 95	> 10000	> 10000	> 10000
c-Abl	280 ± 25	> 10000	3600 ± 1100	> 10000

[a] Data represent the mean ± SEM (n ≥ 3) drug concentrations required to inhibit enzyme activity by 50% (IC_{50} value; nM) at the following ATP concentrations 1.0 µM (c-Kit, CSF-1R, c-Met), 2.0 µM (HER-1), 5.0 µM (c-Abl), 7.5 µM (CDK-1), 8.0 µM (VEGFR-1, KDR, FGFR-1, Tie-2), 10.0 µM (PDGFR-β), 20.0 µM (c-Src), 30.0 µM (IGF-1R), and, unless otherwise indicated, are data generated under the same assay conditions; [b] alternative nomenclature for these kinases is given in parenthesis; [c] published data from reference (28); [d] published data from reference (30).

Quinazolines ZD4190 and ZD6474 (AstraZeneca)

ZD4190 (Figure 1) is an ATP-competitive kinase inhibitor, structurally based on the 4-aminoquinazoline scaffold (25), which has been extensively exploited in the design of EGFR kinase inhibitors (26). Lipophilic *meta-* and bulky *ortho*-substituents in the aniline moiety are optimal for FGFR-1 inhibition, but poorly tolerated by VEGFR-1 and –2, allowing manipulation of these substituents to optimise potency and selectivity towards VEGFR-2. The quinazoline 6- and 7-substituents were also optimized for potency and the triazole-bearing 7-substituent introduced excellent pharmacokinetic properties. However, ZD4190 is not selective for the PDGFR-kinase family, since it also inhibits c-Src, v-Abl and in particular EGFR-kinase (Table 1).

Molecular modelling can help to rationalize this selectivity: The putative binding mode of ZD4190 in our model of the ATP-binding site of VEGFR-2 (Plate 2) is based upon the assumption that ZD4190 adopts the same orientation as seen for a 4-aminoquinazoline analogue bound to cyclin dependent kinase-2 (27). Like the adenine-N1 atom of ATP, the quinazoline-N1 atom accepts an H-bond from the backbone-NH of Cys917, while the aniline moiety projects into a hydrophobic pocket, formed by residues Val914, Val912, Val897, Leu887, Cys1043, Phe1045 and the hydrocarbon part of the side-chain of Lys866, which is not exploited by ATP. The triazole side-chain extends towards solvent, via a hydrophobic channel constituted from residues Leu838 and Gly920. Thus, the lack of selectivity of ZD4190 over EGFR may be attributed to the residues forming the hydrophobic pocket being similar in both kinases (the equivalent residues to Val914, Val912, Val897, Leu887 and Cys1043 in EGFR are Thr, Leu, Cys, Met, Thr respectively, while Phe1045 and Lys866 are conserved).

Consistent with its inhibition of VEGFR-2, ZD4190 inhibits the VEGF-stimulated growth of human umbilical vein endothelial cells (HUVECs) *in vitro* (IC_{50} 50 nM). Following oral administration to mice (100 mg/kg), plasma levels were 13 µM after 24h, compatible with a once-daily oral dosing (25, 28) and accordingly, in human xenograft models in mice, ZD4190 dose-dependently inhibited the growth of established tumors, with good efficacy being reached at 100 mg/kg/day. It is difficult to assess the role of VEGFR-2 inhibition in this cytostatic activity, since this is also consistent with EGFR inhibition, however, it has been shown that acute treatment of mice bearing PC-3 prostate tumors, reduced vascular permeability within the xenografts, probably due to the inhibition of VEGF signaling.

ZD6474 has replaced ZD4190 in development and is a promising clinical candidate. This compound potently inhibits VEGFR-2 (IC_{50} 40 nM), displays high selectivity over VEGFR-1 (IC_{50} 1.6 µM), EGFR (IC_{50} 0.5 µM) and FGFR-1 (IC_{50} 3.6 µM), and possesses a similar selectivity profile to that of ZD4190 against growth factor-stimulated endothelial cell proliferation (29). Following once-daily chronic treatment, at well tolerated doses (12.5 to 100 mg/kg), ZD6474 shows broad spectrum tumoristatic activity in human

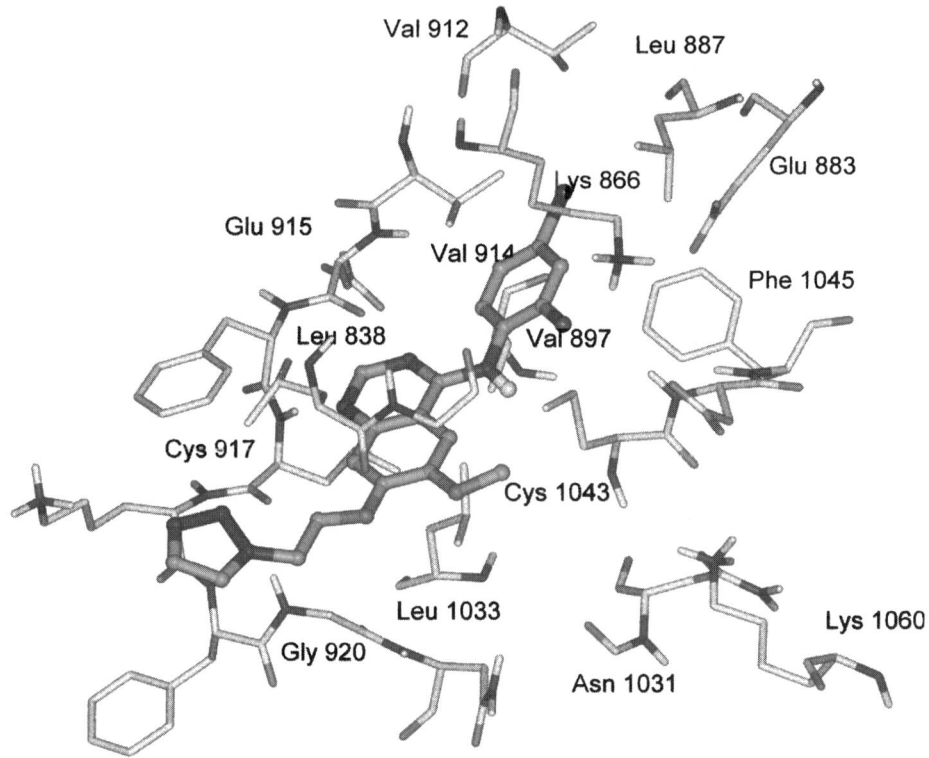

Plate 2. Portrayal of ZD4190 bound to the VEGFR-2 ATP-binding site. (Color figure can be found in color insert.)

tumor xenograft models, with efficacy being independent of tumor size at the start of the experiment.

Indolinones SU5416 and SU6668 (Pharmacia; formerly Sugen).

Workers at Sugen identified the indolin-2-one structure as a central pharmacophore for the design of protein kinase inhibitors and SU5416 and SU6668 (Figure 1) emerged from this work as anti-angiogenic agents (30). With the exception of PDGFR-β, which is potently inhibited by SU6668 but not by SU5416, both indolinones possess similar potency and selectivity for the PDGFR-family of tyrosine kinases (Table 1). Notably, both compounds are more potent inhibitors of VEGFR-1 than of VEGFR-2. Much of the observed selectivity can be rationalized using models of SU5416 and SU6668 bound within the ATP-cleft of the kinases, as represented in Plate 3 for SU6668 ligated to VEGFR-2. These binding models have been derived from the structural information provided by the crystal structures of two related indolinone-type tyrosine kinase inhibitors in complex with the catalytic domain of FGFR-1 (31). The lactam amide group of the inhibitor forms the same bidentate donor-acceptor H-bonds with the backbone of Cys917 and Glu915 as the adenine-N1 and 6-amino group of ATP. In terms of lipophilic interactions, the inhibitors do not interact as extensively as ZD4190 with the previously described hydrophobic pocket, with contacts only being made with three residues, namely Val897, Val914 and Cys1043. However, this does not appear to translate into less potency against the VEGFR kinases, since in comparison to ZD4190 potency against VEGFR-1 is actually increased (Table 1). The propionic acid moiety of SU6668, like the triazole of ZD4190, reaches the solvent with the pyrrole-ring inserting into the channel formed by Leu838 and Gly920. In accordance with this orientation, the enhanced potency of SU6668 against PDGFR-β has been rationalized on the basis of the propionic acid interacting with Arg604, located on the N-lobe at the entrance to the ATP-binding site (32); however in VEGFR-2 the corresponding residue is Lys836 which would not normally be expected to give rise to major differences in binding. The lack of activity of SU5416 and SU6668 towards the EGFR kinase may be explained on the grounds that the Val897, Val914 and Cys1043 triad in the VEGFR kinases, correspond to the less hydrophobic residues Cys, Thr and Thr in the EGFR kinase and these are probably less effective in forming hydrophobic interactions with the proximal indolinone-phenyl ring. (This is not the case for ZD4190 since the lipophilic aniline-ring extends much further into the lipophilic pocket so the these residues play less of a role in determining selectivity).

As anticipated based upon their ability to inhibit kinases, SU5416 (IC$_{50}$ 60 nM) and SU6668 (IC$_{50}$ 410 nM) potently inhibit VEGF-induced mitogenesis in HUVECs and have little activity against FGF-induced proliferation (32 - 34). However, despite inhibiting PDGFR-β, SU6668 only

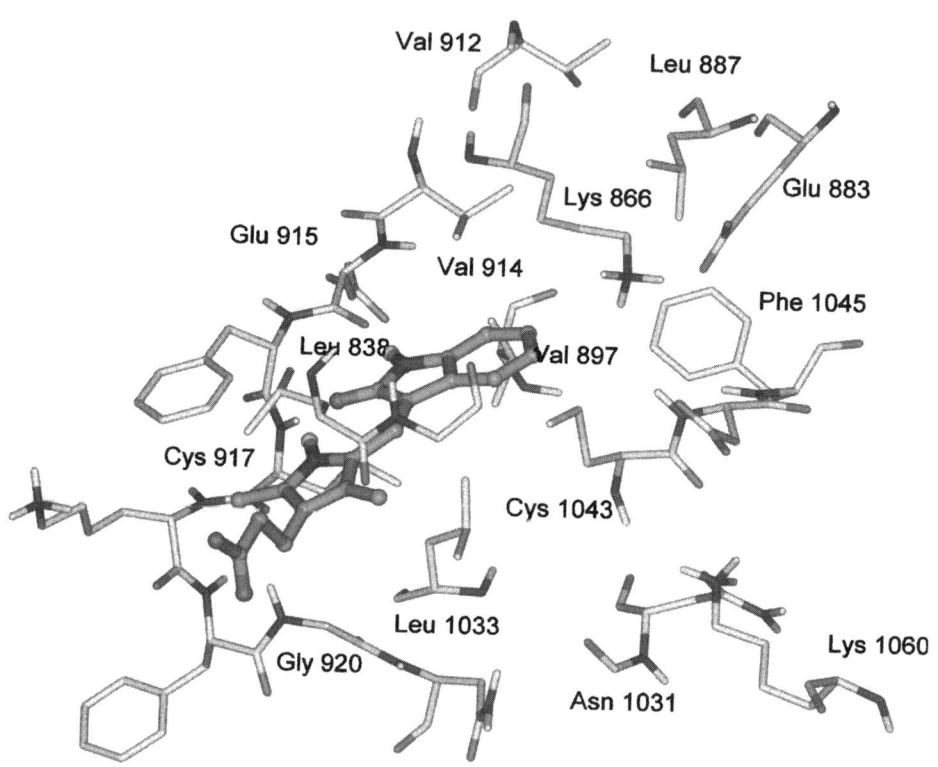

Plate 3. Portrayal of SU6668 docked in the VEGFR-2 ATP-binding site. (Color figure can be found in color insert.)

inhibits PDGF-induced mitogenesis of NIH-3T3 fibroblasts weakly (IC_{50} 16.5 µM) and to a similar extent to that of SU5416 (IC_{50} 4.54 µM).

Both SU5416 (25 mg/kg/day i.p.) and SU6668 (75 – 200 mg/kg/day i.p. or p.o.) showed similar inhibitory effects on the growth of most types of subcutaneous tumor xenografts in athymic mice (*32 - 34*). SU6668, in contrast to SU5416, also inhibited the growth of SF763T glioblastoma and SKOV3TP5 ovarian carcinoma xenografts, which may be a result of its different *in vitro* profile (Table 1). Direct evidence for the anti-angiogenic activity of these compounds comes from multifluorescence videomicroscopy studies, where in addition to inhibiting the growth of subcutaneously implanted glioma cells, both compounds reduced the permeability and vascular density of the newly formed tumor vasculature (*32, 35*). In more challenging models, both SU5416 and SU6668 have been shown to inhibit the growth of established tumors and in some cases induce tumor regression, possibly as a result of the tumors becoming necrotic (*32, 34*).

The potential of these anti-angiogenic drugs is further supported by encouraging findings in clinical studies in cancer patients, where intraperitoneal infusion of SU5416 has shown some objective responses at doses which were well tolerated and produced manageable mild-to-moderate side-effects (*34*).

Phthalazine PTK787 / ZK222584 (Novartis / Schering co-development)

A third structural class of VEGFR-kinase inhibitors is represented by the phthalazine PTK787 (ZK222584; Figure 1). PTK787 potently inhibits recombinant VEGFR kinases and is the most selective VEGFR kinase inhibitor described (Table 1). This profile can be rationalized (*36*) on the basis of the putative binding mode of PTK787 to the ATP-binding site of VEGFR-2 (Plate 4). Unlike ATP (Plate 1) and many reported kinase inhibitors, it does not form direct hydrogen bonds with the peptide backbone of the hinge region (Cys917 and Glu915 in the case of VEGFR-2), but occupies hydrophobic regions within the binding site. The chlorophenyl moiety extends into the same hydrophobic pocket as the anilino moiety of ZD4190, although less deeply, while the phthalazine bicycle makes hydrophobic contacts with Leu1033, Gly920 and Leu838. Although no direct H-bonds with the hinge region are possible in this proposed orientation, the anilino-NH of the inhibitor is located at distances from the backbone of Glu915 and Cys917, which could allow water-mediated H-bonds to be formed. The essential pyridyl-N of the inhibitor is believed to form an H-bond with the amino group Lys1060 located on the activation loop of the kinase and, since Lys1060 is only conserved within tyrosine kinases of the PDGF family this interaction may contribute to the selective recognition of the compound by the kinases of this family. However, based on the model, an alternative H-bond cannot be completely ruled out involving the proximal Asn1031 or a residue of the glycine rich loop which

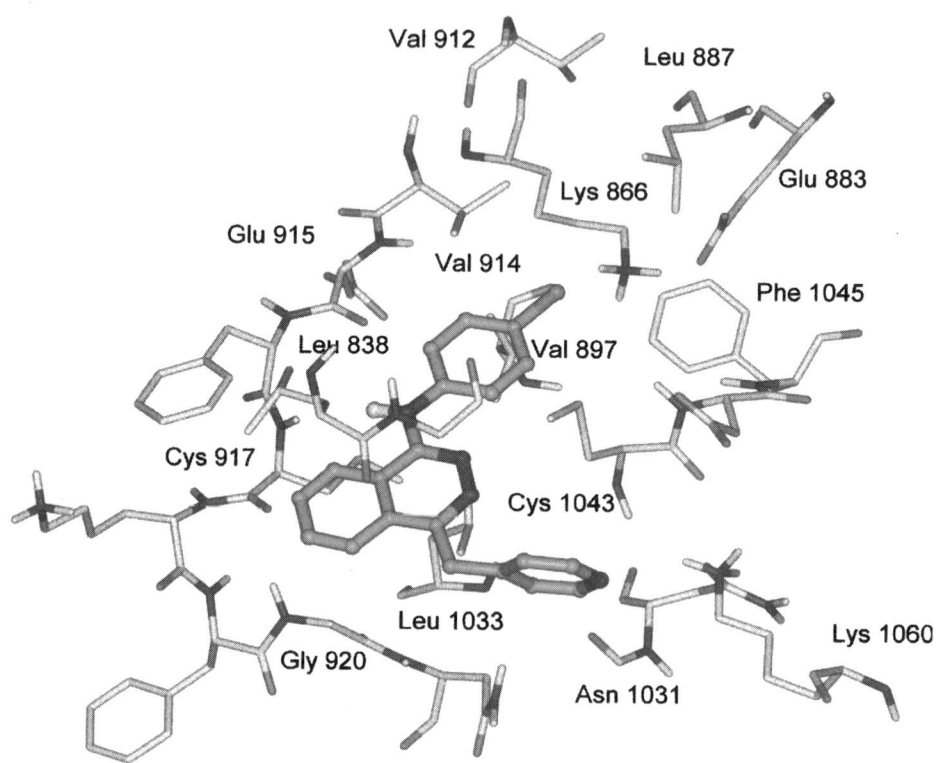

Plate 4. Portrayal of PTK787 docked in the VEGFR-2 ATP-binding site. (Color figure can be found in color insert.)

may be prone to conformational rearrangement (*31*). The lack of activity of PTK787 against the EGFR kinase can be rationalized on the same grounds as that discussed above for the indolinones.

PTK787 readily penetrates cells and inhibits the autophosphorylation of VEGFR-2 in both HUVECs and KDR-transfected chinese hamster ovary cells (*37*). It possesses good functional activity in cellular systems, inhibiting VEGF-mediated cell proliferation (IC_{50} 16 nM), cell survival and cell migration (IC_{50} 58 nM) in HUVECs. As expected for a selective VEGF inhibitor, it neither inhibited FGF-stimulated proliferation of HUVECs, nor the proliferation of cells not expressing VEGF receptors. Anti-angiogenic activity has been demonstrated *in vitro* through inhibition of growth factor-stimulated capillary sprouting from rat aorta (IC_{50} 675 nM).

Following oral administration to mice (50 mg/kg), PTK787 is rapidly absorbed giving peak plasma concentrations of about 30 µM, with concentrations remaining above 1 µM at 8h (*37*). In accordance with this pharmacokinetic profile, the compound displays good efficacy by the oral route. Thus, in a growth factor implant model, PTK787 (12.5 – 50 mg/kg p.o.) inhibited VEGF-induced vascularization of a subcutaneous implant in mice, quantified by measurement of the weight and blood content of the vascularized tissue surrounding the implant. In a model of malignant pleural effusion, a complication of advanced lung cancer, consistent with inhibiting VEGF-mediated events, PTK787 (50 mg/kg/day, p.o.) suppressed tumor vascularization and markedly reduced the volume of pleural effusate in nude mice bearing lung tumors produced by human PC14PE6 adenocarcinoma cells (*38*). In addition, pretreatment of nude mice with PTK787 (50 mg/kg/day, p.o.) inhibited skin vascular permeability induced by either human recombinant $VEGF_{165}$ or by pleural effusate harvested from PE14PE6-treated mice.

Consistent with its anti-angiogenic activity, PTK787 possesses good efficacy in inhibiting the growth of VEGF-producing human carcinomas in a number of murine xenograft models, with CWR-22 human prostate carcinoma being completely inhibited in some mice (*37*). In the case of A431 epithelial carcinoma, as well as inhibiting tumor growth PTK787 (50 mg/kg p.o.) reduced the occurrence of microvessels and induced necrosis within the tumors. PTK787 (50 mg/kg/day p.o.) has also been shown to have anti-metastatic activity in an orthotopic model of murine renal cell carcinoma, where both tumor growth (34%) and the number of metastases (65%) were inhibited in comparison with control animals.

In terms of safety, PTK787 is well tolerated in animals and does not impair wound healing, as has been speculated for anti-angiogenic agents, nor does it have significant effects on leukocyte levels as might result from c-kit inhibition (*37*). On the basis of its promising pre-clinical profile, PTK787 / ZK222584 was selected for clinical evaluation.

Prospects for the design of improved second generation VEGFR inhibitors as anti-angiogenic agents.

In this brief review, we have sought to illustrate the value of structural biology and molecular modelling in the design and rationalization of the activity and selectivity of VEGFR kinase inhibitors. Protein kinases have been optimized via evolution to recognize a signal and subsequently activate a phosphorylation process. In utilizing ATP for phosphorylation, it is not stongly bound and, although residues involved in ATP-binding are highly conserved, the amino-acid residues in close proximity to this binding site are not. Therefore, the future pospects for the rational design of inhibitors to target these proximal regions and impart kinase selectivity holds great promise and will improve as more crystal structures of kinases ligated to ATP-competitive inhibitors become available.

Given the knowledge of the structural biology of VEGF and its receptor signaling pathway, the design principles for VEGFR kinase inhibition and the understanding obtained in patients treated with first generation therapeutic agents, prospects are good for the design of relatively safe, orally active drugs capable of completely abrogating the pro-angiogenic activity of VEGF in cancer. However, although VEGF is amongst the most potent angiogenic factors, it is not the only cytokine to be upregulated in angiogenesis and several such cytokines are sometimes produced by the same tumor. In addition, by blocking one angiogenic pathway, the resulting tumor hypoxia may serve to upregulate other pathways. Consequently, anti-angiogenic therapy directed against a single cytokine might be insufficient and for optimal therapy the clinical oncologist might require a palette of anti-angiogenic drugs.

References

1. Carmeliet, P.; Jain, R. K. *Nature* **2000**, *407*, 249-257.
2. Kerbel, R. S. *Carcinogenesis* **2000**, *21*, 505-515.
3. Folkman, J. *Nature Med.* **1995**, *1*, 27-31.
4. Raleigh, J. A.; Calkins-Adams, D. P.; Rinker, L. H.; Ballenger, C. A.; Weissler, M. C.; Fowler, W. C.; Novotny, D. B.; Varia, M. A. *Cancer Res.* **1998**, *58*, 3765-3768.
5. Weidner, N. *Am. J. Pathol.* **1995**, *147*, 9-19.
6. Yancopoulos, G. D.; Davis, S.; Gale, N. W.; Rudge, J.S.; Wiegand, S.J.; Holash, J. *Nature* **2000**, *407*, 242-248.
7. Neufeld, G.; Cohen, T.; Gengrinovitch, S.; Poltorak, K. *FASEB J.* **1999**, *13*, 9-22.
8. Veikkola, T; Alitalo, K. *Semin. Cancer Biol.* **1999**, *9*, 211-220.
9. Veikkola, T.; Karkkainen, M.; Claesson-Welsh, L.; Alitalo, K. *Cancer Res.* **2000**, *60*, 203-212.

10. Ravi, R.; Mookerjee, B.; Bhujwalla, Z. M.; Sutter, C. H.; Artemov, D.; Zeng, Q.; Dillehay, L. E.; Madan, A.; Semenza, G. L.; Bedi, A. *Genes Dev.* **2000**, *14*, 34-44.
11. Marti, H. J. H.; Bernaudin, M.; Bellail, A.; Schoch, H.; Euler, M.; Petit, E.; Risau, W. *Am. J. Pathol.* **2000**, *156*, 965-976.
12. Semenza, G. L. *Biochem. Pharmacol.* **2000**, *59*, 47-53.
13. Dibbens, J. A.; Miller, D. L.; Damert, A.; Risau, W.; Vadas, M. A.; Goodall, G. J. *Mol. Biol. Cell* **1999**, *10*, 907-919.
14. Muller, Y. A.; Christinger, H. W.; Keyt, B. A.; de Vos, A. M. *Structure* **1997**, *5*, 1325-1338.
15. Hata, S.; Rook, S. L.; Aiello, L.P. *Diabetes* **1999**, *48*, 1145-1155.
16. Kappel, A.; Ronicke, V.; Damert, A.; Flamme, I.; Risau, W.; Breier, G. *Blood* **1999**, *93*, 4284-4292.
17. Weismann, C.; Fuh, G.; Christinger, H.W.; Eigenbrot, C.; de Vos, A.M. *Cell* **1997**, *91*, 695-704.
18. Shalaby, F.; Rossant, J.; Yamaguchi, T. P.; Gertsenstein, M.; Wu, X. F.; Breitman, M. L.; Schuh, A. C. *Nature* **1995**, *376*, 62–66.
19. Hiratsuka, S.; Minowa, O.; Kuno, J.; Noda, T.; Shibuya, M. *Proc. Natl. Acad. Sci. USA* **1998**, *95*, 9349-9354.
20. Hanks, S. K.; Quinn, A. M. *Methods Enzymol.* **1991**, *200*, 38-62.
21. Dougher-Vermazen, M.; Hulmes, J. D.; Boehlen, P.; Terman, B. I. *Biochem. Biophys. Res. Commun.* **1994**, *205*, 728-738.
22. McTigue, M. A.; Wickersham, J. A.; Pinko, C.; Showalter, R. E.; Parast, C. V.; Tempczyk-Russell, A.; Gehring, M. R.; Mroczkowski, B.; Kan, C.-C.; Villafranca, J. E.; Appelt, K. *Structure* **1999**, *7*, 319-330.
23. Mohammadi, M.; Schlessinger, J.; Hubbard, S.R. *Cell* **1996**, *86*, 577-587.
24. Mohammadi, M.; Froum, S.; Hamby, J. M.; Schroeder, M. C.; Panek, R. L.; Lu, G. H.; Eliseenkova A. V.; Green, D.; Schlessinger, J.; Hubbard, S. R. *EMBO J.* **1998**, *17*, 5896-5904.
25. Hennequin, L. F.; Thomas, A. P.; Johnstone, C.; Stokes, E. S. E.; Plé, P. A.; Lohmann, J.-J. M.; Ogilvie, D. J.; Dukes, M.; Wedge, S. R.; Curwen, J. O.; Kendrew, J.; Lambert-van der Brempt, C. *J. Med. Chem.* **1999**, *42*, 5369-5389.
26. Traxler, P. *Exp. Opin. Ther. Patents* **1998**, *8*, 1599-1625.
27. Shewchuk, L.; Hassel, A.; Wisely, B.; Rocque, W.; Holmes, W.; Veal, J.; Kuyper, L. F. *J. Med. Chem.* **2000**, *43*, 133-138.
28. Wedge, S. R.; Oglivie, D. J.; Dukes, M.; Kendrew, J.; Curwen, J. O.; Hennequin, L. F.; Thomas, A. P.; Stokes, E. S. E.; Curry, B.; Richmond, G. H. P.; Wadsworth, P. F. *Cancer Res.* **2000**, *60*, 970-975.
29. Wedge, S. R.; Oglivie, D. J.; Dukes, M.; Kendrew, J.; Hennequin, L. F.; Stokes, E. S. E.; Curry, B. 91st Annual Meeting of the American Association of Cancer Research, San Francisco, CA, March 2000; Poster 3600.

30. Sun, L.; Tran, N.; Liang, C.; Tang, F.; Rice, A.; Schreck, R.; Waltz, K.; Shawver, L. K.; McMahon, G.; Tang, C. *J. Med. Chem.* **1999**, *42*, 5120-5130.
31. Mohammadi, M.; McMahon, G.; Sun, L.; Tang, C.; Hirth, P.; Yeh, B. K.; Hubbard, S. R.; Schlessinger, J. *Science* **1997**, *276*, 955-960.
32. Laird, A. D.; Vajkoczy, P.; Shawver, L. K.; Thurnher, A.; Liang C.; Mohammedi, M.; Schlessinger, J.; Ullrich, A.; Hubbard, S. V.; Blake, R. A.; Fong, T. A. T.; Strawn, L. M.; Sun, L.; Tang, C.; Hawtin, R.; Tang, F.; Shenoy, N.; Hirth, K. P.; McMahon, G.; Cherrington, J. M. *Cancer Res.* **2000**, *60*, 4152-4160.
33. Fong, T. A. T.; Shawver, L. K.; Sun, L.; Tang, C.; App, H.; Powell, T. J.; Kim, Y. H.; Schreck, R.; Wang, X.; Risau, W.; Ullrich, A.; Hirth, K. P.; McMahon, G. *Cancer Res.* **1999**, *59*, 99-106.
34. Mendel, D. B.; Laird, D.; Smolich, B. D.; Blake, R.A.; Liang, C.; Hannah, A. L.; Shaheen, R. M.; Ellis, L. M.; Weitman, S.; Shawver, L. K.; Cherrington, J. M. *Anti-Cancer Drug Design* **2000**, *15*, 29-41.
35. Vajkoczy, P.; Menger, M. D.; Vollmar, B.; Schilling, L.; Schmiedek, P.; Hirth, K. P.; Ullrich, A.; Fong, T. A. T. *Neoplasia* **1999**, *1*, 31-41.
36. Bold, G.; Altmann, K.-H.; Frei, J.; Lang, M.; Manley, P. W.; Traxler, P.; Wietfeld, B.; Buchdunger, E.; Cozens, R.; Ferrari, S.; Furet, P.; Hofmann, F.; Martiny-Baron, G.; Mestan, J.; Rösel, J.; Sills, M.; Stover, D.; Acemoglu, F.; Boss, E.; Emmenegger, R.; Lässer, L.; Masso, E.; Roth, R.; Schlachter, C.; Vetterli, W.; Wyss, D.; Wood, J. M. *J. Med. Chem.* **2000**, *43*, 2310-2323.
37. Wood, J. M.; Bold, G.; Buchdunger, E.; Cozens, R.; Ferrari, S.; Frei, J.; Hofmann, F.; Mestan, J.; Mett, H.; O'Reilly, T.; Persohn, E.; Rösel, J.; Schnell, C.; M.; Stover, D.; Theuer, A.; Towbin, H.; Wenger, W.; Woods-Cook, K.; Menrad, A. Siemeister, M.; Schirner, M.; Thierauch, K.-H.; Schneider, M. R.; Drevs, J.; Martiny-Baron, G.; Totzke, F.; Marmé, D. *Cancer Res.* **2000**, *60*, 2178-2189.
38. Yano, S.; Herbst, R. S.; Shinohara, H.; Knighton, B.; Bucana, C. D.; Killion, J. J.; Wood, J.; Fidler, I. J. *Clin. Cancer Res.* **2000**, *6*, 957-965.

Chapter 18

Carbohydrate-Based Tumor Antigens as Antitumor Vaccine Agents

Jennifer R. Allen[1] and Samuel J. Danishefsky[1,2]

[1]The Laboratory of Bioorganic Chemistry, Sloan-Kettering Institute for Cancer Research, 1275 York Avenue, New York, NY 10021
[2]Department of Chemistry, Columbia University, Havemeyer Hall, New York, NY 10027

Abstract. Previous efforts have shown that synthetically derived tumor-associated antigens and glycoconjugates can be used as targets for cancer immunotherapy. Synthetically derived antigens offer many advantages, including the preparation of rigorously pure material and the production of sufficient material to contemplate pre-clinical and clinical studies. The task of achieving such quantities for clinical investigation, however, requires a highly convergent and reliable synthetic route. In this report, we relate the synthesis of the *n*-pentenyl glycosides (NPG) of the tumor-associated antigens Globo-H and Fucosyl GM1 using a highly convergent synthetic protocol. The elaboration of the NPG glycoside to form protein glycoconjugates and amino acid derivatives is also described.

An important goal of contemporary immunotherapy is that of stimulating an immune response in a tumor-bearing host through the use of a tumor-associated antigen. It has been known for some time that specific types of glycolipids or glycoproteins, which may be detectable in normal tissue types by immunohistology, may be significantly over expressed in tumors of that tissue [1]. Furthermore, high levels of expression on tumor cells can often provoke an antibody response. However, this initial antibody response is apparently ineffective in providing immunoprotection or immunorejection and disease can begin and progress. Possibly, tumor cells are capable of sending a variety of decoys, which disguise their initial growth and allow for the tumor antigens to be seen as "self" by the immune system. Accordingly, the idea of using synthetically derived, cell free glycoconjugates as versions of tumor-associated immunostimmulatory antigens in the development of anti-tumor vaccines to break the tolerance of the immune response is attractive [2]. Investigations using carbohydrate-based vaccines have thus far been conducted with the hope that patients immunized in an adjuvant setting might produce antibodies reactive with cancer cells, and that the production of such antibodies might mitigate against tumor spread, thereby creating a more favorable prognosis [3].

Cancer carbohydrate antigens such as Tn, TF, GD_3, GM_2, KH-1, Le^y, fucosyl GM_1 and Globo-H are suitable targets for both active and passive immunotherapies because they are over-expressed at the surface of malignant cells in a variety of cancers [4]. Obviously, such discoveries can only follow from painstaking structural analysis. Moreover, due to major advances in the areas of monoclonal antibody (mAb) technology and immunohistology, the antigens have been immunocharacterized by suitable mAb's and therefore have relevant serological markers available for immunological studies. However, the required delicate, yet complex, characterizations are even more impressive when it is considered that the availabilities of such carbohydrates is typically limited to sub-milligram levels following isolation and purification of specimens from human cancer tissue collections. Consequently, in the absence of an enzymatically mediated synthesis, the task of delivery of carbohydrate-based antigens in amounts required to study immunotherapy at the pre-clinical and clinical level falls to organic synthesis.

An additional challenge to cancer vaccine strategies is the successful delivery of the antigen in a favorable molecular context for eliciting a therapeutically useful immunological response [5]. Here again, chemical synthesis can play a pivotal role. A variety of approaches have been adopted to elicit a more sustained immune response beyond that triggered by exposure to native synthetic antigens. These attempts include chemical modification of carbohydrates, additional administration of immunological adjuvants and covalent attachment of carbohydrates to immunogenic protein carriers [6]. More specifically, studies have shown that a coherent vaccine can be achieved through covalent bioconjugation to carrier protein keyhole limpet hemocyanin (KLH) [7].

Our laboratory has been interested in the development of anti-cancer vaccines using vaccine constructs prepared by total chemical synthesis for some time [8]. For example, a KLH construct containing the MBr1 antigen globo-H, **1**, is one of our most clinically promising anti-cancer vaccines (Figure 1). It has been demonstrated in phase I clinical trials that the globo-H based vaccine, in concert with a non-synthetic immunoadjuvant (i.e. QS-21), is safe in humans as regards to autoimmunity problems and induces specific antibodies against tumor cells carrying on their cell surface the same antigenic structure contained in the vaccine. As a result, this glycoconjugate has now advanced to phase II and phase III human clinical trials. In addition, we have synthesized vaccine glycoconjugates containing the Lewis[y] antigen, and the α-O-linked antigens Tn and TF. Pre-clinical and human clinical studies with these vaccines, as well as those of other laboratories [9], suggest that the use of cancer vaccines containing carbohydrate antigens of fully synthetic origin to induce an anti-cancer immune response is in fact showing considerable promise for eventual therapy.

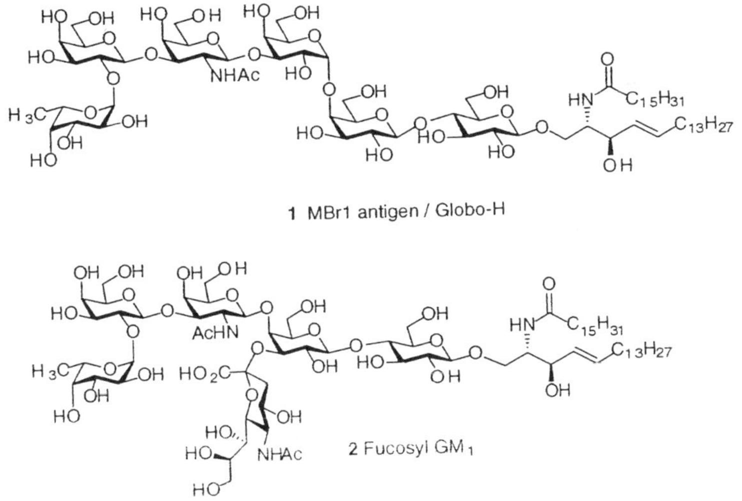

Figure 1. Tumor-associated antigens Globo-H and Fucosyl GM_1.

In this paper, we relate a summary of our synthetic efforts involving n-pentenyl glycosides (NPGs) of complex oligosaccharide tumor-associated antigens. These studies began with the synthesis of fucosyl GM_1 as its n-pentenyl glycoside and were extended to include a second-generation synthesis of the MBr1 antigen, globo-H. Lastly, the incorporation of n-pentenyl glycosides into glycopeptides is a recent advance in our laboratory which we briefly describe herein.

Synthesis of Fucosyl GM_1 NPG

The introduction of NPG's into our synthetic arsenal originated as a solution to a problem encountered during the synthesis of a bioconjugatable precursor of fucosyl GM_1 [10]. We became interested in the fucosylated GM_1 ganglioside, **2**, for a number of reasons. Nilsson *et al.* identified fucosyl GM_1 as a specific marker associated with small cell lung cancer (SCLC) cells [11]. Furthermore, monoclonal antibodies (F12) to the antigen serve to detect fucosyl GM_1 in tissues and serum of SCLC patients [12]. Immunohistochemistry studies have suggested that, due to its highly restricted distribution in normal tissues, fucosyl GM_1 could be an excellent target for immune attack against SCLC [13]. Thus, in anticipation of building an effective anti-tumor vaccine, our program was directed to the preparation of the antigen so functionalized as to secure the need for its conjugation to carrier protein.

Initial synthetic efforts were aimed at the hexasaccharide contained in fucosyl GM$_1$ as its corresponding glycal. A [3+3] coupling of donor **3** and glycal acceptor **4** produced the hexasaccharide **5**, albeit in low yield, as shown in Scheme 1. However, we were unsettled to find that under no conditions were we able to accomplish the removal of the protecting groups contained in **5** to achieve peracetate **6**.

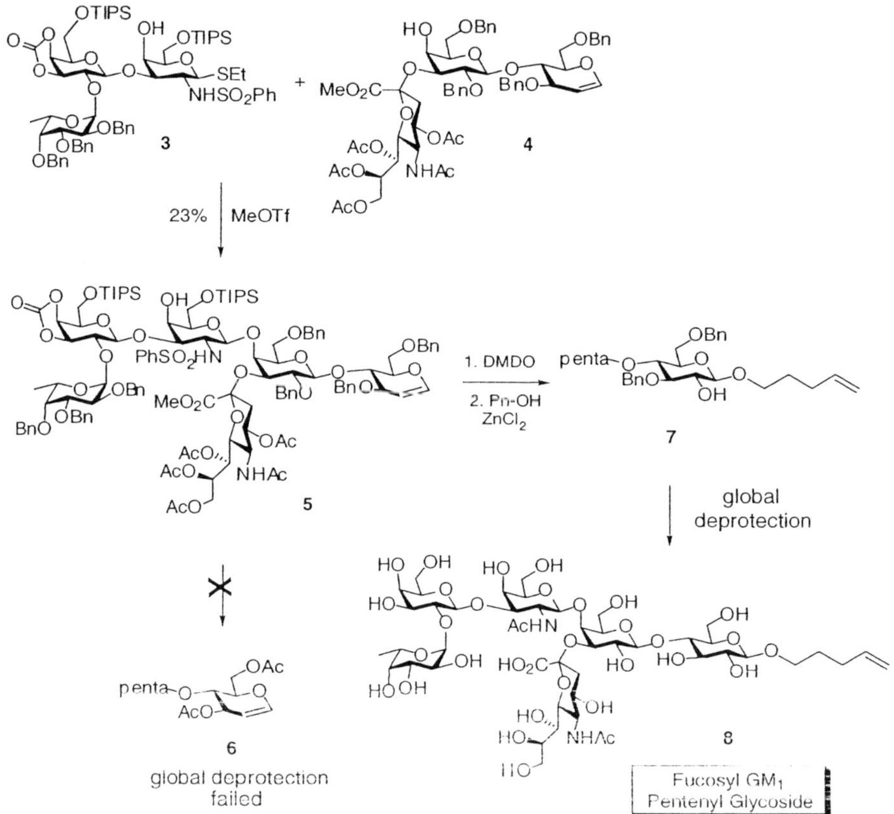

Scheme 1.

In an effort to find a hexasaccharide which was suitable for global deprotection, we considered replacing the reducing end glycal. For vaccine development we required, in any case, a linker that could be modified to allow for conjugation to KLH. Previous syntheses in our carbohydrate-based vaccine program utilized the terminal olefin contained in an allyl glycoside as a linker for bioconjugation. As depicted in the bottom of Scheme 2, the allyl glycoside in those constructs was installed in a late stage glycosidation by solvolysis of a glycal epoxide. Typically, the glycal was maintained through the global deprotection to the peracetate stage. In those investigations, it was demonstrated that epoxidation of the peracetylated glycal, followed by treatment with allyl

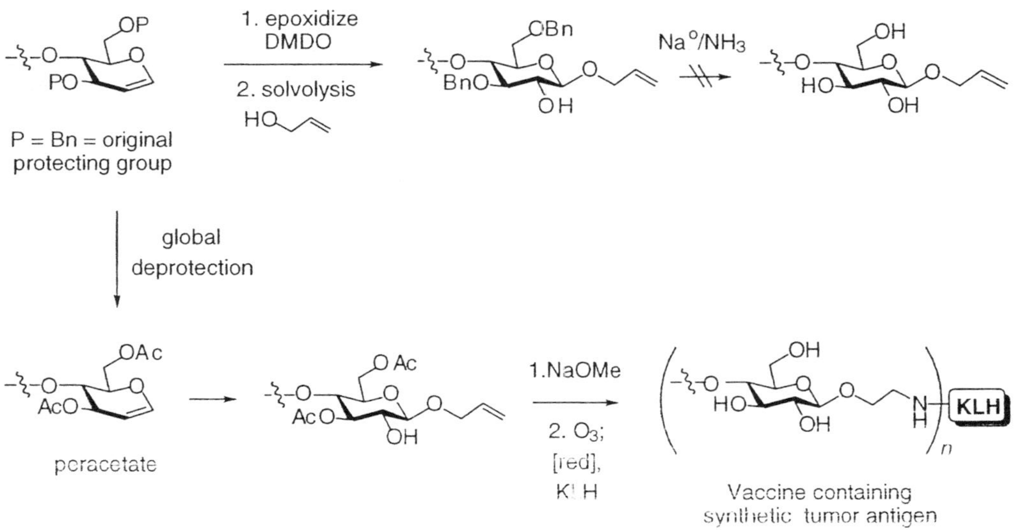

Scheme 2. Solvolysis of glycal epoxides with allyl alcohol en route to KLH conjugation.

alcohol gave the corresponding allyl glycoside. Finally, removal of the ester protecting groups yielded fully deprotected allyl glycoside poised for bioconjugation to KLH. Through these previous studies, it was also determined that pre-functionalization to the allylic ether at the stage of original protection was not possible due to its instability under the necessary dissolving metal reduction conditions in the global deprotection sequence (top of Scheme 2). The inability to affect the global deprotection of hexasaccharide **5** to peracetate **6** (Scheme 1) undermined application of this methodology to the synthesis of fucosyl GM_1 allyl glycoside for vaccine development.

Following the difficulties we encountered, it was recognized that that the well-known pentenyl glycoside linkage, thoroughly and elegantly developed by Fraser-Reid and associates, might be used to advantage in this context [14]. *n*-Pentenyl glycosides (NPG) have been widely employed as substrates for a variety of reactions occurring at the anomeric center of oligosaccharides. NPG's are stable to a range of reaction conditions and reagents, but are readily activated for glycosidation reactions by treatment with a halogen oxidant. As a result of their stability and the neutral conditions required for their activation, pentenyl glycosides have been demonstrated to be valuable linkages for mechanistic and synthetic studies. Therefore, in the context of our studies directed at fucosyl GM_1, epoxidation of **5** under standard conditions utilizing DMDO (dimethyl dioxirane) as the oxidant followed by reaction with pentenyl alcohol (Pn-OH) gave the corresponding NPG **7**. Fortunately, subsequent experiments determined that conversion of the hexasaccharide glycal to **7** allowed for smooth global deprotection to **8** (Scheme 1). Importantly, the pendant olefin contained in **8** was used to bioconjugate fucosyl GM_1 to carrier protein KLH, via reductive amination [15], to afford vaccine glycoconjugate **9** (Scheme 3).

Following the conversion of **7** to **8** and the successful conjugation to give glycoconjugate **9**, we reasoned that incorporation of the reducing end glycoside at an earlier stage in the synthesis could result in a more convergent assembly of fucosyl GM_1. In addition, by this route we could explore the possibility that the pivotal [3+3] coupling step might be more effective with a glycoside rather than with a glycal at the "reducing end" of the trisaccharide acceptor. As shown in Scheme 4, we first focused on pentenyl lactoside, **10**. The C3' and C4' hydroxyl groups we engaged as the dimethyl ketal **11**. Perbenzylation of **11** to give **12**, followed by acetonide removal with aqueous acetic acid, yielded the desired acceptor **13**. Sialylation using phosphite donor **14** proceeded in good yield to give the trisaccharide acceptor, **15**.

In the subsequent coupling event, reaction of donor **3** with a 2.0 molar excess of the acceptor **15** containing the pentenyl linker proceeded with MeOTf promotion in 70% yield (see Scheme 5). It will be recalled that the previous [3+3] coupling using the glycal acceptor **4** proceeded in low yield (23%, Scheme 1). Moreover, subsequent transformations were required to reach **8**. Formation of compound **16** by the direct [3+3] coupling route presented in Scheme 4 represents a significant improvement in overall chemical yield and processing. Thus, these results demonstrate that in this particular merger, replacement of the glycal functionality by a more stable glycosidic protecting group does indeed improve the efficiency of the coupling step. Global deprotection of **16** yielded the characterized hexasaccharide, **8**.

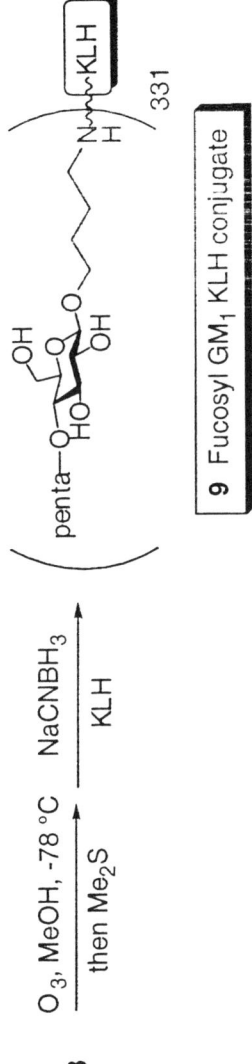

Scheme 3.

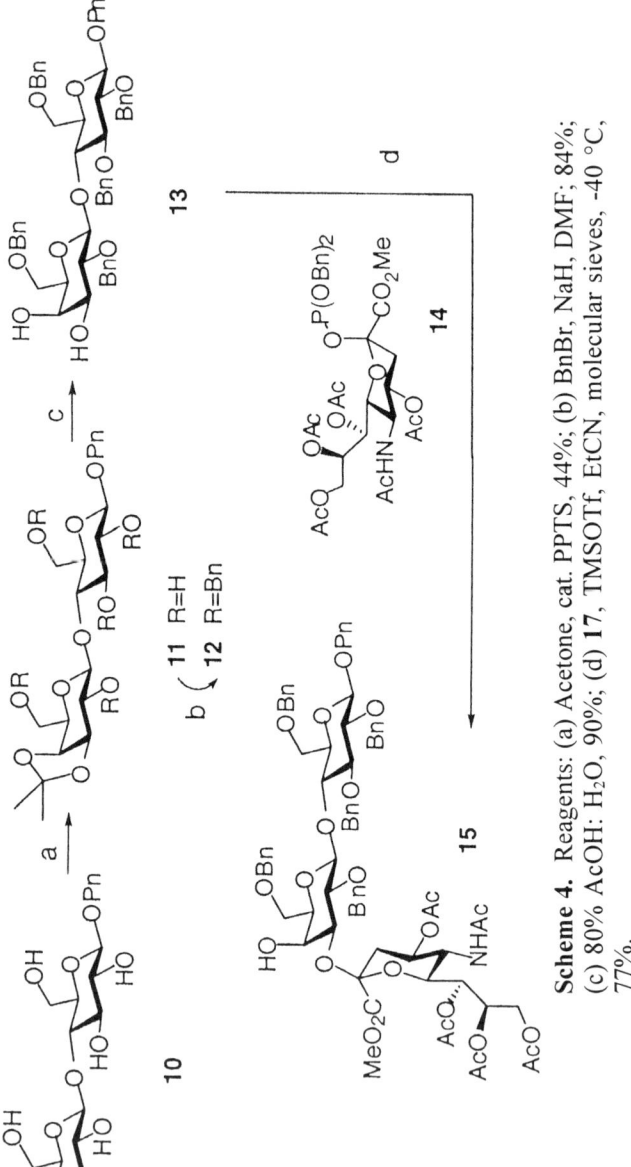

Scheme 4. Reagents: (a) Acetone, cat. PPTS, 44%; (b) BnBr, NaH, DMF; 84%; (c) 80% AcOH: H$_2$O, 90%; (d) **17**, TMSOTf, EtCN, molecular sieves, -40 °C, 77%.

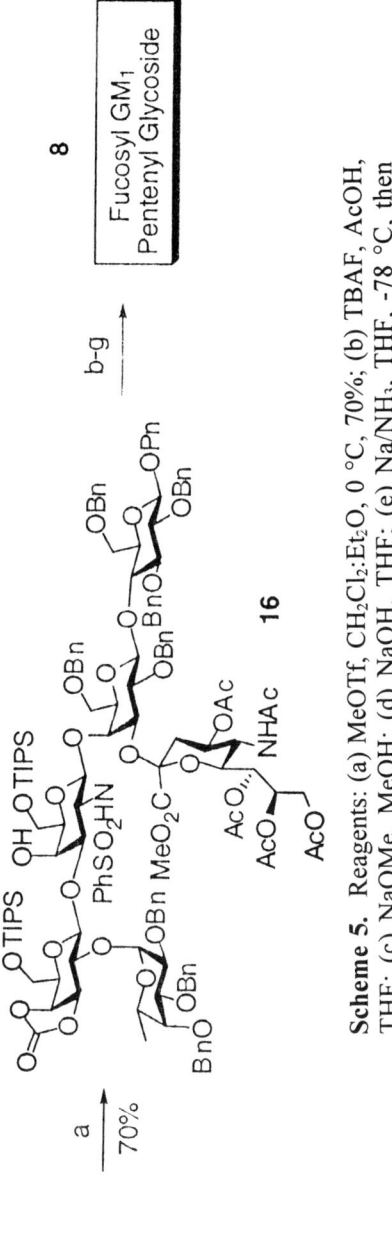

Scheme 5. Reagents: (a) MeOTf, CH$_2$Cl$_2$:Et$_2$O, 0 °C, 70%; (b) TBAF, AcOH, THF; (c) NaOMe, MeOH; (d) NaOH, THF; (e) Na/NH$_3$, THF, -78 °C, then MeOH; (f) Ac$_2$O, pyridine, DMAP, CH$_2$Cl$_2$, 45% 5 steps, (g) steps c-d, 96%.

With the successful synthesis of **8**, we secured the goals of chemical synthesis and bioconjugation of the SCLC specific fucosyl GM_1. The approach to this hexasaccharide relies on a highly convergent [3+3] coupling reaction between the known donor **3** and acceptor **15** which carries with it a pentenyl linker for bioconjugation. The use of a pentenyl glycoside protecting group at the reducing end of **8** offers clear advantages. We observed an increase in chemical yield in the crucial coupling using acceptor **15**, with the more stable glycosidic linkage, as compared to trisaccharide glycal **4**. Also in these studies, it was demonstrated that the pentenyl linker serves well as a linker for conjugation purposes. Clearly, the applicability for incorporating reducing end NPG's into complex tumor antigen targets has been established.

Synthesis of Globo-H NPG

The (mAb) MBr1 antigen globo-H (**1**, Figure 1) was originally isolated and characterized by Hakomori *et al* from the human breast cancer cell line MCF-7 [16], and has subsequently been further characterized as being over-expressed in other types of carcinomas including colon, prostate, lung, ovary and small cell lung cancers [17]. The first generation total synthesis of this antigen by our laboratory [18] has culminated in phase II and III human clinical trials using a fully synthetic globo-H vaccine against prostate and breast cancer [19].

The previous synthesis of the MBr1 antigen from our laboratory was as its allyl glycoside according to the chemistry outlined in Scheme 2. Utilizing all glycal building blocks, a flexible terminal glycal was maintained throughout the hexasaccharide construction. The glycal was then used to install the allyl glycoside, required for immunoconjugation to biocarier proteins, or the ceramide side chain present en route to globo-H glycolipid **1** for proof of structure.

We recognized, however, that while the previous protocols were effective in producing adequate quantities of synthetic material for proof of structure, immunocharacterization, conjugation, mouse vaccinations and phase I human clinical trials, a single target formation protocol required for advanced clinical trials could benefit from a more convergent synthetic approach. The obvious similarity of the globo-H hexasaccharide to the hexasaccharide core contained in fucosyl GM_1 directed us to think in terms of a reducing end *n*-pentenyl glycoside that could be incorporated at an early stage of the synthesis. Furthermore, although the direct implications of such a linkage had not been realized with fucosyl GM_1, an NPG moiety might serve as useful donors for alternative glycosylations.

To explore this possibility [20], we focused our attention on a plan involving the same trisaccharide donor **3** sector used in our previous syntheses (see Scheme 1), in preparation for a [3+3] coupling to produce the hexasaccharide core. In the case of globo-H, as well as fucosyl GM_1, the synthesis of the requisite trisaccharide acceptor began with the *n*-pentenyl lactoside, **10** (see Scheme 4). Subsequent steps were designed to generate a free acceptor site at C4' of **18** for an eventual [2+1] coupling to give the trisaccharide.

We began with a stannane mediated monobenzylation of **10** to selectively give the C3' benzyl ether, as shown in Scheme 6. This transformation was followed by benzylidene acetal formation between the C4' and C6' hydroxyls to give **17**. Perbenzylation of the remaining hydroxyl groups in **17** and regioselective reductive cleavage of the benzylidene with sodium cyanoborohydride and anhydrous HCl gave the acceptor **18**. In preparation for the [2+1] coupling, the required differentially protected α-fluoro donor **20** was prepared from glycal **19** in good yield. Gratifyingly, reaction of **18** and **20** under strongly fluorophilic $Cp_2Zr(OTf)_2$ promotion provided trisaccharide **21** in 80% yield with excellent anomeric selectivity. Finally, discharge of the lone PMB group provided the necessary trisaccharide acceptor **22**, now poised for hexasaccharide assembly.

In the [3+3]coupling event, treatment of donor **3** and acceptor **22** with MeOTf gave hexasaccharide **23** in 60% yield (Scheme 7). The [3+3] coupling yield using trisaccharide acceptor **22** was comparable to the [3+3] procedure using the glycal-based acceptor in our earlier synthesis. However, the tremendous advantage of using **22**, is manifested in the deprotection steps which follow. As in earlier steps, the NPG linkage proved highly reliable under the listed deprotection conditions and the fully deprotected pentenyl glycoside of globo-H, **25**, prominently poised for bioconjugation, was isolated in the yields shown in Scheme 7. Importantly, in the second generation NPG variation, progress toward **25** from hexasaccharide construct **23** was greatly simplified because the need for additional functionalization to allow for conjugation had been eliminated.

Glycoamino Acids Derived From *n*-Pentenyl Glycosides

Native glycoproteins and glycopeptides are typically glycosylated through either serine or threonine with an α-glycosidic linkage. A particularly striking example is that of mucin glycoproteins. Mucins possess amino acid sequences with a very high percentage of serine and threonine residues wherein the first carbohydrate moiety is prevailingly α-*O*-linked *N*-acetylgalactosamine [21]. In addition, Tn, TF, STn, ST and glycophorin, comprise a class of tumor-associated antigens carrying this α-*O*-linkage [22]. Such compounds have received much synthetic and immunological attention [23]. Primary synthetic effort directed at *O*-linked glycopeptides has centered on the stereoselective construction of the naturally required α-glycosidic linkage and has provided a framework for organic chemists to continue to progress toward a general and processable synthesis of native glycopeptides.

With NPG **8** and **25** in hand [10, 20], the contemplated next step in our progression to a more effective presentation of antigens in anti-tumor vaccines required an efficient and general way to incorporate these two antigens into glycopeptides. Initial efforts were directed at performing a direct glycosylation of NPG donors [24] with acceptors containing the required serine or threonine linkage. Unfortunately, these attempted couplings failed and this approach was abandoned.

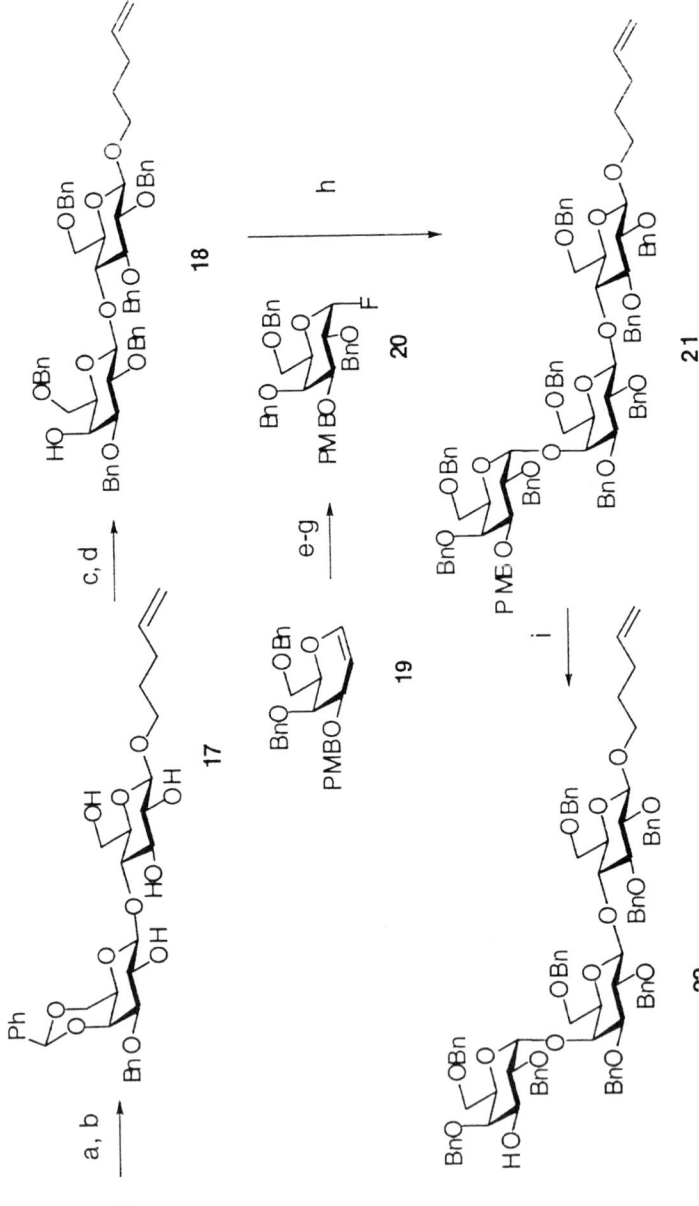

Scheme 6. Reagents: (a) BnBr, Bu$_2$SnO, Bu$_4$NI, C$_6$H$_6$, 53%; (b) PhCH(OMe)$_2$, CSA, CH$_3$CN, 72%; (c) BnBr, NaH, DMF, Et$_4$NI, 97%; (d) NaCNBH$_3$, HCl, Et$_2$O, THF, 75%; (e) DMDO, CH$_2$Cl$_2$; (f) HF·pyridine; (g) BnBr, NaH, DMF, 76%, 3 steps; (h) Cp$_2$Zr(OTf)$_2$, toluene:THF (5:1), 80% (α), α:β 10:1; (i) DDQ, CH$_3$CN, H$_2$O, 92%.

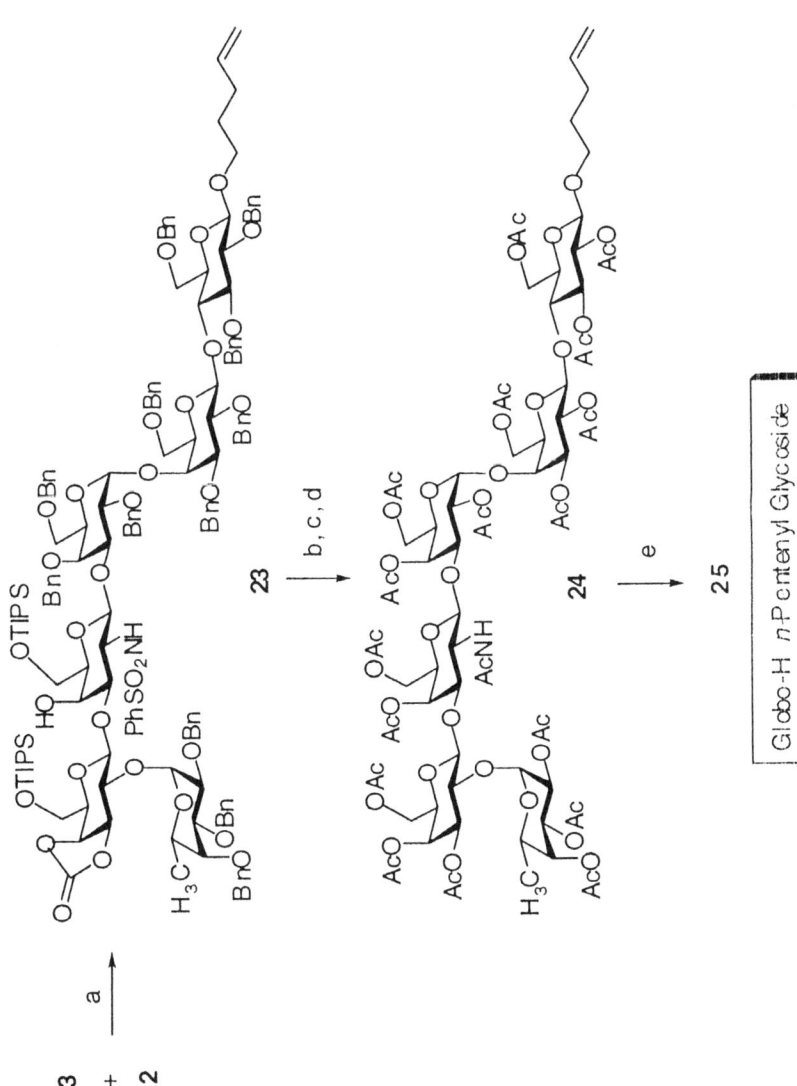

Scheme 7. Reagents: (a) MeOTf, $CH_2Cl_2:Et_2O$, 60-70% (1:2), 0 °C; (b) TBAF, THF; then NaOMe, MeOH; (c) Na/NH_3, THF, -78 °C; then MeOH; (d) Ac_2O, pyridine, DMAP, 42%; (e) NaOMe, MeOH, quantitative.

The transformation of the olefinic unit contained in *n*-pentenyl glycosides to various spacer functionalities has been described [25]. Using this unsaturation to achieve an amino acid attachment, would provide access to glycoamino acids derived from *n*-pentenyl glycosides such as fucosyl GM_1 and Globo-H. Fortunately, precedent from Toone and co-workers suggested a plausible approach. Their work detailed methodology based on catalytic asymmetric hydrogenation of *C*-allyl glycosylated enamide esters to produce carbon linked glycosyl serines [26].

Application of an analogous approach is detailed in Scheme 8. As a starting point, we chose to examine peracetylated lactose derivatives as models [27]. The required lactose derived enamide ester substrate was prepared by ozonolysis of the NPG **26** to give the corresponding aldehyde derivative. The crude aldehyde was then subjected to Horner-Emmons olefination using tetramethylguanidine and phosphonate **27**. Phosphonate **27**, with *N*-Boc and 2-(trimethylsilyl)ethyl ester (TSE) protection, was chosen because of the need for the resulting glycoamino acids to be orthogonally suitable for peptide couplings in the presence of acetate carbohydrate protecting groups. The enamide ester **28** was obtained as a single geometric isomer in 88% yield for the 2-step procedure.

In the hydrogenation of similar *C*-glycosylated enamides, Toone had reported the use of chiral DuPHOS ligands as catalyst precursors [28] and had optimized the resulting diastereomeric excesses of the monosaccharide products with respect to ligand and solvent. Following their report, we found optimal conditions for asymmetric hydrogenation of enamide ester **28** to be as shown. The protected glycoamino acid **29** was obtained in 98% yield and was determined to have been formed with a diastereomeric ratio (dr) of >20:1. Remarkably, the *t*-Boc protons are nearly baseline resolved and, in the asymmetric reaction, the minor isomer could not be detected by ^{1}H NMR. ^{13}C analysis also supports the conclusion that the minor isomer is not formed within the limit of NMR detection. Hydrogenation of **28** with an achiral catalyst (Pd/C, MeOH) produced a 1:1 mixture of R and S configured **29**, providing a comparison for diastereomeric ratio determination. This reaction also indicates that chirality transfer to yield **29** occurs from the chiral ligand and not carbohydrate derived substrate control. A final step to be performed prior to moving to synthesis and assembly of tumor antigens was that of demonstrating deprotectability of the blocking groups contained in the amino acid side chain. In the event, reaction of **29** with TBAF in THF gave acid **30**, suitably prepared for peptide coupling, in 93% isolated yield.

Following this application, the resulting glycoamino acids would represent a non-natural β-glycosidic linkage between oligosaccharide and amino acid. However, maturation of methodology to achieve such a linkage would provide an exciting synthetic possibility. We envision the study of the effect of these types of non-natural linkages in artificial glycopeptides as mimics of tumor cell surfaces [29]. The strategy described herein is now being applied to the complex tumor-associated antigens fucosyl GM_1 and globo-H, as well as other interesting NPG derived antigens. The incorporation of these non-natural glycoamino acids into glycopeptides is

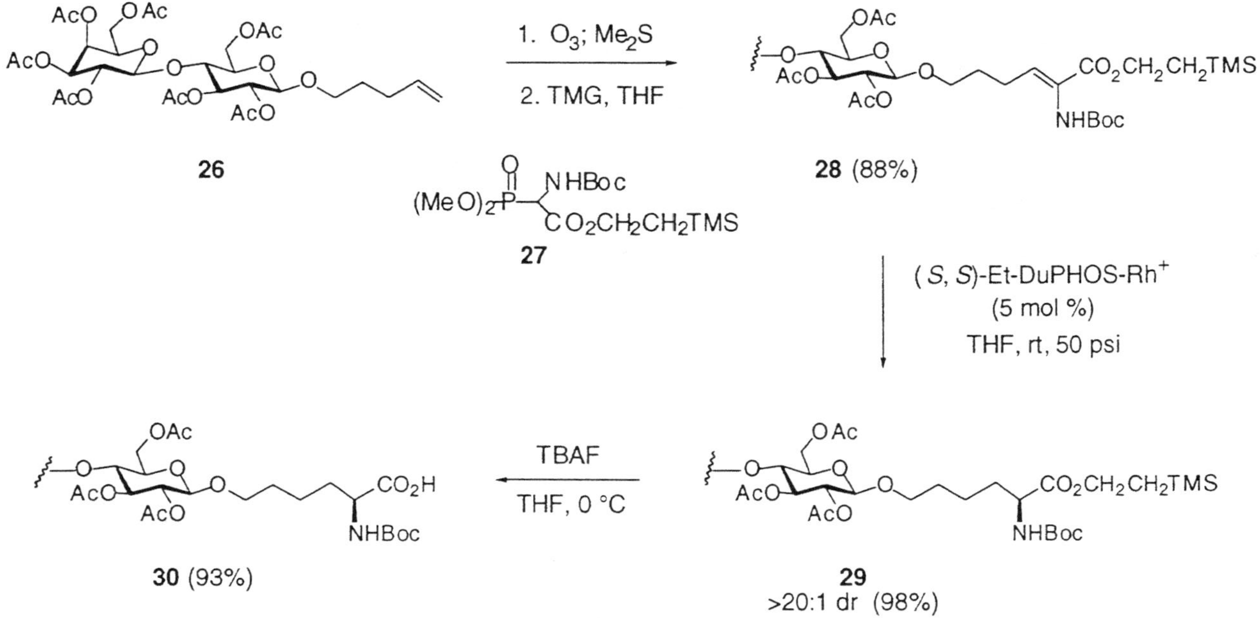

Scheme 8.

also being pursued [30]. The synthesis and immunological studies of these constructs are well underway and will be reported in due course.

Literature cited.

[1] (a) S. Hakomori *Cancer Res.* **1985**, *45*, 2405; (b) T. Feizi *Cancer Surveys* **1985**, *4*, 245; (c) K. O. Lloyd *Am. J. Clin. Pathol.* **1987**, *87*, 129; (d) K. O. Lloyd *Cancer Biol.* **1991**, *2*, 421.

[2] (a) A. Lanzavecchia *Science* **1993**, *260*, 937-944; (b) D. M. Pardoll *Curr. Opin. Immunol.* **1993**, *5*, 719-725; (c) P. O. Livingston *Curr. Opin. Immunol.* **1992**, *4*, 2; (d) G. Dranoff, E. Jaffee, A. Lazenby, P. Golumbek, H. Levitsky, K. Brose, V. Jackson, H. Hamada, D. Paradoll, R. Mulligan *Proc. Natl. Acad. Sci. USA*, **1993**, *90*, 3538; (e) M. -H. Tao, R. Levy, R. *Nature*, **1993**, *362*, 755; (f) T. Boon *Int. J. Cancer*, **1993**, *54*, 177.

[3] (a) T. Toyokuni, A. K. Singhal *Chem. Soc. Rev.* **1995**, *24*, 23; (b) R. Lo-Man, S. Bay, S. Vicher-Guerre, E. Deriaud, D. Cantacuzene, C. Leclerc *Cancer Res.* **1999**, *59*, 1520; (c) G. D. MacLean, M. B. Bowen-Yacyshyn, J. Samuel, A. Miekle, G. Stuart, J. Nation, S. Poppema, M. Jerry, R. Kaganty, T. Wont, B. M. Longnecker *J. Immunol.* **1992**, *11*, 292; (d) P. Y. S. Fung, M. Made, R. R. Koganty, B. M. Longnecker *Cancer Res.* **1990**, *50*, 4308; (e) P. O. Livingston, G. Ritter, P. Srivastava, M. Padavan, M. J. Calves, H. F. Oettgen, L. J. Old *Cancer Res.* **1989**, *49*, 7045; (f) Y. Haraday, M. Sakatsume, G. A. Nores, S. Hakomori, M. Taniguchi *Jpn. J. Cancer Res.* **1989**, *80*, 988 and references therein.

[4] For a review see S. Hakomori, Y. Zhang *Chem. Biol.* **1997**, *4*, 97.

[5] (a) Stroud, M. R.; Levery, S. B.; Martensson, S.; Salyan, M. E. K.; Clausen, H.; Hakomori, S. *Biochemistry*, **1994**, *33*, 10672; (b) Yuen, C.-T.; Bezouska, K.; O'Brien, J.; Stoll, M.; Lemoine, R.; Lubineau, A.; Kiso, M.; Hasegawa, A.; Bockovich, N. J.; Nicolaou, K. C.; Feizi, T. *J. Biochem.* **1994**, *269*, 1595; (c) Stroud, M. R.; Levery, S. B.; Nudelman, E.; Salyan, M. E. K.; Towell, J. A.; Roberts, C. E.; Watanabe, M.; Hakomori, S. *J. Biol. Chem.* **1991**, *266*, 8439.

[6] (a) Livingston, P. O.; Koganty, R. R.; Longenecker, B. M.; Lloyd, K. O.; Calves, M. J. *Vaccine Res.* **1991**, *1*, 99; (b) Kensil, C. R.; Patel, U.; Lennick, M.; Marciani, D. *J. Immunol.* **1991**, *146*, 431.

[7] See (a) Livingston, P. O.; Zhang, S.; Lloyd, K. O. *Cancer Immunol. Immunother.* **1997**, *45*, 1; (b) Ragupathi, G. *Cancer Immunol. Immunother.* **1996**, *43*, 152 and references therein.

[8] For a review of efforts from our laboratory see Allen, J. R.; Danishefsky, S. J. *Angew. Chem. Int Ed. Eng.*, **2000**, *39*, 836

[9] (a) Cappello, S.; Liu, N. X.; Musselli, C.; Brezicka, F.-T.; Livingston, P. O.; Ragupathi, G. *Cancer Immunol. Immunother.* **1999**, *48*, 483; (b) Helling, F.; Shang, Y.; Calves, M.; Oettgen, H. F.; Livingston, P. O. *Cancer Res.* **1994**, *54*, 197; (c) Helling, F.; Zhang, S.; Shang, A.; Adluri, S.; Calves, M.; Koganty, R.; Longenecker, B. M.; Yao, T-J.; Oettgen, H. F.; Livingston, P. O. *Cancer Res.* **1995**, *55*, 2783; (d) Reddish, M.A.; Jackson, L.; Koganty, R.; Qiu, D.; Hong, W.; Longenecker, B. M. *Glycoconjugate J.* **1997**, *14*, 549.

[10] Allen, J. R.; Ragupathi, G.; Livingston, G.; Danishefsky, S. J. *J. Am. Chem. Soc.* **1999**, *121*, 10875.
[11] (a) Nilsson, O.; Mansson, J.-E.; Brezicka, T.; Holmgren, J.; Lindholm, L.; Sorenson, S.; Yngvason, F; Svennerholm, L. *Glycocojugate J.* **1984**, *1*, 43; (b) Brezicka, F. T.; Olling, S.; Nilsson, O.; Bergh, J.; Holmgren, J.; Sorenson, S.; Yngvason, F. *Cancer Res.* **1989**, *49*, 1300.
[12] (a) Nilsson, O.; Brezicka, T. F.; Holmgren, J.; Sorenson, S.; Svennerholm, L.; Yngvason, F; Lindholm, L. *Cancer Res.* **1986**, *46*, 1403; (b) Vangsted, A. J.; Clausen, H.; Kjeldsen, T. B.; White, T.; Sweeney, B.; Hakomori, S.; Drivsholm, L.; Zeuthen, J. *Cancer Res.* **1991**, *51*, 2879.
[13] Zhang, S.; Cordon-Cardo, C.; Zhang, H. S.; Reutter, V. E.; Adluri, S.; Hamilton, W. B.; Lloyd, K. O.; Livingston, P. O. *Int. J. Cancer* **1997**, *73*, 42.
[14] Fraser-Reid, B. O.; Udodong, U. E.; Zufan W.; Ottosson, H.; Merritt, R.; Rao, S.; Roberts, C.; Madsen R. *Synlett*, **1992**, 927
[15] (a) Bernstein, M. A.; Hall, L. D. *Carb. Res.* **1980**, *78*, C1; (b) Lemieux, R. U. *Chem. Soc. Rev.*, **1978**, *7*, 423.
[16] (a) Kannagi, R.; Levery, S. B.; Ishijamik, F.; Hakomori, S.; Schevinsky, L. H.; Knowles, B. B.; Solter, D. *J. Biol. Chem.* **1983**, *258*, 8934; (b) Bremer, E. G.; Levery, S. B.; Sonnino, S.; Ghidoni, R.; Canevari, S.; Kannagi, R.; Hakomori, S. *J. Biol. Chem.* **1984**, *259*, 14773.
[17] (a) Livingston, P.O. *Cancer Biol.* **1995**, *6*, 357-366; (b) Zhang, S., Cordon-Cardo, C., Zhang, H.S., Reuter, V.E., Adluri, S., Hamilton, W.B., Lloyd, K.O. & Livingston, P.O. *Int. J. Cancer* **1997**, *3*, 42-49.
[18] Park, T. K.; Kim, I. J.; Hu, S.; Bilodeau, M. T.; Randolph, J. T.; Kwon, O.; Danishefsky, S. J. *J. Am. Chem. Soc.* **1996**, *118*, 11488.
[19] (a) Ragupathi, G.; Park, T. K.; Zhang, S. Kim, I. J.; Graber, L.; Adluri, S.; Lloyd, K. O.; Danishefsky, S. J.; Livingston, P. O. *Angew. Chem. Int. Ed. Eng.* **1997**, *36*, 125; (b) Ragupathi, G.; Slovin, S. F.; Adluri, S.; Sames, D.; Kim, I. J.; Kim, H.; Spassova, M.; Bornmann, W. G.; Lloyd, K. O.; Scher, H. I.; Livingston, P. O.; Danishefsky, S. J. *Angew. Chem. Int. Ed. Engl.* **1999**, *38*, 563; (c) Slovin, S. F.; Ragupathi, G.; Adluri, S.; Ungers, G.; Terry, K.; Kim, S.; Spassova, M.; Bornmann, W. G.; Fazzari, M.; Dantis, L.; Olkiewicz, K.; Lloyd, K. O.; Livingston, P. O.; Danishefsky, S. J.; Scher, H. I. *Proc. Natl. Acad. Sci. U.S.A.* **1999**, *96*, 5710.
[20] Allen, J. R.; Allen, J. G.; Williams, L. J.; Zhang, X-F.; Zatorski, A.; Ragupathi, G.; Livingston, P. O.; Danishefsky, S. J. *Chem. Eur. J.* **2000**, *6*, 1366.
[21] (a) Podolsky, D. K. *J. Biol. Chem.* **1985**, *260*; (b) Carlstedt, I.; Davies, J. R. *Biochem. Soc. Trans.* **1997**, *25*, 214; (c) Poland, P.A.; Kinlough, C. L.; Rokaw, M. D.; Magarian-Blander, J.; Finn, O J.; Hughey, R. P. *Glycoconjugate J.* **1997**, 14, 89.
[22] (a) Springer, G. F. *Science* **1984**, *224*, 1198; (b) Campbell, B. J.; Finnie, E. F.; Hounsell, E. F.; Rhodes, J. M. *J. Clin. Invest.* **1995**, *95*, 571.
[23] (a) Schultheiss-Reimann, P.; Kunz, H. *Angew. Int. Ed. Eng.* **1983**, *22*, 63; (b) Kunz, H.; Waldmann, H. *Angew. Int. Ed. Eng.* **1984**, *23*, 71; (c) Kunz, H.; Birnbach, S. *Angew. Int. Ed. Eng.* **1986**, *25*, 360; (d) Paulsen, H.

Angew. Int. Ed. Eng. **1982**, *21*, 155; (e) Kunz, H. *Pure Appl. Chem.* **1993**, *65*, 1223; (f) Schmidt, R. R. *Angew. Int. Ed. Eng.* **1986**, *25*, 212; (g) Komba, S.; Meldal, M.; Werdelin, O.; Jensen, T.; Bock, K. *J. Chem. Soc. Perkin Trans I* **1999**,415; (h) Winterfield, G. A.; Ito, Y.; Ogawa, T.; Schmidt, R. R. *Eur. J. Org. Chem.* **1999**, 1167 and references therein.

[24] (a) Udodong, U. E.; Madsen, R.; Roberts, C.; Fraser-Reid B. O. *J. Am. Chem. Soc.* **1993**, *115*, 7886; (b) Merritt, J. R.; Fraser-Reid B. O. *J. Am. Chem. Soc.* **1994**, *116*, 8334; (c) Fraser-Reid, B. O.; Wu, Z.; Udodong, U. E.; Ottosson H. *J. Org. Chem.* **1990**, *55*, 6068; (d) Mootoo, D. R.; Date, V.; Fraser-Reid, B. O. *J. Am. Chem. Soc.* **1988**, *110*, 2662; (e) Mootoo, D. R.; Konradsson, P.; Fraser-Reid B. O. *J. Am. Chem. Soc.* **1989**, *111*, 8540.

[25] Buskas, T.; Soderberg, E.; Konradsson, P.; Fraser-Reid, B. O. *J. Org. Chem.* **2000**, *65*, 958.

[26] (a) Debenham, S. D.; Debenham, J. S.; Burk, M. J.; Toone, E. J. *J. Am. Chem. Soc.* **1997**, *119*, 9897; (b) Debenham, S. D.; Cossrow, J.; Toone, E. J. *J. Org. Chem.* **2000**,*64*, 9153.

[27] Allen, J. R.; Danishefsky, S. J. *J. Prakt. Chem.* submitted for publication.

[28] (a) Burk, M. J. *Accts. Chem . Res.* **2000**, *33*, 3631; (b) Burk, M. J.; Gross, M. F.; Harper, T. G. P.; Kalberg, C. S.; Lee, J. R.; Martinez, J. P. *Pure & Appl. Chem.* **1996**, *68*, 37.

[29] For references on glycopepide mimetics see (a) Marcaurelle, L. A.; Bertozzi, C. R. *Chem. Eur. J.* **1999**, *5*, 1384; (b) Bertozzi, C. B ; Hoeprich, P. D.; Bednarski, M. D. *J. Org. Chem.* **1992**, *57*, 6092; (c) Kessler, H.; Wittman, V. Kock, M.; Kottenhahn, M. *Angew. Chem., Int. Ed. Eng.* **1992**, *31*, 902; (d) Burkhart, F.; Hoffmann, M.; Kessler, H. *Angew. Chem., Int. Ed. Eng.* **1997**, *36*, 1191; (e) Fuchss, T.; Schmidt, R. R. *Synthesis*, **1998**, 753; (f) Ben, R. N.; Orellana, A.; Arya, P. *J. Org. Chem.* **1998**, *63*, 4817; (g) Arya, P.; Ben, R. N.; Qin, H. *Tetrahedron Lett.* **1998**, *39*, 6131; (h) Dondoni, A. Marra, A.; Massi, A. *J. Org. Chem.* **1999**, *64*, 933; (i) Vincent, S. P.; Schleyer, A.; Wong, C.-H. *J. Org. Chem.* **2000**, *65*, 4440 and references therein.

[30] Allen, J. R.; Danishefsky, S. J. *J. Am. Chem. Soc.* **2000**, submitted for publication.

Chapter 19

Drugs to Enhance the Therapeutic Potency of Anticancer Antibodies: Antibody–Drug Conjugates as Tumor-Activated Prodrugs

Walter A. Blättler and Ravi V. J. Chari

ImmunoGen, Inc., 128 Sidney Street, Cambridge, MA 02139

The development of disease-fighting agents will be increasingly based on information from the genomic analysis of healthy and diseased cells. In cancer this information will eventually lead to the discovery of somatic mutations that led to the generation of tumorigenic cells and the appearance of neoplasias. This will spur the development of treatments directed towards these fundamental tumorigenic mechanisms. Today, genomic analysis of cancer tissues has already yielded biological markers for cancers, such as proteins that are preferentially or exclusively expressed on cancer cells. Monoclonal antibodies can be generated to recognize and specifically bind to these cancer markers. If the markers are expressed on the surface of the cancer cells, monoclonal antibodies administered to a patient will travel to the tumor and accumulate at the tumor through their binding. This can be exploited for the specific delivery of cell-killing drugs in form of covalently linked antibody-drug conjugates. Such antibody-drug conjugates are the most rapid way of converting today's genomic information into specific anti-cancer treatments. The principles guiding the preparation of antibody-drug conjugates, the generation of effective conjugates, and the current development status are discussed.

Design of Cytotoxic Drugs for Delivery by Monoclonal Antibodies

Background and Principles of Antibody-Drug Conjugate Preparation

Multidrug therapy is a standard modality for the treatment of most cancers. However, despite the large number of chemotherapeutic drugs available, long-term remissions or cures are achieved only in a small set of cancers. The therapeutic efficacy of the cytotoxic drugs is limited by their general toxicity to proliferating cells, resulting in a small therapeutic window (the therapeutic window is defined as the therapeutic dose range up to the maximum tolerated dose). Indeed, many drugs have to be used near their maximum tolerated dose for a clinically meaningful, therapeutic effect.

The recognition that the efficacy of current agents was not limited by a lack of cytotoxic potency but rather by a narrow therapeutic window caused by the lack of tumor selectivity gave impetus to the development of drugs linked to tumor-specific monoclonal antibodies. It was perceived that the antibody-targeted delivery of the cytotoxic agents would lead to enhanced accumulation at the tumor sites, thereby increasing the selectivity and therapeutic window, and ultimately, the therapeutic benefit. However, the first generation of antibody-drug conjugates, which included clinically used cytotoxic agents such as methotrexate, *Vinca* alkaloids and doxorubicin (for a review see *1*), failed to live-up to these expectations. The pre-clinical and clinical results were disappointing. In general, less activity was observed than with the free cytotoxic agents while the toxicities were very similar. The therapeutic window had been narrowed.

During this initial development it was not appreciated sufficiently that the conjugation of the small molecular cytotoxic agents to a large protein, the 150 kD monoclonal antibody, totally altered the pharmacodynamic and pharmacokinetic behavior if the drug *in vivo*. The typically short serum half-life, measured in minutes, for a chemical drug was transformed to that of a humanized or human monoclonal antibody, which is days or even weeks. The hugely increased area under the curve (AUC) influenced the toxicity. The pharmacodynamics, specifically the delivery to the tumors and the tumor penetration, was also that of a large protein and is characterized by a limited and relatively slow tumor penetration through a diffusion mechanism (*2*). Since these key pharmacological factors were largely a characteristic of the monoclonal antibody and could not be easily changed[1], the antibody-drug conjugate problem became a drug design task for chemists, which can be paraphrased as follows: new cytotoxic agents had to be found or designed that were non-toxic during the long circulation in form of an antibody-drug conjugate but were cytotoxic at the concentration achievable by antibodies at the tumor site.

Delivery and binding of monoclonal antibodies to tumors had been studied quantitatively in patients using antibodies labeled with radioisotopes (*3*). Antibody levels at tumors were achieved which were commensurate with near antigen saturation for antibodies that had a high binding activity (apparent dissociation constants, K_D, of about 10^{-10} M or lower). Therefore, cytotoxic agents had to be identified that were active in the concentration range of antibody binding. Figure 1a) shows the binding curve of a typical humanized monoclonal antibody from the authors' laboratory. The antibody huC242 binds to colon tumor cells with an apparent K_D of about 5×10^{-11} M and saturation binding is approached at an antibody concentration of about 3×10^{-10} M. The binding curve of the corresponding drug-conjugate, huC242-DM1, is also shown to indicate that conjugation does not affect the binding of the antibody. The binding curve is juxtaposed with the results from an *in vitro* cytotoxicity assay [Figure 1 b)]. COLO 205 cells were treated with different cytotoxic drugs under the same conditions using a wide range of concentrations and the cytotoxic effect was assayed by determining the percentage of surviving cells. For each drug a typical killing curve was obtained. The clinical cancer agents tested, doxorubicin, etoposide (VP-16), 5-fluorouracil (5-FU), and cisplatin, show activity only at concentrations greater than 10^{-8} M, which is outside the antibody concentrations achievable at tumors, as discussed above. With this data in mind, one would have predicted that the first generation of conjugates, which were based on clinical anti-cancer agents, would fail. Figure 1b) also includes the toxicity curves for two new cytotoxic drugs, called DM1 and DC1, which are

---

[1] Although several types of antibody fragments and different recombinant forms of antibodies were prepared that displayed different pharmacokinetic behaviors from that of intact antibodies (for a recent review, see Hudson, P.J., *Current Opinion Immunol.* **1999**, *11*, 548-557. However, careful studies in animal models showed (see for example Hu, S.Z.; Raubitschek, A.; Sherman, M.; Williams, L.E.; Wong, J.J.C.; Shively, S.E.; Wu, A.M. *Cancer Res.* **1996**, *56*, 3055-3061) that maximal total accumulation at the tumor was achieved with intact IgG antibodies. For a more thorough discussion of intact antibodies *vs.* antibody fragments in the context of immunoconjugates the reader is referred to Chapter 1 in *Monoclonal Antibody-Based Therapy of Cancer*, Michael L. Grossbard, Ed.; Marcel Dekker, Inc., NY, **1998**, pp. 1-22.

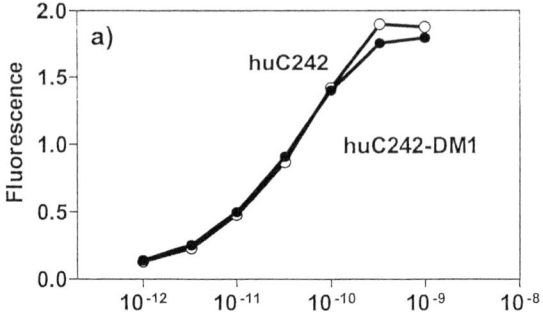

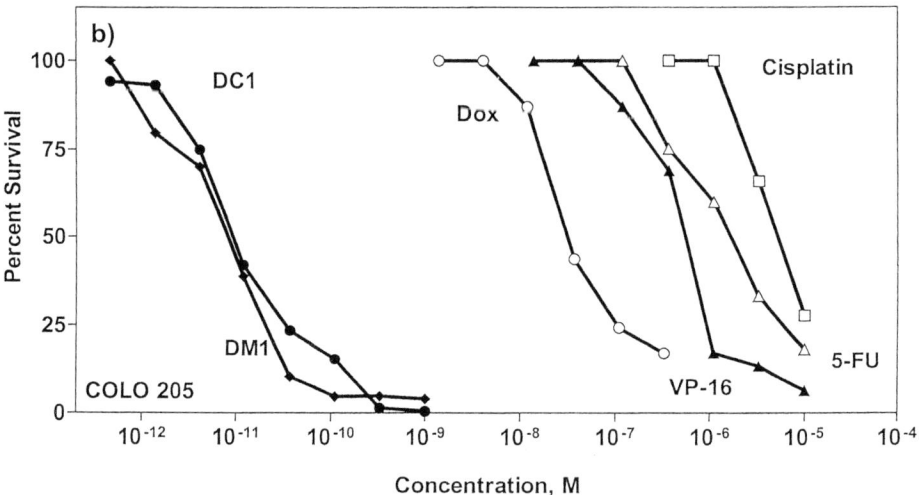

Figure 1a). Binding of the monoclonal antibody huC242 and its maytansinoid conjugate, huC242-DM1, to COLO 205 colon tumor cells as measured by indirect immunofluorescence using a fluorescence activated cell analyzer; 1b). Cytotoxicity measured for six different cytotoxic agents towards COLO 205 colon tumor cells. All agents were tested in the same fashion: cells were exposed to the agents on time 0 and surviving cells were assayed after 24 hours.

active at the concentration range necessary for the generation of active antibody-drug conjugates. Their chemistry and use will be described below.

As a means to explain the title, it should be mentioned that some monoclonal antibodies are active against certain tumors in "naked" form, i.e., the inherent effector functions of the antibody are sufficient to have an anti-cancer effect. However, it is the authors' belief that most anti-cancer antibodies to solid tumors are not sufficiently active in naked form. Their effector function has to be complemented by a cytotoxic agent, hence the title: *Drugs to Enhance the therapeutic Potency of Anti-Cancer Antibodies*. It also allows us to highlight the important, dual character of antibody-drug conjugates. In the view of biochemists and immunologists, they are proteins that carry a few small, chemical molecules, the drugs. In the chemist's view they are cytotoxic drugs, which were converted into prodrugs through modification with an antibody.
The prodrug is then biologically activated at the tumor. Therefore, in the authors' laboratory we named such antibody-drug conjugates *Tumor-Activated Prodrugs (TAPs)*. It is this dual character inherent in antibody-drug conjugates which must guide the development of such agents for the treatment of cancer.

Novel Cytotoxic Drugs for Antibody-Drug Conjugate Preparation

The success of an antibody-drug conjugate depends on the characteristics of the antibody, the potency of the drug and the method of linkage of the antibody to the drug. The antibody should be carefully selected such that it binds specifically to tumor tissue, and has little cross-reactivity with healthy tissues. It should bind to the tumor cells with high avidity (K_D around 1×10^{-10} M) and be internalized into the cell at a reasonable rate upon binding to its antigen. The linker between the antibody and drug has to be designed in a manner that ensures stability in plasma, but upon internalization of the antibody-drug conjugate, allows for the rapid release of the cytotoxic drug in its fully active form inside the tumor cell. Furthermore, the conjugate must remain intact during storage in aqueous solution for an extended period of time to allow formulations for convenient intravenous administration. In an attempt to meet these criteria, several types of cleavable linkers have been evaluated, notably acid-labile and peptidase-labile linkers (*1*). However, studies in the authors' laboratories led to the conclusion that disulfide linkers are the ideal choice. Antibody-drug conjugates prepared using disulfide bonds were found to be stable upon storage in aqueous, buffered solutions for up to two years. Pharmacologically, these conjugates have the characteristic of prodrugs: the cytotoxic drug, when modified with the antibody (in other words, when the drug is linked to the antibody) is non-toxic, presumably because it can not reach its intracellular killing site. Only at the tumor, where the antibody component of the conjugate binds to the specific antigen on the surface of tumor cells, is the conjugate transported to the interior of the cell, presumably through the cellular mechanism of receptor-mediated endocytosis. Conversion of the prodrug to the active agent is effected by release of the drug from the antibody, likely by cleavage of the linker disulfide bond *via* disulfide exchange with an intracellular thiol, such as glutathione. Thus, drugs that are appropriate for use in antibody-drug conjugates should not only be highly potent ($IC_{50} < 1 \times 10^{-10}$ M), but also possess a disulfide or thiol substituent. Since there are few known drugs that meet these criteria, it becomes necessary to first identify drugs that are 100 to 1000-fold more cytotoxic than clinically used anti-cancer drugs, and then incorporate a disulfide-containing substituent at a site that will maintain the potency of the parent drug. Examples of potent drugs with diverse mechanisms of action, and their conversion into analogs bearing a disulfide-containing substituent are given below.

Maytansinoids: maytansinoids are anti-mitotic agents that exert their cytotoxic effect by inhibiting tubulin polymerization. Although their mechanism of action is similar to that of vincristine, they are about 1000-fold more cytotoxic *in vitro* than vincristine. The parent drug maytansine [**1**, see Figure 2a)] was first isolated from an Ethiopian shrub by Kupchan et al (*4*). The anti-tumor activity of maytansine was extensively evaluated in human clinical trials. However, results of these trials were disappointing as maytansine, although potent in vitro, displayed a poor therapeutic window *in vivo* (*5*).

Is it possible then to harness the high potency of maytansine for the selective destruction of tumor cells? In principle, this can be achieved by linking maytansine to tumor-specific antibodies via disulfide bonds. The first step then was to synthesize analogs of maytansine that bear a thiol-containing substituent. The ester functionality at the C-3 position is essential for the cytotoxic activity of maytansine. However, the chain length of the acyl group on the *N*-methyl-*L*-alanine moiety at C-3 can be varied without affecting potency, suggesting that the acyl group might tolerate the addition of an uncharged substituent (*6*). With this information in mind, the thiol-containing maytansinoid DM1 [**2**, see Figure 2a)] was designed. The synthesis of DM1 [Figure 2b)] starts from Ansamitocin P-3 (**3**), which can be readily produced by microbial fermentation (*7*). Reductive cleavage of the C-3 ester with lithium trimethoxyaluminum hydride under carefully controlled conditions (- 35 to - 40 °C, 1 h) provided maytansinol (**4**). Maytansinol was esterified with *N*-methyl-*N*-(3-methyldithiopropanoyl)-*L*-alanine (**5**) to give the disulfide-containing maytansinoid ester, *L*-DM1SMe (**6**). Reduction of the disulfide bond with dithiothreitol (DTT) provided the thiol-containing maytansinoid DM1 (**2**).

a)

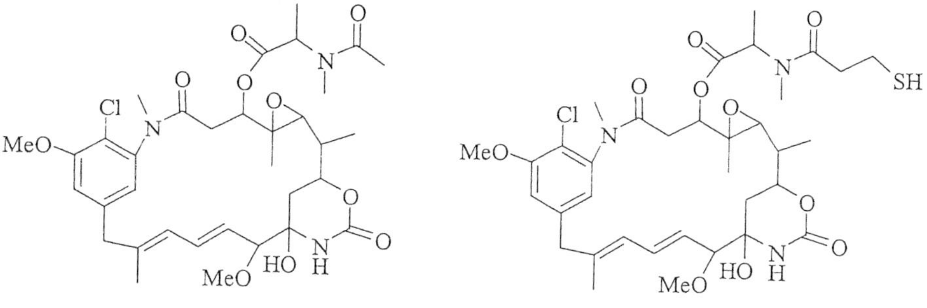

1

2

*Figure 2a). Structures of Maytansine (**1**) and DM1(**2**).*

b)

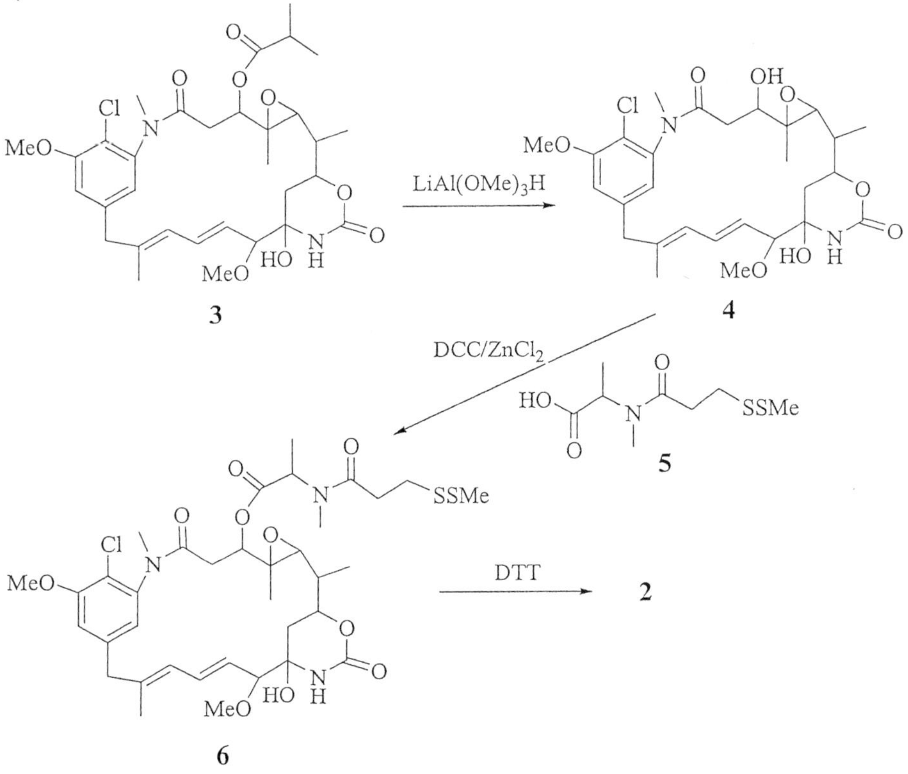

Figure 2b). Synthesis of DM1.

Thiol-containing compounds, such as DM1, are unstable under cell culture conditions, as they are prone to oxidation. Thus, the more stable disulfide-containing maytansinoid L-DM1SMe (**6**) was evaluated for *in vitro* cytotoxicity towards a panel of human tumor cell lines. In a direct comparison (*8*), DM1SMe was about 3 to 10-fold more potent than the parent drug maytansine (IC$_{50}$ values of 3×10^{-12} to 1×10^{-11} M for DM1SMe versus 1×10^{-11} to 1×10^{-10} M for maytansine).

CC-1065 Analogs: CC-1065 (**7**, see Figure 3) is a structurally novel anti-tumor antibiotic whose mode of action is sequence-selective binding to the minor groove of DNA, followed by alkylation of an adenine base on the DNA. *In vitro*, CC-1065 is 1000-fold more cytotoxic than other DNA-interacting agents, such as doxorubicin and cisplatin. Adozelesin (**8**) is a structurally simpler synthetic analog of CC-1065, which retains the high potency of the parent compound (*9*), and is, therefore, a good candidate for use in targeted delivery in the form of an antibody-drug conjugate. In order to prepare an antibody conjugate, the disulfide-containing analogs **9a,b** were designed (Figure 3). In these compounds (also known as DC1 derivatives), the alkylating cyclopropapyrroloindole (CPI) portion of adozelesin was replaced by the cyclopropabenzindole (CBI) unit. Bis-indolyl-CBI compounds have been shown to have the same potency as the corresponding CPI compounds (*10*). In addition CBI is more stable in aqueous solution and easier to synthesize than CPI (*11*). In order to enable linkage to an antibody *via* a disulfide bond, the terminal, un-substituted benzofuran ring of adozelesin was replaced by an indole ring bearing a disulfide-containing side chain at the C-5 position.

DC1 derivative **9a** was evaluated for *in vitro* cytotoxicity towards human tumor cell lines. This compound had an IC$_{50}$ value of 2×10^{-11} M, which is similar to that reported for adozelesin and for the un-substituted bis-indolyl-CBI (*12*).

Calicheamicins: calicheamicins are anti-tumor antibiotics that bind to the minor groove of DNA and produce site-specific double strand DNA breaks. They are reported to be potent at sub-picomolar concentrations (*13*). The core calicheamicin structure includes a trisulfide functionality that undergoes reduction, followed by rearrangement of the enediyne to give an active diradical intermediate that produces strand breaks in the DNA. Calicheamicin γ_1 (**10**, see Figure 4) has been converted to its corresponding disulfide derivative **11**, which also bears a hydrazide functionality allowing for linkage to oxidized sugar residues on an antibody (*13*).

Taxoids: Paclitaxel and its semisynthetic analog docetaxel are two of the most active agents in the treatment of various cancers. In contrast to the

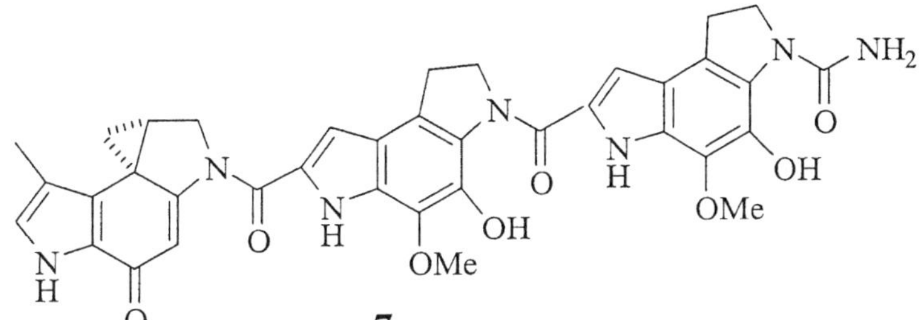

*Figure 3. Structures of CC-1065 (**7**), adozelesin (**8**), and DC1 (**9**).*

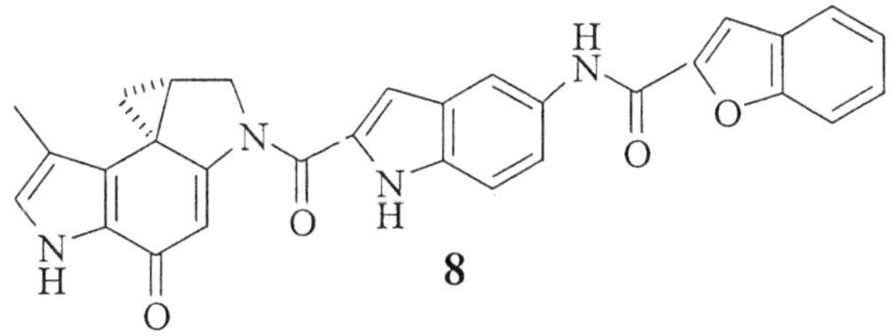

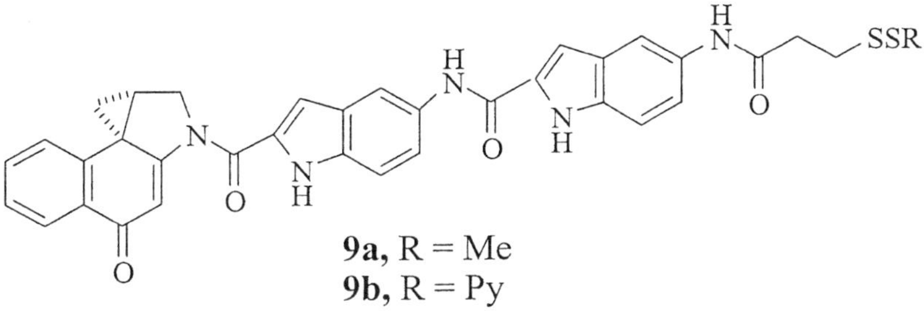

9a, R = Me
9b, R = Py

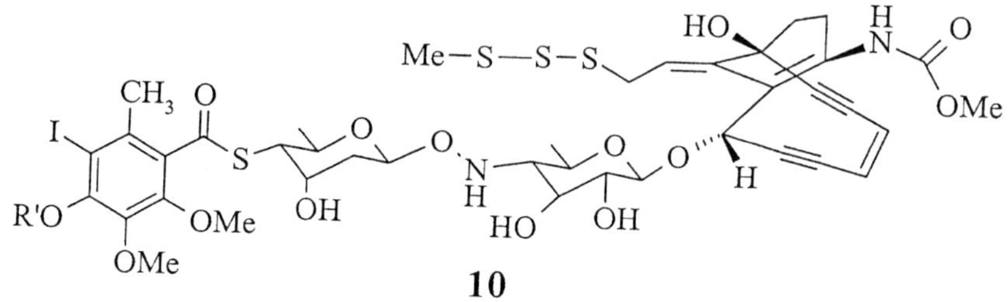

10

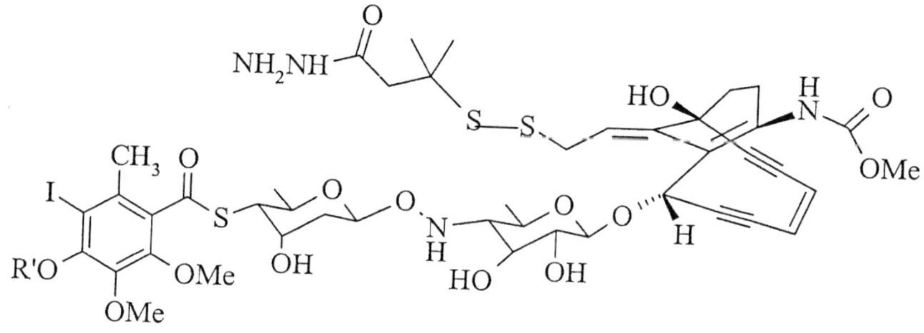

11

*Figure 4. Structures of calicheamicin γ_1 (**10**) and calicheamicin hydrazide (**11**).*

mechanism of action of the *Vinca* alkaloids and maytansine, taxoids enhance the polymerization of tubulin and stabilize microtubules resulting in cell death. Paclitaxel and docetaxel display *in vitro* activities in the same concentration range as other chemotherapeutic agents [see Figure 1b)], thus are not sufficiently potent for use in antibody conjugates. However, two different research groups have reported the synthesis of new taxoids that are up to 100-fold more cytotoxic than the parent taxoids. While these new compounds may not be useful by themselves, their higher potency could be exploited in the form of antibody conjugates. Compounds **12** and **13** (Figure 5) are representative examples of more potent paclitaxel (*14*) and docetaxel (*15*) analogs, respectively. These compounds have to be further modified to introduce a disulfide-containing substituent, without altering their potency, to enable linking to antibodies.

Other agents: of the many anthracycline analogs prepared, 3'-(3-cyano-4-morpholinyl)-3'-desaminodoxorubicin (**14**, see Figure 5) is the most potent with *in vitro* activities at concentrations lower than 10^{-10} M (*16*). This potency coupled with the broad spectrum of activity of the anthracyclines would make it an attractive candidate for antibody conjugate formation. However its poor stability has precluded such a development. Despite this fact, the doxorubicin derivative **14** shows that anthracycline agents can be prepared with sufficient cytotoxic potency to make them good candidate drugs for delivery through antibody conjugates.

The four drug classes mentioned: maytansinoids, CC-1065 analogs, taxoids, and anthracyclines, utilize four different mechanisms for cell killing. The armamentarium of cytotoxic agents for the treatment of cancer today includes many agents with many more different killing mechanisms. If antibody-drug conjugates will be shown to be superior to unconjugated cytotoxics, then the number and type of "linkable" drugs should be increased to include members of additional drug classes. Today, noticeably absent are drugs that are inhibiting nucleotide biosynthetic pathways, such as the anti-folates and nucleoside analogs, or inhibitors of topoisomerases. The challenge is to prepare drugs with increased *in vitro* cytotoxic potency

Antibody-Drug Conjugates in Development

Antibody-Calicheamicin Conjugates

Calicheamicin has been linked to the antibodies CTMO1 directed against ovarian cancer, and hP67.6 directed against the CD33 antigen expressed on leukemic cells of patients with acute myeloid leukemia (AML). A structural

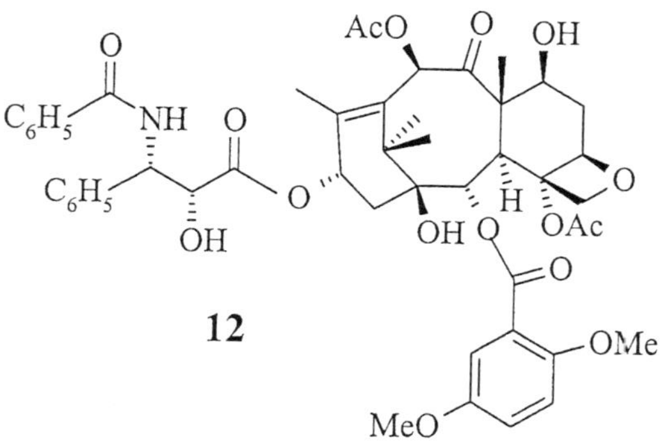

12

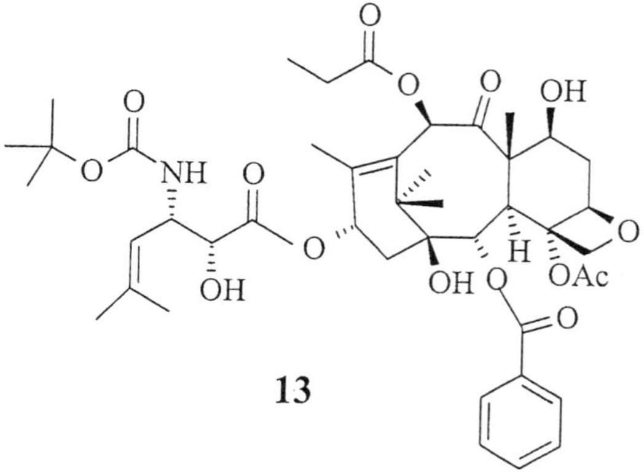

13

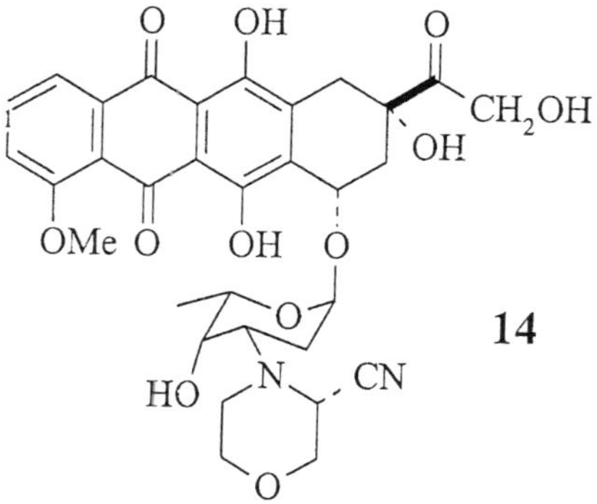

14

Figure 5. Highly potent analogs of paclitaxel, docetaxel and doxorubicin.

representation of antibody-calicheamicin conjugates is given in Figure 6. In tumor xenograft models, the CTM01 conjugate of calicheamicin displayed good anti-tumor efficacy resulting in complete tumor regressions (13). The anti-CD33-calicheamicin conjugate has been evaluated in clinical trials in humans with AML. The conjugate was found to selectively ablate malignant hematopoiesis in some patients with relapsed AML (17). Importantly, the therapy has been reported to be associated with few toxic side effects. This drug, also known as ozogamicin gemtuzumab and Mylotarg, has been recently approved for the treatment of AML in patients who are 60 years and older.

Antibody-Maytansinoid Conjugates

C242-DM1: in huC242-DM1, the potent maytansinoid DM1 (**2**) is linked to the humanized monoclonal antibody huC242. The C242 antibody binds with high affinity to the carbohydrate antigen, CanAg, which is expressed on the surface of human colorectal and pancreatic cancer cells and on a majority of non-small cell lung cancer samples (18). The antibody has minimal cross-reactivity with normal human tissues. The murine C242 antibody that was originally produced was humanized by variable domain resurfacing using the method previously described (19). Several humanized antibodies have now been evaluated in human clinical trials and have been found to be non-immunogenic. Thus, it is expected that huC242-DM1 will also not be immunogenic, allowing for repeated cycles of therapy in humans.

In order to link DM1 to C242, the antibody was first reacted with N-succinimidyl 4-(2'-pyridyldithio)pentanoate (SPP, **15**). Displacement of the succinimidyl group by a nucleophile on the antibody, such as the ε-amino group on a lysine residue, results in the incorporation of a 2-pyridyldithio substituent on the antibody (Figure 7). Typically, on the average, 4 to 5 of these active disulfide groups are introduced per molecule of antibody. Reaction of the modified antibody with the thiol-containing maytansinoid DM1 results in disulfide exchange yielding the antibody-maytansinoid conjugate huC242-DM1. A conjugate typically contains three to four maytansinoid molecules linked per molecule of antibody.

Evaluation for *in vitro* cytotoxicity towards target and non-target cells showed that C242-DM1 is very effective in killing antigen-positive human colon cancer COLO 205 cells, with an IC_{50} value of 3.2×10^{-11} M. Exposure of these cells to a higher concentration of conjugate (5×10^{-9} M) eliminates greater than 99.999% of the cells. The conjugate is about 1000-fold less cytotoxic (IC_{50} = 3.6×10^{-8} M) towards the antigen-negative human melanoma cell line A-375 [Figure 8a)]. Thus, the non-target cells, to which the antibody does not bind, are

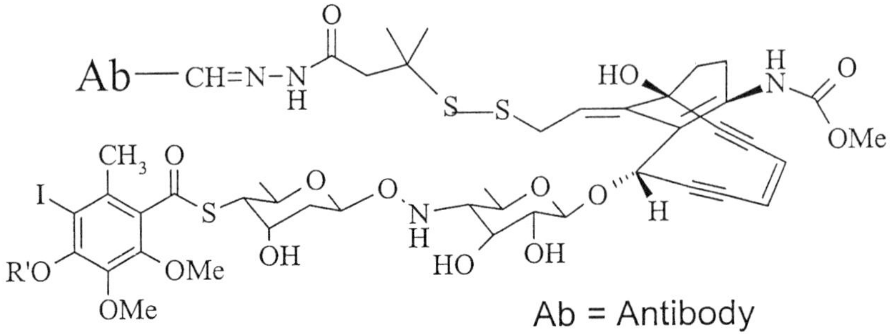

Figure 6. Structural representation of Antibody-calicheamicin conjugates.

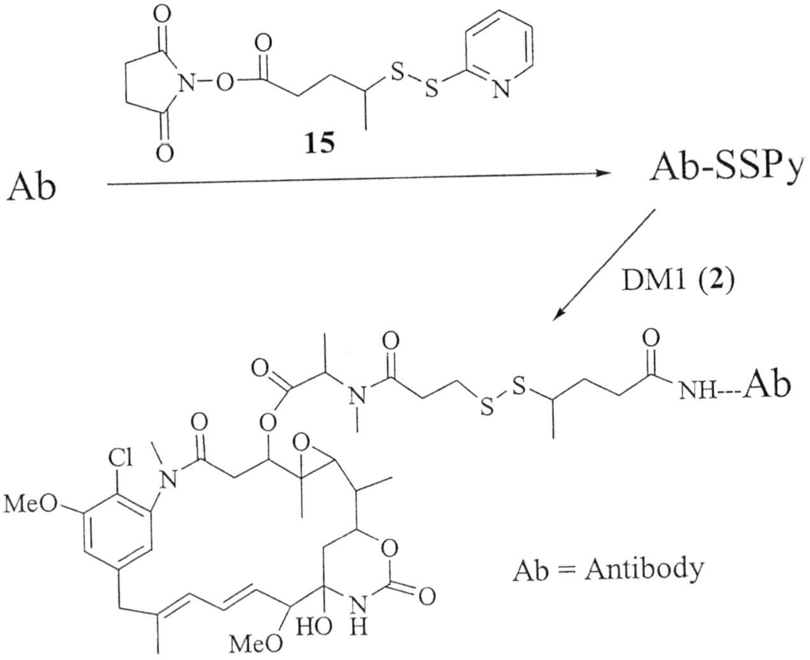

Figure 7. Preparation of Antibody-DM1 conjugates.

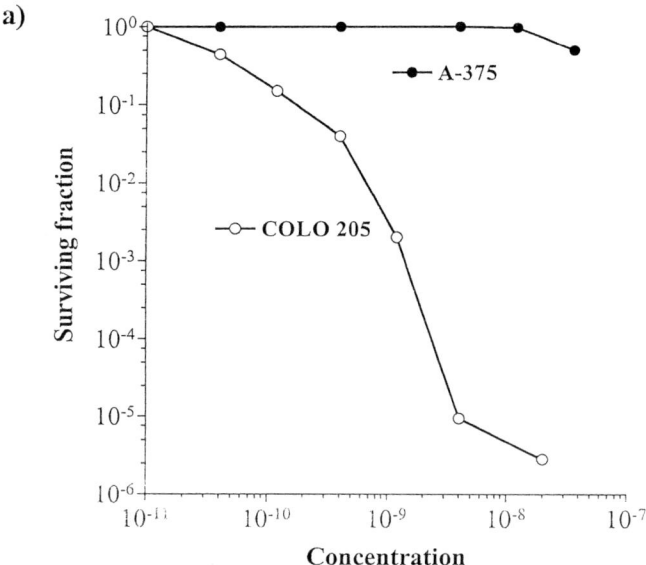

Figure 8a). In vitro cytotoxicity and specificity of C242-DM1. Surviving fractions of cells were determined using clonogenic assays after a continuous exposure to the conjugate (Reprinted from Chari, R. V. J. *Adv. Drug Delivery Revs.* **1998,** *31,* 89-104 *with permission from Elsevier Science).*

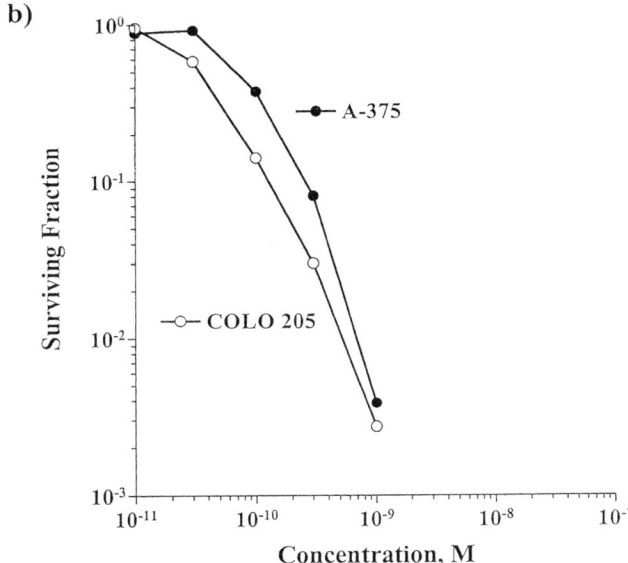

Figure 8b). In vitro potency of free DM1 analog **6** *Surviving fractions of cells were determined using clonogenic assays after a continuous exposure to the drug.*

unaffected at a concentration where the conjugate can kill 5 logs of target cells, demonstrating the antigen-specificity of the cytotoxic effect (20). In contrast, the non-conjugated maytansinoid is equally cytotoxic towards both cell lines [Figure 8b)] indicating that linking to the C242 antibody converts the maytansinoid into a prodrug which is only activated by specific binding and internalization into the target cell.

The anti-tumor efficacy of C242-DM1 was compared to that of unconjugated maytansine and to 5-FU/Leukovorin and irinotecan (CPT-11), the only drugs approved in the U.S.A. for use against colorectal cancer. Different groups of immunodeficient mice bearing established, subcutaneous HT-29 human colon tumor xenografts were treated with one of these agents and the tumor sizes were measured. Unconjugated maytansine had little anti-tumor effect in this model, while treatment with 5-FU/Leukovorin or CPT-11, used at their respective maximum tolerated doses, resulted in modest delays in tumor growth. In contrast, treatment with C242-DM1 resulted in complete eradication of the tumor [Figure 9a)]. Importantly, C242-DM1 was non-toxic to the animals. C242-DM1 was then tested in mice bearing very large (average size 500 mm^3) colon tumor xenografts (20). Even in this challenging model, the conjugate displayed exceptional activity resulting in cures of all the animals [Figure 9b)]. Thus, targeted delivery using the C242 antibody greatly improves the therapeutic window of the maytansinoid. This approach opens the door to the therapeutic use of highly potent compounds that were previously too toxic to be useful.

C242-DM1 was further evaluated in cynomolgus monkeys where the antigen for C242 is expressed, like in humans, on a limited number of normal tissues. Animals were treated with C242-DM1 daily for five consecutive days, and the level of biologically active conjugate in circulation in the blood was measured. The circulating serum concentration of active C242-DM1 (0.4 – 1.0 x 10^{-6} M) was about 100-fold higher than that required to kill 5 logs of tumor cells in an *in vitro* assay. Importantly, this level of active conjugate was maintained for 5 days without signs of toxicity, suggesting that C242-DM1 behaves like a prodrug.

HuN901-DM1: the second maytansinoid conjugate in development, consists of DM1 linked to the humanized antibody huN901, which binds to the CD56 antigen (21) that is expressed on all small cell lung cancers (SCLCs) and neuroblastomas. The conjugate was prepared by the method outlined earlier in Figure 7. HuN901-DM1 was evaluated for anti-tumor efficacy in a human small cell lung cancer xenograft model. Immunodeficient mice bearing xenografts established subcutaneously with SW2 cells were treated either with VP-16 or cisplatin, the standard agents used clinically in SCLC, or with huN901-DM1.

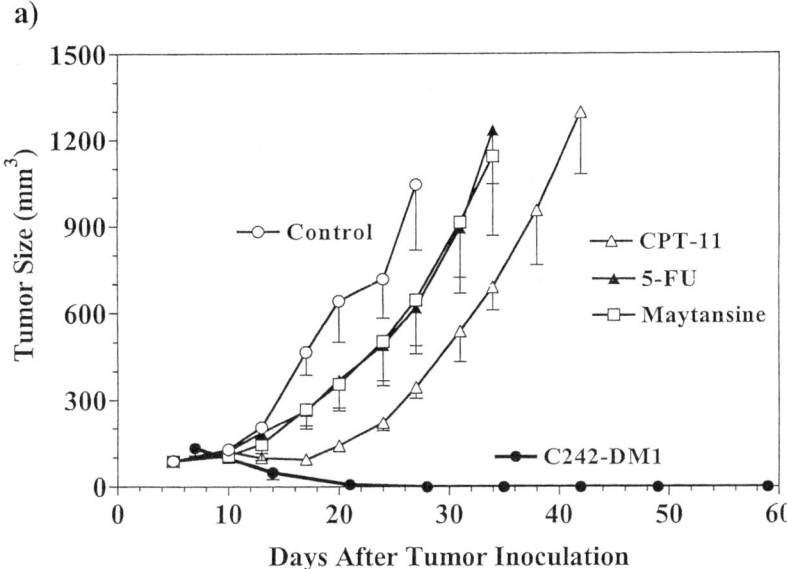

Figure 9a). Comparison of the anti-tumor efficacy of C242-DM1 and chemotherapeutic drugs in a human colon tumor xenograft (HT-29) model.

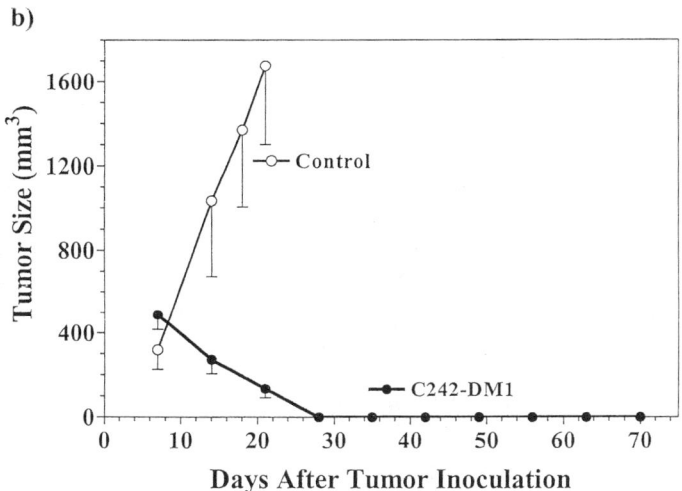

Figure 9b). Efficacy of C242-DM1 against large colon tumor (COLO 205) xenografts (Reprinted from Chari, R. V. J. *Adv. Drug Delivery Revs.* **1998,** *31,* 89-104 *with permission from Elsevier Science).*

Another group of animals received a mixture of unconjugated DM1 and free huN901 antibody. Tumors in animals treated with the maximum tolerated dose of VP-16 or cisplatin showed a modest growth delay as compared to the tumors in the untreated control group of animals. The mixture of free drug and antibody had no effect on tumor growth, while treatment with huN901-DM1 resulted in cures lasting greater than 200 days in all the animals (Figure 10). Thus, huN901-DM1 is a promising new agent with exceptional anti-tumor activity in an animal model of SCLC.

Antibody-CC-1065 Analog Conjugates: the CC-1065 analog, DC1, has been conjugated to antibodies as outlined in Figure 11. The antibody was reacted with the modifying agent 16 to introduce, on the average, 4 to 5 thiol groups per antibody molecule. Disulfide exchange between the modified antibody and the DC1 analog 9b provided a conjugate with about 4 molecules of DC1 linked per antibody molecule. Anti-B4-DC1 was prepared by linking DC1 to the anti-B4 antibody, which binds to the CD19 antigen expressed in B cell lymphoma.

Anti-B4-DC1 was evaluated for *in vitro* cytotoxicity towards the antigen-positive, human lymphoma cell line Namalwa and towards antigen-negative MOLT 4 cells (*12*). Anti-B4-DC1 was extremely cytotoxic for the target Namalwa cells with an IC_{50} value of 5×10^{-11} M, after a 72 hour exposure to the drug. The killing curve is very steep, with greater than 99.999% of cells killed at a conjugate concentration of 1×10^{-9} M. Anti-B4-DC1 is at least 1000-fold less cytotoxic to the non-target MOLT 4 cells ($IC_{50} > 5 \times 10^{-8}$ M), demonstrating the target specificity of the cytotoxic effect (Figure 12).

The anti-tumor activity of Anti-B4-DC1 was evaluated in an aggressive, metastatic tumor model of human B cell lymphoma (*12*). Anti-B4-DC1 showed remarkable anti-tumor efficacy, resulting in a median increase in life span of 265%, with 30% of the animals being cured. In comparison, the standard chemotherapeutic drugs, doxorubicin, vincristine, etoposide or cyclophosphamide, used clinically for the treatment of lymphoma had only a modest effect (median increase in life span of 22-91%).

Clinical Status and Conclusions

The new generation of antibody-drug conjugates, incorporating drugs that are considerably more potent than standard anti-cancer chemotherapeutics, are now in various stages of clinical evaluation. The most advanced of these agents is the humanized anti-CD33 antibody-calicheamicin conjugate, Mylotarg, which has been approved recently for the treatment of acute myeloid leukemia in

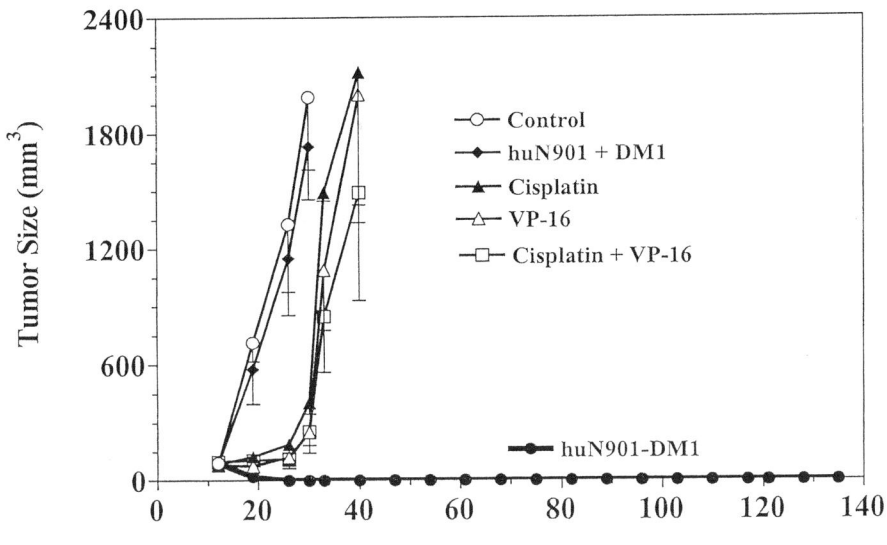

Figure 10. Comparison of the anti-tumor efficacy of huN901-DM1 and chemotherapeutic drugs in a human SCLC xenograft (SW2) model.

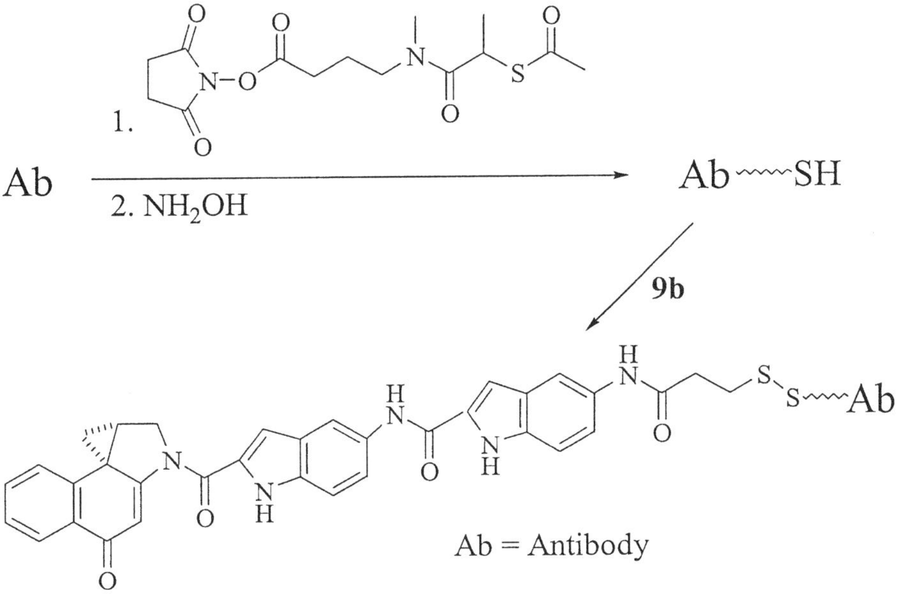

Figure 11. Preparation of Antibody-DC1 conjugates.

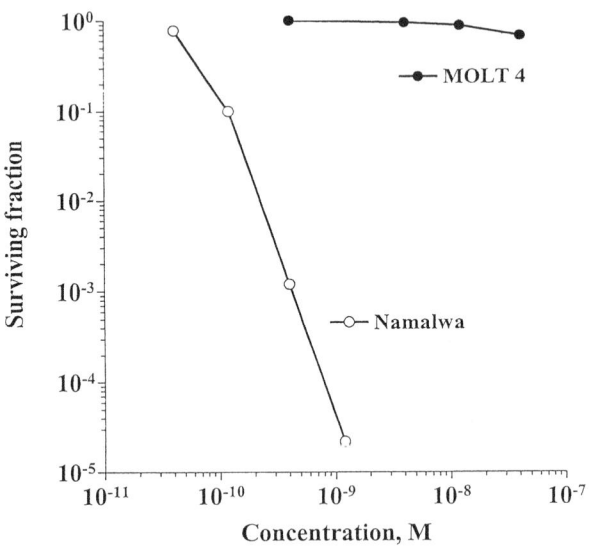

Figure 12. In vitro cytotoxicity and specificity of Anti-B4-DC1. Surviving fractions of cells were determined using clonogenic assays after a 72 h exposure to the conjugate.

certain patients. The antibody-maytansinoid conjugate, huC242-DM1 (also known as SB-408075) is in Phase I clinical evaluation in patients with colorectal and pancreatic cancers. Based on the impressive efficacy data in animal tumor models, it is expected that several more of these agents will be evaluated in the clinic in the near future. The clinical results and the approval of Mylotarg, opens the door to future developments of antibody conjugates of highly potent drugs that are too toxic to be useful as unmodified cytotoxics. In addition, it may be now possible to enhance the potency of naked antibodies that have some anti-tumor activity on their own, by linking them with highly cytotoxic drugs like DM1 and calicheamicin. Information from genomics research, coupled with the ability to generate monoclonal antibodies to any identified target and chemical creativity for the generation of potent, "linkable" cytotoxic drugs, may yet lead to more effective cancer treatments. In addition, such treatments may preserve a better quality of live for the patients than the currently available treatments.

References

1. Chari, R. V. J. *Adv. Drug Delivery Revs.* **1998**, *31*, 89-104.
2. Hobbs, S. K.; Monsky, W. L.; Yuan, F.; Roberts, W. G.; Griffith, L.; Torchilin, V. P.; Jain, R. K. *Proc. Natl. Acad. Sci. USA.* **1998**, *95*, 4607-4612.
3. *Antibodies as carriers of cytotoxicity*; Sedlacek, H.-H.; Seemann, D.; Hoffmann, D,; Czech, J.; Lorenz, P.; Kolar, C.; Bosslet, B., Eds.; Contributions to Oncology Beiträge zur Onkologie; Karger: Basel, 1992; Vol. 43, pp 29-97.
4. Kupchan, S. M.; Komoda, Y.; Branfman, A. R.; Sneden, A. T.; Court, W. A.; Thomas, G. J.; Hintz, H. P. J.; Smith, R. M.; Karim, A.; Howie, G. A.; Verma, A. K.; Nagao, Y.; Dailey, R. G.; Zimmerly, V. A.; Summer, W. C. *J. Org. Chem.* **1977**, *42*, 2349-2347.
5. Issell, B. F.; Crooke, S. T.; *Cancer Treatment Revs.* **1978**, *5*, 199-207.
6. Kupchan, S. M.; Sneden, A. T.; Branfman, A.R.; Howie, G. A.; Rebhun, L. I.; McIvor, W. E.; Wang, R. W.; Schnaitman, T.C. *J. Med. Chem.* **1978**, *21*, 31-37.
7. Hatano, K.; Higashide, E.; Akiyama, S.; Yoneda, Y. *Agric. Biol. Chem.* **1984**, *48*, 1721-1729.
8. Chari, R. V. J.; Martell, B. A.; Gross, J. L.; Cook, S. B.; Shah, S. A.; Blättler, W. A.; McKenzie, S. J.; Goldmacher, V. S. *Cancer Res.* **1992**, *52*, 127-131.
9. Li. L. H.; Kelly, R. C.; Warpehoski, M. A.; McGovren, J.P.; Gebgard, L.; DeKoning, T. F. *Invest. New Drugs.* **1987**, *9*, 137-148.

10. Boger, D. L.; Ishizaki, T.; Kito, P. A.; Suntonwart, O. *J. Org. Chem.* **1990,** *55,* 5823-5833.
11. Boger, D. L.; Yun, W.; Teegarden, B. R. *J. Org. Chem.* **1992,** *57,* 2873-2876.
12. Chari, R. V. J.; Jackel, K. A.; Bourret, L. A.; Derr, S. M.; Tadayoni, B. M.; Mattocks, K. M.; Shah, S. A.; Liu, C. Blättler, W. A.; Goldmacher, V. S. *Cancer Res.,* **1995,** *55,* 4079-4084.
13. Hinman, L. M.; Hamaan, P. R.; Wallace, R.; Menendez, A. T.; Durr, F. E., Upeslacis, J. *Cancer Res.* **1993,** *53,* 3536-3542.
14. Kingston, D. G. I.; Choudhary, A. G.; Chordia, M. D.; Gharpure, M.; Gunatilaka, A. A. L.; Higgs, P. I.; Rimoldi, J. M.; Samala, L.; Jagtap, P. G.; Jiang, Y. Q.; Lin, C. M.; Hamel, E.; Long, B. H.; Fairchild, C. R.; Johnston, K. A. *J. Med. Chem.* **1998,** *41,* 3715-3726.
15. Ojima, I.; Slater, J. C.; Michaud, E.; Kuduk, S. D.; Bounaud, P-Y.; Vrignaud, P.; Bissery, M-C.; Veith, J. M.; Pera, P.; Bernacki, R. J. *J. Med. Chem.* **1996,** *39,* 3889-3896.
16. Acton, E. M.; Tong, G. L.; Mosher, C. W.; Wolgemuth, R. L. *J. Med. Chem.* **1984,** *27,* 638-645.
17. Sievers, E. L.; Applebaum, F. R.; Spielberger, R. T.; Forman, S. J.; Flowers, D.; Smith, F. O.; Shannon-Dorcy, K.; Berger, M. S.; Bernstein, I. D. *Blood,* **1999,** *93,* 3678-3684.
18. Baeckstrom, D.; Hansson, G. C.; Nilsson, O.; Johansson, C.; Gendler, S. J.; Lindholm, L. *J. Biol. Chem.* **1991,** *266,* 21537-21547.
19. Roguska, M. A.; Pedersen, J. T.; Keddy, C. A.; Henry, A. H.; Searle, S. J.; Lambert, J. M.; Goldmacher, V. S.; Blättler, W. A.; Rees, A. R.; Guild, B. C. *Proc. Natl. Acad. Sci. (USA).* **1993,** *91,* 969-973.
20. Liu, C.; Tadayoni, B. M.; Bourret, L. A.; Mattocks, K. M.; Derr, S. M.; Widdison, W. C.; Kedersha, N. L.; Ariniello, P. D.; Goldmacher, V. S.; Lambert, J. M., Blättler, W. A.; Chari, R. V. J. *Proc. Natl. Acad. Sci. USA.* **1996,** *93,* 8618-8623.
21. Roy, D.C.; Ouellet, S.; Houillier, L.; Ariniello, P. D.; Perreault, C.; Lambert, J. M. *J. Natl. Cancer. Inst.* **1996,** *88,* 1136-1145.

Author Index

Afonso, A., 214
Aki, C. J., 214
Al-Awar, Rima S., 171
Allen, Jennifer R., 299
Altmann, Karl-Heinz, 112
Altstadt, Thomas, 43
Alvarez, C. S., 214
Anthony, N. J., 190
Baxter, Andy, 260
Bell, I., 190
Bernacki, Ralph J., 59
Bishop, W. R., 214
Blättler, Walter A., 317
Blommers, Marcel J. J., 112
Bombardelli, Ezio, 59
Borella, Christopher P., 59
Borzilleri, Robert M., 97
Bounaud, Pierre-Yves, 59
Brüggen, J., 231
Bryant, M. S., 214
Buchdunger, Elisabeth, 231, 245
Buchmann, Bernd, 131
Bunte, Thomas, 131
Buser, C. A., 190
Caravatti, Giorgio, 112, 231
Carr, D., 214
Chakravarty, Subrata, 59
Chao, J., 214
Chari, Ravi V. J., 317
Chen, Shu-Hui, 43
Connolly, M., 214
Cooper, A. B., 214
Corbett, Thomas H., 171
Cozens, R., 231

Cragg, G. M., 1
Danishefsky, Samuel J., 59, 299
Davide, J. P., 190
Decker, S., 1
del Rosario, J., 214
Desai, J., 214
Deskus, J.., 214
deSolms, S. J., 190
Dextraze, Pierre, 43
Dinsmore, C., 190
Doll, R. J., 214
Du, Karen, 43
Fairchild, Craig, 43
Ferlini, Cristiano, 59
Flörsheimer, Andreas, 112
Fray, Andrew H., 171
Furet, Pascal, 231, 245, 282
Ganguly, A. K., 214
Geng, Xudong, 59
Gibbs, J. B., 190
Girijavallabhan, V., 214
Golik, Jerzy, 43
Graham, S. L., 190
Hansel, Steven, 43
Hartman, G. D., 190
He, Lifeng, 59
Heimbrook, D. C., 190
Hoffmann, Jens, 131
Horwitz, Susan Band, 59, 81
Huber, H., 190
Humphreys, W. Griffith, 97
Hunt, John T., 199
Inoue, Tadashi, 59
James, L., 214

Johnson, J. I., 1
Johnston, Kathy A., 43
Kadow, John F., 43
Kelly, J., 214
Kim, Soong-Hoon, 97
Kirschmeier, P., 214
Klar, Ulrich, 131
Koblan, K. S., 190
Kohl, N. E., 190
Kramer, Robert A., 43
Kuduk, Scott D., 59
Langley, David R., 43
Lee, Francis Y. F., 97
Lee, Frank, 43
Lichtner, Rosemarie B., 131
Lin, Songnian, 59
Liu, M., 214
Liu, Y.-T., 214
Lobell, R. B., 190
Long, Byron, 43
Lydon, N., 231
Mallams, A. K., 214
Manley, Paul W., 282
Martello, Laura A., 81
Martinelli, Michael J., 171
Mastalerz, Harold, 43
McDaid, Hayley M., 81
Michaud, Evelyne, 59
Miller, Michael L., 59
Moher, Eric D., 171
Montana, John, 260
Moore, Richard E., 171
Nicolaou, Kyriacos C., 112, 148
Njoroge, F. G., 214
Nomeir, A. A., 214
Norman, Bryan H., 171
Ojima, Iwao, 59
O'Reilly, Terrence, 112, 231
Orr, George A., 59

Ouellet, Carl, 43
Park, Young Hoon, 59
Patel, Vinod F., 171
Patton, R., 214
Pera, Paula, 59
Perrone, Robert, 43
Pettit, George R., 16
Pinto, P., 214
Rane, D. F., 214
Rao, Srinivasa, 59
Regueiro-Ren, Alicia, 97
Remiszewski, S., 214
Riva, Antonella, 59
Rose, William C., 43
Rossman, R. R., 214
Sausville, E. A., 1
Scambia, Giovanni, 59
Schinzer, Dieter, 112
Schmidt, Alfred, 112
Schultz, Richard M., 171
Schulze, Gene, 43
Schwede, Wolfgang, 131
Scola, Paul, 43
Shih, Chuan, 171
Skuballa, Werner, 131
Slater, John C., 59
Smith, Amos B., III, 81
Staab, Andrew, 43
Strickland, C., 214
Sun, Chung-Ming, 59
Taveras, A. G., 214
Toth, John E., 171
Traxler, P., 231
Varie, David L., 171
Veith, Jean M., 59
Vibulbhan, B., 214
Vite, Gregory D., 97
Walker, Michael, 43
Walsh, John J., 59

Wang, J., 214
Wang, L., 214
Wang, Tao, 59
Wartmann, Markus, 112
Weber, P., 214
Wei, Jen-Mei, 43
Williams, T. M., 190
Winssinger, Nicolas, 148
Wittman, Mark, 43
Wolin, R., 214
Wright, J. J. Kim, 43
Xue, Quifen May, 43
Yang, Chia-Ping H., 81
Yvas, Dolatrai M., 43
Zimmermann, Jürg, 245
Zoeckler, Mary, 43

Subject Index

A

Abl kinase
  low molecular weight inhibitors, 246
  See also STI571
ABT-770
  enzyme inhibitory profile, 274f
  matrix metalloproteinase (MMP) inhibitor, 272, 274
  traditional non-substrate based template, 274
Actinomycetales, industry focus, 6
Actinomycin, antitumor antibiotics, 6
ADAM and ADAM-TS gene families, angiogenesis and bone resorption, 277
Adenosine 5'-triphosphate (ATP)
  ATP-competitive inhibitors of epidermal growth factor (EGF) receptor tyrosine kinase, 233–234
  interaction with vascular endothelial growth factor receptor (VEGFR-2) kinase, 285
  model of ATP binding site of EGF receptor tyrosine kinase, 237f
  portrayal of PTK787 bound to VEGFR-2 ATP-binding site, 294
  portrayal of SU6668 bound to VEGFR-2 ATP-binding site, 292
  portrayal of ZD4190 bound to VEGFR-2 ATP-binding site, 290
  proposed and alternative binding mode of PK1166 in ATP binding site of EGF receptor tyrosine kinase, 238f
  representation of ATP docked in binding site in VEGFR-2, 287
  structure, 286f
  structure of VEGFR-kinase and interaction with, 285

See also Vascular endothelial growth factors (VEGFs)
African bush willow, combretastatins, 17–18
Alkylation, potential of sarcodictyins, 165, 168f
17-Allylamino-17-demethoxygeldanamycin (17-AAG), Phase I trials, 9
Amaryllidaceae species, pancratistatin, 22
Amide-containing, epothilone analogs, 119, 120f, 121
Angiogenesis
  ADAM and ADAM-TS gene family, 277
  process in cancer, 282–283
  production of cytokines from effector cells, 282–283
  role of vascular endothelial growth factors (VEGFs) in tumor, 283–284
  See also Vascular endothelial growth factors (VEGFs)
Animals, potential for biosynthetic products for cancer treatment, 17
Anthracycline
  antitumor antibiotics, 6
  preparation of analogs, 327
  See also Antibody-drug conjugates
Antibody-drug conjugates
  active antibodies in "naked" form, 321
  anthracycline analogs, 327
  antibody–calicheamicin conjugates, 327, 329
  antibody–CC-1065 analog (DC1) conjugates, 334
  antibody–maytansinoid conjugates, 329, 332, 334

342

background and principles of,
preparation, 318–321
binding curve of typical humanized
monoclonal antibody, 320f
C242-DM1 antibody–maytansinoid
conjugate, 329, 332
calicheamicins, 324, 326f
CC-1065 and analogs adozelesin and
DC1, 324, 325f
clinical status, 334, 337
comparing anti-tumor efficacy of
C242-DM1 and chemotherapeutic
drugs in human colon tumor
xenograft model, 332, 333f
comparison of anti-tumor efficacy of
huN901-DM1 and
chemotherapeutic drugs in human
SCLC xenograft model, 335f
conjugates characteristic of prodrugs,
321–322
delivery and binding of monoclonal
antibodies to tumors, 319
efficacy of C242-DM1 against large
colon tumor xenografts, 332, 333f
highly potent analogs of paclitaxel,
docetaxel, and doxorubicin, 324,
327, 328f
in vitro cytotoxicity and specificity
of Anti-B4-DC1, 334, 336f
in vitro cytotoxicity and specificity
of C242-DM1, 331f
in vitro potency of free DM1 analog,
331f
maytansinoids, 322, 324
measuring cytotoxicity for six
cytotoxic agents towards colon
tumor cells, 320f
preparation of antibody-DC1
conjugates, 335f
preparation of antibody-DM1
conjugates, 330f
structural representation of antibody–
calicheamicin conjugates, 330f
structures of parent drug maytansine
and thiol-containing DM1, 323f

success of antibody-drug conjugates,
321
synthesis of thiol-containing
maytansinoid DM1, 323f
taxoids, 324, 327, 328f
Anti-cancer drug screen
COMPARE algorithm, 7–8
interpreting results, 7–8
in vitro, 7
value of characterizing expression of
molecular targets, 8
Anti-cancer drugs. *See* Biosynthesis of
anticancer drugs
Antigens. *See* Carbohydrate-based
tumor antigens
Anti-neoplastic activity,
phyllanthoside, 20
Anti-tumor activity
BMS-214662 [(R)-7-cyano-2,3,4,5-
tetrahydro-1-(1H-imidazol-4-
ylmethyl)-3-phenylmethyl)-4-(2-
thienylsulfonyl)-1H-1,4-
benzodiazepine)], 209f
cryptophycin 52 (LY355703), 185–
187
discovery of, in cryptophycins, 172–
174
epothilone B, 114, 115f
pyrrolo-pyrimidine PKI166 against
tumors growing in mice, 240–241
Anti-tumor antibiotics, natural
products, 6
Anti-tumor vaccine agents. *See*
Carbohydrate-based tumor antigens
Arthritis
Ro31-9790, 264
Ro113-0830 rheumatoid arthritis
model, 271
Trocade, 266–267
Astrocyte reversal (9ASK) system,
combretastatins, 17
Aureolic acid, antitumor antibiotics, 6
Autoradiography, farnesyltransferase
inhibitors, 192, 193f
Aziridinyl epothilones

conversion of epothilone A to, 104
in vitro data, 105t
synthesis, 103–104

B

Batimastat
  development as anti-cancer agent, 264–265
  structure, 264
BAY12-9566
  carboxylic acid system, 272
  clinical trials, 272
  enzyme inhibitory profile, 273f
  See also Matrix metalloproteinase (MMP) inhibitors
BB-3644, second generation matrix metalloproteinase (MMP) inhibitor, 265
Benzocycloheptapyridine FTIs. See Farnesyltransferase inhibitors (FTI)
Benzoquinone ansamycins, anticancer potential, 9
Bioactive compounds, marine environment, 5–6
Biological activity. See Epothilones
Biology
  activities of selected designed epothilone analogs, 159t
  epothilones, 149
  evaluation of epothilone libraries, 152–153, 157, 159
  evaluations of sarcodictyin libraries, 164–165, 167t
  farnesyltransferase inhibitors (FTI), 192, 193f
  FTI in animals, 192, 194
  sarcodictyins, 149
Biosynthesis of anticancer drugs
  bryostatins, 22, 25–26
  cephalostatins, 32, 34
  combretastatins, 17–18
  dolastatins, 29, 32
  focusing on evolutionary biosynthetic anticancer substituents, 34
  pancratistatin, 21–22
  phyllanthoside/phyllanthostatin 1, 18, 20–21
  potential for discovering from animal, plant, and microorganisms, 16–17
  spongistatins, 26–29
  structures of bryostatins, 23, 24
  structures of cephalostatins, 33
  structures of combretastatins, 19
  structures of dolastatins, 30, 31
  structures of phyllanthosides/phyllanthostatins, 19
  structures of spongistatins, 27
Bleomycin, antitumor antibiotics, 6
BMS-247550
  one pot synthesis, 108
  tumor growth delay for BMS-247550 versus paclitaxel in clinically derived tumor xenograft model, 110f
BMS-275291
  clinical trials, 267
  enzyme inhibitory profile, 268f
  See also Matrix metalloproteinase (MMP) inhibitors
Bone metastatis, matrix metalloproteinase (MMP) inhibitors, 276–277
Bone resorption, ADAM and ADAM-TS gene family, 277
Bristol–Myers, farnesyltransferase program, 211
Bryostatins
  Bugula neritina, 25
  clinical development, 26
  curative levels against variety of murine experimental cancer systems, 25–26
  marine source, 22, 25
  P388 lymphocytic leukemia, 25
  potential anticancer agent, 5

structures, 23, 24
yields, 25
Bugula neritina
　bryostatin, 25
　potential anticancer agent from, 5

C

Calicheamicins
　antibody–calicheamicin conjugates, 327, 329
　anti-CD33 antibody–calicheamicin conjugate (Mylotarg), 334, 337
　anti-tumor antibiotics, 324
　structural representation of antibody-drug conjugates, 330f
　structures of, and calicheamicin hydrazide, 326f
　See also Antibody-drug conjugates
Camptothecins, naturally derived agents, 5
Cancer clinical trials. See Human cancer clinical trials
Cancer drug discovery
　diversity generation, 3–6
　empirical, 2f
　empirical versus rational paradigms, 1–2
　examples of rationally improved anti-cancer agents, 8–12
　farnesyltransferase inhibitors, 11–12
　heat shock protein 90 (Hsp90) interactors, 8–9
　iterations of classical targets, 12
　kinase antagonists, 9–11
　National Cancer Institute (NCI) 60 cell line screen, 6–8
　natural products, 4–6
　rational, 3f
　synthetic compounds and combinatorial chemistry, 3–4
CaNT murine colon tumor
　combretastatin, 18
　pancratistatin, 22

Carbohydrate-based tumor antigens
　active and passive immunotherapies, 300
　challenge of successful delivery of antigen, 300
　development of anti-cancer vaccines using vaccine constructs, 300–301
　fucosyl GM_1 KLH conjugate, 304, 305
　global deprotection for fucosyl GM_1 *n*-pentenyl glycoside (NPG), 302
　glycoamino acids from NPGs, 309, 312, 313
　reagents and conditions for fucosyl GM_1 NPG, 306, 307
　reagents and conditions for Globo-H NPG, 310, 311
　solvolysis of glycal epoxides with allyl alcohol en route to KLH conjugation, 303
　synthesis of fucosyl GM_1 NPG, 301–308
　synthesis of Globo-H NPG, 308–309
　tumor-associated antigens Globo-H and fucosyl GM_1, 301f
CC-1065 analog
　antibody-drug conjugates, 334
　in vitro cytotoxicity and specificity of anti-B4-DC1, 336f
Cell mass, reduction, 12
Cephalostatins
　P388 murine lymphocytic leukemia, 32
　possibilities for modes of action, 32, 34
　potent activity against human cancer cells, 32
　preclinical development, 34
　structures, 33
CGS 27023A
　clinical trials, 269
　enzyme inhibitory profile, 269f
　non-substrate based matrix metalloproteinase (MMP) inhibitor, 268

See also Matrix metalloproteinase (MMP) inhibitors
Chlorohydrin derivative
  cryptophycin, 185
  macrocyclization and preparation of cryptophycin and, 175–176
  See also Cryptophycins
Chronic myeloid leukemia (CML)
  Bcr-Abl protein as major factor in pathophysiology, 246
  description, 245–246
  See also Phenylamino-pyrimidines (PAPs); STI571
CisPlatin
  antitumor activity against human tumor xenografts, 186t
  use with farnesyltransferase inhibitors (FTI), 192
Clinical trials
  matrix metalloproteinase (MMP) inhibitors, 276
  monitoring development of farnesyltransferase inhibitors (FTI), 196–197
  See also Human cancer clinical trials
Collagenases, group of matrix metalloproteinases (MMPs), 261
Combinatorial chemistry
  library synthesis, 152
  methods, 3–4
  peptoids, 4
  pin technology, 3
  screening metalloproteinase libraries, 274
  solid phase synthesis of epothilone A analogs, 154, 155
  split-pool method, 4
  tea-bag method, 4
Combretastatins
  A-4 and its phosphate prodrug, 18
  activity in astrocyte reversal (9ASK) system and P388 lymphocytic leukemia cell line, 17
  development of A-4 for human cancer clinical trials, 18

  isolation and synthesis of A-1, A-2, A-3, and A-4, 18
  isolation from African bush willow, 17–18
  structures, 19
COMPARE computer algorithm
  correlations of activity patterns, 7–8
  insight into mechanism of drug action, 8
  pancratistatin, 21–22
Conjugates. *See* Antibody-drug conjugates
Convergent synthetic approach, seco-cryptophycin, 175f
Cross resistance, discodermolide in Taxol-resistant cell lines, 87, 90
Cryptophycins
  analog preparation, 178
  antitumor activity of LY355703 (cryptophycin 52) against human tumor xenografts, 186t
  chemical structure of cryptophycin 1 and its four fragments, 173f
  comparison of epoxide and chlorohydrin derivatives, 185
  convergent synthetic approach to "seco-cryptophycin", 175f
  description, 171–172
  discovery of antitumor activity, 172–174
  in vivo antitumor activity of cryptophycin 52 (LY355703), 185–187
  key structure-activity relationship (SAR) findings, 179f
  macrocyclization and preparation of cryptophycin 52 and its chlorohydrin derivatives, 176f
  mechanism of action, 171–172
  pharmacological property, 172
  retrosynthetic analysis of cryptophycin 52 (LY355703), 174–175
  SAR of β-epoxide region of fragment A, 182, 183f

SAR of C/D ester bond and C6,C7-
positions, 180, 181*f*
SAR of D-chlorotyrosine of fragment
B, 184
SAR of phenyl group of fragment A,
182, 184
SAR regions of interest, 178, 180
SAR studies, 178, 180
sensitivity to assays, 178
Sharpless Asymmetric Epoxidation
(SAE) route to preparing fragment
A, 177*f*
stereochemistry, 173
total analysis of LY355703, 175–176
Cyclin-dependent kinases (CDKs)
flavopiridol, 10
modulators, 9–11
9-nitropaullone, 10–11
regulating cell cycle progression, 9
UCN-01 (7-hydroxystaurosporine),
9–10
Cyclopropyl derivatives
in vitro data for cyclopropyl
epothilones, 107*t*
stereoselective synthesis of 12,13-
cyclopropyl epothilones, 106
synthesis of cyclopropyl epothilones,
105–106
Cys-Aliphatic1-Aliphatic2-Xaa
(CAAX)
CAAX-based peptidic inhibitors,
202–205
C-terminal tetrapeptide sequence,
201–202
See also Farnesyltransferase
inhibitors (FTI)
Cytotoxicity
antimitotic agents in drug-resistant
cell line, 88*t*, 89*t*
discodermolide in Taxol-resistant
cell lines, 87, 90
in vitro cytotoxicity and specificity
of Anti-B4-DC1, 334, 336*f*
in vitro cytotoxicity and specificity
of maytansinoid C242–DM1, 331*f*

measuring cytotoxicity for six
cytotoxic agents towards colon
tumor cells, 320*f*
Taxol® enhancing cytotoxicity of
discodermolide in A549-T12 cells,
92*f*
Cytoxan, antitumor activity against
human tumor xenografts, 186*t*

D

Daunomycin-related agents, antitumor
antibiotics, 6
Deep-sea vents, untapped resources of
bioactive compounds, 6
Degradation, relay synthesis strategy,
101*f*
Degradation of natural product,
epothilone derivatives, 100–102
Deoxycoformycin, clinically active
agent, 6
Deoxygenation, oxiranyl epothilones,
101
Developmental Therapeutics Program
(DTP), National Cancer Institute
(NCI), 4
Didemnin B, first marine-derived
compound to enter clinical trials, 5
Dipeptide-based farnesyltransferase
inhibitors, 205–206
Direct semisynthesis. *See* Epothilone
derivatives
Discodermolide
cross resistance and cytotoxicity in
Taxol-resistant cell lines, 87, 90
cytotoxicity of antimitotic agents in
drug-resistant cell line not
expressing P-glycoprotein (Pgp),
89*t*
cytotoxicity of antimitotic agents in
drug-resistant cell line
overexpressing Pgp, 88*t*
structure, 68*f*
tubulin polymerization and

microtubule stabilization
properties, 150f
Discodermolide and Taxol®
chemical structures of Taxol,
epothilone A and B, deleutherobin,
and discodermolide, 84f
differences in bundle formation after
incubation in, 86f
enhancing microtubule bundle
formation, 83, 87
in vitro polymerization of
microtubules and bundle
formation, 83, 87
potentiation of apoptosis by, 95f
synergism in human carcinoma cell
lines, 93f, 94f
synergistic drug combination,
90, 92
Taxol® enhancing cytotoxicity of
discodermolide in A549-T12 cells,
92f
Docetaxel (TAXOTERE®)
evolution of taxoids, 61–68
highly potent analogs of paclitaxel,
docetaxel, and doxorubicin, 328f
structure, 44, 61
See also TAXOL® (paclitaxel)
Dolabella auricularia, dolastatins, 29
Dolastatins
human cancer clinical trials, 32
marine sea hare, 29
preclinical development, 29, 32
structures, 30, 31
synthesis of dolastatin 10, 29
yield, 25
Doxorubicin
antitumor activity against human
tumor xenografts, 186t
highly potent analogs of paclitaxel,
docetaxel, and, 328f
use with farnesyltransferase
inhibitors (FTI), 192
Drug. See Antibody-drug conjugates
Drug discovery. See Cancer drug
discovery

Drugs. See Biosynthesis of anticancer
drugs
Dysoxylum binectariferum, structural
basis for flavopiridol, 5

E

Eleutherobin
cytotoxicity of antimitotic agents in
drug-resistant cell line not
expressing P-glycoprotein (Pgp),
89t
cytotoxicity of antimitotic agents in
drug-resistant cell line
overexpressing Pgp, 88t
structure, 68f
tubulin polymerization and
microtubule stabilization
properties, 150f
Empirical paradigm
drug discovery, 2f
versus rational, 1–2
Epidermal growth factor (EGF)
receptor tyrosine kinase
ATP-competitive inhibitors of, 233–
234
model of ATP binding site of, 237f
4-phenylamino-quinazolines, 233f
proposed and alternative binding
mode of PK1166 in ATP binding
site of, 238f
rational for inhibitors of, in cancer,
232
See also Pyrrolo[2,3-d]pyrimidine
and pyrazolo[3,4-d]pyrimidine
derivatives
Epothilone A and B
cytotoxicity of antimitotic agents in
drug-resistant cell line not
expressing P-glycoprotein (Pgp),
89t
cytotoxicity of antimitotic agents in
drug-resistant cell line
overexpressing Pgp, 88t

structures, 68f
tubulin polymerization and microtubule stabilization properties, 150f
See also Epothilones
Epothilone derivatives
advantages over prototypical taxanes, 98
conversion of epothilone A to 12,13-aziridyinyl epothilones, 104
12,13-cyclopropyl epothilones, 105–106
degradation of epothilone C to C1–C12 fragment and conversion to known precursor, 102
degradation of natural product, 100–102
degradation/relay synthesis strategy, 101f
deoxygenation of oxiranyl epothilones, 101
direct semisynthesis, 103–109
discovery, 98
epothilone B-derived π-allylpalladium complex, 107f
evaluation of Nicolaou's total synthesis route, 99
feasibility of relay synthesis, 102
flaws of epothilone B hampering clinical development, 98–99
in vitro data for aziridinyl epothilones, 105t
in vitro data for cyclopropyl epothilones, 107t
in vitro data for lactam analogues, 109t
mechanism of action, 97–98
one pot synthesis of clinical candidate BMS-247550, 108
potency of epothilone B lactam, 108–109
replacement of 12,13-oxirane with C–N bond, 103
stereoselective synthesis of 12,13-cyclopropyl epothilones, 106
structures of representative, 98f
susceptibility to palladium-catalyzed ring opening to intermediate π-allylpalladium complex, 106, 107f
synthesis of aziridinyl analogues, 103–104
synthesis of epothilone B lactam, 107–108
three-step semisynthesis of epothilone B lactam, 108
total synthesis, 99–100
total synthesis of lactams, 100
toxicity of epothilone B, 99
tumor growth delay for BMS-247550 versus paclitaxel in clinically derived tumor xenograft model, 110f
Epothilone syntheses
A, B, C, and D analogs, 132
B and D analogs, 132–139
building block A (C1–C6), 133, 134
building block B (C7–C12), 133, 135
building block C (C13–C17), 136–137
construction of framework, 137–138
epothilone fragments, 116
epothilones and analogs–generalized retrosynthesis, 117
library synthesis, 152
N-methyl amide analog, 120
optimized synthesis of B for designed focused libraries, 157
radiolabelled B and D analogs, 144, 145
requisite building blocks in retro approach, 116, 118, 119
retrosynthetic analysis of A using ring-closing olefin metathesis, 151f
retrosynthetic analysis of epothilone library by solid phase olefin metathesis, 152f
retrosynthetic analysis of focused libraries of B analogs allowing for diversity of R^1 and late stage

introduction of heterocycle R^2, 156f
retrosynthetic approach, 149, 151
schematic of total synthesis, 138–139
side-chain modified epothilone B analogs, 126
solid phase synthesis of A analogs by combinatorial approach, 154, 155
stereoselective synthesis of B analogs, 158
trans-epothilones A–epoxidation selectivity, 123

Epothilones
analogs from epoxide ring opening, 125
antiproliferative effects in multi-drug resistant (MDR) cells, 132
antitumor activity of epothilone B in human models, 114
antitumor effects of epothilone B and Taxol® against subcutaneously transplanted human epidermoid carcinoma, 115f
bioactive conformation of side chain, 142–143
biological activities of selected designed analogs, 159t
biological evaluation of epothilone library, 159
biology, 149
C12/C13 amide- and imidazole-containing epothilone analogs, 119, 121
enhanced overall activity in panel of tumor cell lines, 142f
epothilone analogs incorporating 12,13 *N*-alkyl amide and imidazole groups as potential deoxyepothilone B mimetics, 120f
heterocycles modifications, 125–128
IC50-values for growth inhibition of human carcinoma cell lines by C12/C13-modified epothilone A analogs, 126t
IC50-values for net growth inhibition of human carcinoma cell lines by epothilone A and B and paclitaxel, 114t
improvement in activity and profile, 143, 144f
improvement of activity, 141–142
indications of different mode of action between B and D, 146
induction of tubulin polymerization and growth inhibition of human carcinoma cell lines by C12/C13 *trans*-epothilones A, 124t
induction of tubulin polymerization and growth inhibition of human carcinoma cell lines by C12/C13 amide- and imidazole-modified epothilone analogs, 122t
in vitro characterization of radiolabelled, 145t, 146
mechanism of action, 113
paralleling biological activity of paclitaxel, 131
pyridine and pyrimidine-based analogs of epothilone B–importance of nitrogen positioning, 127t
pyridine-based analogs of epothilone B–biological activity, 127t
radiolabelled B and D analogs, 144–146
reactions of epoxide moiety of epothilone A, 125
stereoselective synthesis of *trans*-deoxyepothilone A, 123
structure and multidrug resistant (MDR) sensitivity, 140–141
structures of epothilones A and B, 113f
summary of structure-activity relationships of library, 156f
toxicity and overcoming P-glycoprotein (Pgp) efflux, 141f
trans-epothilones A, 121, 124

use with farnesyltransferase
inhibitors (FTI), 192
Epoxide derivative, cryptophycin, 185
Epoxide ring opening, epothilone
analogs by, 125
Epstein–Barr virus (EBV),
phyllanthoside, 21
Etoposide, antitumor activity against
human tumor xenografts, 186*t*

F

Farnesyl protein transferase (FPT)
active site cavity, 216
benzocycloheptapyridyl Sch-66336, 216
binding interactions with inhibitors, 220–222
library screening at Schering–Plough Research Institute (SPRI), 216
structure-activity relationship (SAR) studies of benzocycloheptapyridine FTIs, 217–224
structure of Sch-66336, 216
X-ray crystal structure determination, 215
X-ray crystal structure of 2-FPT, 218*f*, 219*f*
See also Farnesyltransferase inhibitors (FTI)
Farnesyl pyrophosphate (FPP), use as substrate, 200
Farnesyltransferase inhibitors (FTI)
anti-proliferative activity in combination with cytotoxic agents, 192
attention of pharmaceutical industry for identification of farnesyl-protein transferase (FTase), 191
binding for 3,8,10- and 3,7,8-trihalobenzocycloheptapyridine FTIs, 223, 224*f*
biology in animals, 192, 194
biology in vitro, 192
clinical trials with 23 [(R)-7-cyano-2,3,4,5-tetrahydro-1-(1H-imidazol-4-ylmethyl)-3-phenylmethyl)-4-(2-thienylsulfonyl)-1H-1,4-benzodiazepine)] (BMS-214662), 210
C-terminal tetrapeptide sequence (CAAX) box, 201–202
Cys-Aliphatic1-Aliphatic2-Xaa (CAAX)-based peptidic inhibitors, 202–205
2-dimensional gel electrophoresis and autoradiography after incubation of cells, 193*f*
dipeptide-based inhibitors, 205–206
enzyme ligation of zinc atoms, 203–204
farnesyl pyrophosphate (FPP-) and bisubstrate-based inhibitors, 200–202
FPT binding pockets surrounding tricyclic structure of FTI 7, 220, 221*f*
FPT potency and pharmacokinetic profile of piperidylacetyl FTIs, 226*t*
in vivo antitumor activity of 23 in HCT-116 xenograft model, 209*f*
intensive investigation of tetrahydrobenzodiazepine 17, 206, 208–211
mechanism of action of Ras proteins, 199–200
monitoring clinical development, 196–197
nonpeptidomimetic inhibitors, 194, 196
optimization of 3-(R)-phenylmethyl series, 208
optimization of 3-(R)-phenylmethyl-7-CN series, 208
optimizing cellular and pharmacokinetic properties leading to Sch-66336, 225–227

pharmacokinetics of 23 in rats, 210f
potency enhancements with
benzocycloheptapyridine FTI, 220
potential liability of thiol tetrapeptide
inhibitor, 203
rational drug design program, 211
rationally improved anti-cancer
agents, 11–12
reaction catalyzed by FT, 201f
relationship of binding orientation
and chirality, 223
Sch-66336 for further development
in treating human cancers, 226
Sch-66336 inhibiting tumor colony
formation, 226
soft agar assay, 192
structure-activity relationships (SAR)
of benzocycloheptapyridine FTIs,
217–225
structure of peptidomimetic FTI L-
744,832, 193f
structure of
tetrahydrobenzodiazepine 17, 207
structures of non-peptide FTIs and
biochemical properties, 195f
synergism of Sch-66336 with other
anticancer agents, 227
tetrahydrobenzodiazepine inhibitors,
206–211
tetrapeptide-based inhibitors, 202–
205
tumor growth inhibition due to
inhibition of farnesylation of Ras,
227
X-ray crystal structure of 2-FPT,
218f, 219f
X-ray structure for 3,8,10- and 3,7,8-
trihalobenzocycloheptapyridine
FTIs, 224f
X-ray structure of 7-FPT, 221f, 222f
Flavopiridol
base structure of product from
Dysoxylum binectariferum, 5
cyclin-dependent kinase (CDK)
modulator, 10

Flora, potential for biosynthetic
products for cancer treatment, 17
5-Flurouracil, antitumor activity
against human tumor xenografts,
186t

G

Gelatinases, group of matrix
metalloproteinases (MMPs), 261
Geldanamycin (GA)
in vivo toxicity and lack of stability, 9
synthesis of dimer, 9
Gel electrophoresis,
farnesyltransferase inhibitors, 192,
193f

H

Halihondria okadai, halichondrin B
from, 5–6
Heat shock protein 90 (Hsp90),
rationally improved anti-cancer
agent, 8–9
Herbimycin A, in vivo toxicity and
lack of stability, 9
Herpes viruses, phyllanthoside, 21
Heterocycle modifications, epothilone
analogs, 125–128
Human cancer clinical trials
antibody–maytansinoid conjugate
huC242-DM1, 337
Batimastat, 264–265
BMS-214662, 210
BMS-275291, 267
bryostatin, 26
combretastatins, 18
dolastatins, 32
Marimastat, 265
phyllanthoside, 20–21
second generation matrix
metalloproteinase (MMP) inhibitor
BB-3644, 265

STI571, 256
Human cancers, spongistatins, 28–29
Human carcinoma cell lines,
  synergism of Taxol® and
  discodermolide, 93f, 94f
Human lung carcinoma cell line
  (A549-T12 cells)
  requirement for Taxol®, 90
  resistance to Taxol®, 87, 90
Human tumor cell lines, 60-cell
  in vitro screening model, 6–8
  pancratistatin, 21–22
Human tumors, Ras mutations, 215
Hybrids. See Taxoids and hybrids

I

IDN5109
  antitumor activity against drug-
    resistant tumors expressing P-
    glycoprotein (Pgp), 66–67
  formerly SB-T-101131, 65
  in vivo antitumor activity, 66f
  oral application after human colon
    carcinoma SW-620 application,
    67f
  potent and orally active taxoid, 64–
    67
  superior activity to paclitaxel and
    docetaxel, 65, 66
  See also Taxoids
Imidazole-containing epothilone
  analogs, 119, 120f, 121
Immunotherapy. See Carbohydrate-
  based tumor antigens
Indolinones, vascular endothelial
  growth factor receptor (VEGFR)-
  kinase inhibitors, 291, 293
Inhibitors
  ultra-tight or irreversible binding, 12
  See also Farnesyltransferase
    inhibitors (FTI); Matrix
    metalloproteinase (MMP)
    inhibitors; Pyrrolo[2,3-
    d]pyrimidine and pyrazolo[3,4-
    d]pyrimidine derivatives; STI571;
    Vascular endothelial growth
    factors (VEGFs)
Irreversible binding inhibitors,
  reduction of cell mass, 12

K

Kinase antagonists
  cyclin-dependent kinases (CDKs), 9–
    11
  flavopiridol, 10
  9-nitropaullone, 10–11
  rationally improved anti-cancer
    agent, 9–11
  UCN-01 (7-hydroxystaurosporine),
    9–10

L

Lactams
  in vitro data for lactam analogues,
    109t
  one pot synthesis of clinical
    candidate BMS-247550, 108
  potency of epothilone B lactam, 108–
    109
  susceptibility of lactones to
    palladium-catalyzed ring opening,
    106, 107f
  synthesis of epothilone B lactam, 108
  synthesis of epothilone lactams, 106–
    108
  three-step semisynthesis of
    epothilone B lactam, 108
Laulimalide, tubulin polymerization
  and microtubule stabilization
  properties, 150f
Leukemia
  Bcr-Abl protein as major factor in
    pathophysiology of chronic
    myeloid leukemia (CML), 246

bryostatin, 25
cephalostatins, 32
chronic myeloid leukaemia, 245–246
combretastatins, 17
dolastatins, 29
phyllanthoside, 20
spongistatins, 26
See also P388 lymphocytic leukemia cell line; Phenylaminopyrimidines (PAPs); STI571
Libraries
biological evaluation of epothilone, 152–153, 157, 159
biological evaluation of sarcodictyin, 164–165, 167*t*
optimized synthesis of epothilone B for designed focused, 157
summary of structure-activity relationships of epothilone, 156*f*
synthesis of epothilones using Radiofrequency Encoded Combinatorial (REC™) chemistry, 152
synthesis of sarcodictyin, 164
synthesis of sarcodictyin analogs, 162–163
See also Epothilones

M

Marimastat
clinical trials, 265
enzyme inhibitory profile, 262*f*
See also Matrix metalloproteinase (MMP) inhibitors
Marine animals
bryostatin, 22, 25
potential for biosynthetic products for cancer treatment, 17
Marine environment, rich source of bioactive compounds, 5–6
Marine sea hare, dolastatins, 29
Matrix metalloproteinase (MMP) inhibitors

ABT-770, 272, 274
Batimastat for development as anti-cancer agent, 264
BAY12-9566 with carboxylic acid, 272
BB-3644 second generation compound, 265
BMS-275291, 267
CGS 27023A, 268–269
chain extension to phenethylamine derivatives, 275
clinical trials of CGS27023A and Prinomastat, 269–270
combinatorial libraries, 274
converting hydroxamic acid to carboxylic acid, 272
design, 263
design of more selective, 270–276
design of substrate based MMPIs around substrate cleavage site, 264*f*
discovery of ADAM and ADAM-TS gene families, 277
drug discovery paradigm, 262–263
early, 264–265
enzyme inhibitory profile of ABT-770, 274*f*
enzyme inhibitory profile of BAY12-9566, 273*f*
enzyme inhibitory profile of BMS-275291, 268*f*
enzyme inhibitory profile of CGS27023A and Prinomastat, 269*f*
enzyme inhibitory profile of Marimastat, 262*f*
enzyme inhibitory profile of Ro113-0830, 271*f*
enzyme inhibitory profile of Trocade, 267*f*
future directions, 276–277
lack of published structure-based drug design data on non-substrate based inhibitors, 270
Marimastat, 265

non-substrate based, 268–270
Prinomastat, 268–270
replacing hydroxamic acid with alternative zinc-binding groups, 267
reverse hydroxamates, 272, 274
Ro31-9790 for development for arthritis, 264
Ro113-0830, 270–271
schematic of key interactions of first generation MMPI and MMP8, 266f
second generation, 266–267
'sheddase' enzymes, 262–263
structure-activity relationship of Pharmacia Upjohn thiadiazoles, 275f
thiadiazoles providing unprimed site binding interactions, 276
thiadiazole ureas, 274
Trocade, 267
tumor metastasis to bone, 276–277
X-ray crystal structure of fluorine-substituted inhibitor, 275–276
Matrix metalloproteinases (MMPs)
activity regulation by tissue inhibitors of metalloproteinases (TIMPs), 261
division into classes, 261
turnover and remodeling of extracellular matrix (ECM), 261
X-ray crystal structures, 277
Maytansinoids
antibody–maytansinoid conjugates, 329, 332, 334
C242–DM1, 329, 332
comparing anti-tumor efficacy of huN901–DM1 and chemotherapeutic drugs, 335f
huN901–DM1, 332, 334
in vitro cytotoxicity and specificity of C242–DM1, 331f
in vitro potency of free DM1 analog, 331f

preparation of antibody–DC1 conjugates, 335f
preparation of antibody–DM1 conjugates, 330f
structure of maytansine and thiol containing DM1, 323f
synthesis of thiol containing DM1, 323f
See also Antibody-drug conjugates
Microorganisms, potential for biosynthetic products for cancer treatment, 17
Microtubule polymer
differences in bundle formation after incubation with Taxol® and discodermolide, 86f
in vitro activity of microtubule agents in tubulin polymerization assay, 85f
in vitro polymerization and bundle formation, 83, 87
Taxol® and discodermolide enhancing bundle formation, 83, 86f
Microtubule stabilization, natural products with, 150f
Microtubule-stabilizing anticancer agents
design and synthesis of series of N-linked macrocyclic hybrids, 71
development of common pharmacophore model, 68–71
lower energy conformation of nonataxel arising from restrained molecular dynamics (RMD) calculations, 69f
macrocyclic hybrids connecting C-2 and C-3' positions of taxoid structure, 70–71
model for designing third generation taxoids, 70
probing drug binding domains, 72–75
structures of epothilones A and B,

eleutherobin, and discodermolide, 68f
three-dimensional pharmacophore common to structurally diverse agents, 69
See also Taxoids
Mitomycin, antitumor antibiotics, 6
Model. *See* Microtubule-stabilizing anticancer agents
MT-MMPs, group of matrix metalloproteinases (MMPs), 261
Multidrug resistance (MDR), description, 67
Mylotarg
humanized anti-CD33 antibody–calicheamicin conjugate, 334, 337
See also Antibody-drug conjugates

N

National Cancer Institute (NCI)
60 cell line screen, 6–8
Developmental Therapeutics Program (DTP), 4
interpreting results from 60 cell line screen, 7–8
in vitro anti-cancer drug screen, 7
research programs for biosynthetic anticancer substituents, 16–17
Natural product degradation, epothilone derivatives, 100–102
Natural products
antitumor antibiotics, 6
cancer treatment, 4–6
deep-sea vents, 6
marine environment, 5–6
plants, 4–5
sponges, 5–6
New generation taxoids. *See* Taxoids and hybrids
9-Nitropaullone, cyclin-dependent kinase (CDK) modulator, 10–11
Nonataxel
lower energy conformation, 69f

structure, 68f
Nonpeptidomimetric inhibitors, farnesyltransferase inhibitors (FTI), 194, 196

O

Osteolysis
osteoclast activity in vicinity of secondary tumor, 276
utility of matrix metalloproteinase (MMP) inhibitors, 276–277
Oxiranyl epothilones, deoxygenation, 101

P

P388 lymphocytic leukemia cell line
bryostatin, 25
cephalostatins, 32
combretastatins, 17
dolastatins, 29
phyllanthoside, 20
spongistatins, 26
Pacific yew tree, paclitaxel (TAXOL®), 60
Paclitaxel
antitumor activity against human tumor xenografts, 186t
evolution of taxoids, 61–68
highly potent analogs of, docetaxel, and doxorubicin, 328f
oral application after human colon carcinoma SW-620 application, 67f
use with farnesyltransferase inhibitors (FTI), 192
See also TAXOL® (paclitaxel)
Pancratistatin
Amaryllidaceae species, 22
anticancer activity, 21–22
prodrug, 21
structure and synthesis, 21

syntheses, 22
(-)-Pantolactone
  starting material for epothilones, 133,
    134
  See also Epothilone syntheses
Paradigms
  empirical drug discovery, 2f
  empirical versus rational, 1–2
  rational drug discovery, 3f
n-Pentenyl glycosides (NPG)
  glycoamino acids from, 309, 312
  See also Carbohydrate-based tumor
    antigens
Peptoids, peptide-like molecules, 4
P-glycoprotein (Pgp)
  epothilones and discodermolide not
    reversing inhibition of Taxol®
    accumulation in Pgp expressing
    cell line, 91f
  photolabeling taxoids for mapping
    Pgp, 75
  probing drug binding domains, 72–
    75
Pharmacophore model. See
    Microtubule-stabilizing anticancer
    agents
Phenylamino-pyrimidines (PAPs)
  antiproliferative activity on factor-
    independent cells, 255
  assay in cellular systems to evaluate
    specificity and mechanism of
    action, 254–255
  assaying tyrosine and
    serine/threonine kinases, 249
  binding model consistent with
    structure-activity relationships,
    251
  compounds prepared, 248t
  docking STI571 into model, 252
  efficacy and specificity of STI571,
    255–256
  general synthesis, 247
  inhibition of tyrosine and
    serine/threonine kinases, 250t

leading to development of selective
  protein kinase inhibitors with high
  potency, 249
pharmacological profile of SIT571,
  254–256
preclinical profile, 256
preparation, 246
proposed binding model of STI571 in
  model of activated form of Abl
  kinase, 253
protein kinase inhibitors, 246
relative orientations of STI571 and
  analog in activated kinase model,
  252, 253
screening for selectivity, 249
selective inhibition of Bcr-Abl-
  positive progenitor cells from
  CML patients, 255
solubility and acidity, 246
STI571 analog binding to inactive
  conformation of kinase, 252
STI571 preferentially binding to
  inactive form of kinase, 252, 254
structural basis of inhibition of Abl
  kinase by STI571, 251–254
structural formulas, 247
See also STI571
Phyllanthoside/phyllanthostatin
  antineoplastic activity against B16
    melanoma and P388 leukemia, 20
  discovery, 20
  phase 1 human trial, 20–21
  source, 18, 20
  strong activity against human
    neoplastic cell lines, 20
  structures, 19
Pin technology, synthesizing new
  compounds, 3
PK1166
  antiproliferative activity, 240t
  antitumor effects against mice, 241f
  binding model, 236–238
  enzymatic profile, 240t
  preclinical studies, 239–242

358

proposed and alternative binding mode of, in ATP binding site of epidermal growth factor receptor (EGFR) kinase, 238f
structure-activity relationships (SAR) leading to, 234–236
synthesis, 239f
See also Pyrrolo[2,3-d]pyrimidine and pyrazolo[3,4-d]pyrimidine derivatives
Plants
cancer treatment, 4–6
potential for biosynthetic products for cancer treatment, 17
Potential anticancer agents, bryostatin 1, 5
Prinomastat
clinical trials, 269–270
enzyme inhibitory profile, 269f
non-substrate based matrix metalloproteinase (MMP) inhibitor, 268
See also Matrix metalloproteinase (MMP) inhibitors
Protein tyrosine kinases
role in intracellular signaling, 245
See also STI571
PTK787/ZK222584
class of vascular endothelial growth factor receptor (VEGFR)-kinase inhibitors, 293–295
efficacy in inhibiting, 295
inhibitory profiles, 288t
oral administration, 295
portrayal of, docked in VEGFR-2 ATP-binding site, 294
safety, 295
structure, 286f
See also Vascular endothelial growth factors (VEGFs)
Pyridine-based analogs, epothilones, 126, 127t
Pyrimidine-based analogs, epothilone B, 127t
Pyrrolo[2,3-d]pyrimidine and pyrazolo[3,4-d]pyrimidine derivatives
antiproliferative activity of PK1166, 240t
antitumor effects of PK1166 in mice, 241f
ATP-competitive inhibitors of epidermal growth factor receptor (EGFR) tyrosine kinase, 233–234
binding model of PK1166, 236–238
effect of 5 and 6 substituents of pyrrolo-pyrimidine, 235t
effects of substituents at 6-phenyl group of pyrrolo-pyrimidine, 235t
enzymatic profile of PK1166, 240t
extensive structure-activity relationship (SAR) determinations, 234–236
model of ATP binding site of EGFR kinase, 237f
4-phenylamino-quinazolines as inhibitors of EGFR tyrosine kinase, 233f
PK1166 inhibiting EGF-stimulated autophosphorylation, 241
preclinical studies with PK1166, 239–242
proposed and alternative binding mode of PK1166 in ATP binding site of EGFR kinase, 238f
pyrazolo[3,4-d]pyrimidines, 236t
rational for EGFR tyrosine kinase inhibitors in cancer, 232
replacements of anilino group, 235t
SAR leading to discovery of potent inhibitor PK1166, 234–236
synthesis of PK1166, 239f
synthetic access to pyrrolo[2,3-d]pyrimidines, 239

Q

Quinazolines, vascular endothelial growth factor receptor (VEGFR)-kinase inhibitors, 289, 291

R

Radiofrequency encoded combinatorial (REC™) chemistry, library synthesis, 152
Ras
 farnesyltransferase inhibitors, 11–12
 mutations in human tumors, 215
 target for anticancer drug development, 11
Rational paradigm
 cyclin-dependent kinase (CDK) modulators, 9–11
 drug discovery, 3f
 examples of rationally improved anti-cancer agents, 8–12
 farnesyltransferase inhibitors, 11–12
 farnesyltransferase program at Bristol–Myers, 211
 flavopiridol, 10
 heat shock protein 90 (Hsp90) interactors, 8–9
 iterations of classical targets, 12
 kinase antagonists, 9–11
 9-nitropaullone, 10–11
 UCN-01 (7-hydroxystaurosporine), 9–10
 versus empirical, 1–2
Reduction, cell mass, 12
Relay synthesis, epothilone derivatives, 101–102
Restrained molecular dynamics (RMD), lower energy conformation of nonataxel, 69
Retrosynthetic approach
 convergent approach to "seco-cryptophycin", 175f
 cryptophycin, 174–176
 epothilone A using ring-closing olefin metathesis, 151f
 epothilone library using solid phase olefin metathesis, 152f
 epothilones, 149, 152
 focused libraries of epothilone B analogs, 156f
 sarcodictyin library, 159–160
 See also Epothilone syntheses
Ro113-0830
 enzyme inhibitory profile, 271f
 preclinical trials, 270–271
 rheumatoid arthritis model, 271
 See also Matrix metalloproteinase (MMP) inhibitors
Ro 31-9790
 arthritis, 264
 structure, 264
 See also Matrix metalloproteinase (MMP) inhibitors

S

Sarcodictyins
 alkylating potential, 168f
 biological activity of selected analogs, 167t
 biological evaluation of libraries, 164–165
 biology, 149
 library synthesis, 164
 loading and cleavage of, scaffold onto solid phase, 161, 164
 retrosynthetic analysis of, library, 160f
 summary of structure-activity, 166f
 synthesis of core structure, 161f
 synthesis of library of analogs, 162–163
 tubulin polymerization and microtubule stabilization properties, 150f
Sch-66336
 benzocycloheptapyridine farnesyltransferase inhibitor, 227
 further development, 226
 inhibition of tumor colony formation, 226
 optimizing cellular and pharmacokinetic properties, 225–227

result of screening compound libraries, 216
structure, 216
synergism with other anticancer agents, 227
See also Farnesyltransferase inhibitors (FTI)
Screen, National Cancer Institute (NCI) 60 cell line, 6–8
Seco-cryptophycin, convergent synthetic approach, 175f
Second generation taxoids. See Taxoids and hybrids
Semisynthesis, direct epothilone derivatives, 103–109
See also Epothilone derivatives
Sharpless Asymmetric Epoxidation (SAE)
controlling two of four asymmetric centers, 174
route to preparation of fragment A for cryptophycin, 176, 177f
See also Cryptophycins
Sheddase enzymes, metalloproteinases cleaving extracellular proteins, 262–263
SMART Microreactors™, library synthesis, 152
Split-pool method, synthesizing new compounds, 4
Sponges, rich source of bioactive compounds, 5–6
Spongistatins
isolation, 26, 28
objective of efficient total synthesis, 29
P388 lymphocytic leukemia, 26
profile against human cancer cell line panel, 28–29
structural determination, 28
structures, 27
yield, 25
Stereoselective synthesis, epothilone B analogs, 157–158
STI571

analog binding to inactive conformation of kinase, 252
antiproliferative activity, 255
assay in cellular systems to evaluation specificity and mechanism of action, 254–255
binding model, 251
docking into model, 252
efficacy and specificity, 255–256
pharmacological profile, 254–256
preclinical trials, 256
preferential binding to inactive kinase, 252, 254
proposed binding mode in model of activated form of Abl kinase, 253
selective inhibition of Bcr-Abl-positive progenitor cells from CML patients, 255
structural basis of inhibition of Abl kinase, 251–254
See also Phenylamino-pyrimidines (PAPs)
Streptomyces, clinically active agents from, 6
Streptozocin, clinically active agent, 6
Stromelysins
group of matrix metalloproteinases (MMPs), 261
thiadiazole ureas as inhibitors, 274
Structure-activity relationships (SAR)
bioactive conformation of side chain of epothilones, 142–143
epothilone library, 153, 156f
improvement in activity and profile of epothilones, 143, 144f
improving activity of epothilones, 141–142
investigation of tetrahydrobenzodiazepine inhibitors, 206, 208–210
sarcodictyins, 164, 166f
SAR studies of cryptophycins, 178, 180–184
structure and multi-drug resistant

(MRD) sensitivity of epothilones, 140–141
studies of benzocycloheptapyridine FTIs, 217–224
See also Cryptophycins; Farnesyltransferase inhibitors (FTI); Taxanes
SU5416
 inhibitory profiles, 288*t*
 structure, 286*f*
 vascular endothelial growth factor receptor (VEGFR)-kinase inhibitors, 291, 293
 See also Vascular endothelial growth factors (VEGFs)
SU6668
 inhibitory profiles, 288*t*
 portrayal of, docked in vascular endothelial growth factor receptor (VEGFR)-2 ATP binding site, 292
 structure, 286*f*
 VEGFR-kinase inhibitors, 291, 293
 See also Vascular endothelial growth factors (VEGFs)
Synergism. *See* Discodermolide and Taxol®
Synthesis, total, epothilone derivatives, 99–100
Synthetic compounds, synthesizing new, 3–4

T

Targets
 iterations of classical, rational improvement, 12
 selection of candidate drug molecules, 2
Taxanes
 ability to overcome resistance in human colon tumor (HCT) 116(VM)46 cell line, 45–46
 biological activity of paclitaxel 7-methyl thio ethers, 50*t*
 BMS-184476, 7-methyl thiomethyl (MTM) analog of paclitaxel, 50–52
 BMS-188797, C-4 methyl carbonate analog of paclitaxel, 48
 C-4 analogs overcoming resistance versus HCT 116(VM)46, 46*t*
 C-4 analogs with modified side chains, 46–48
 C-4 analogs with paclitaxel side chain, 45–46
 C-7 sulfur analogs, 52
 7-deoxy-6 substituted analogs, 54–55
 distal tumor activity of BMS-188797 and BMS-184476, 49*t*
 in vitro biological activity for new, 55*t*
 in vitro profile of C-4 ester and carbonate analogs, 45*t*
 in vitro profiles of two active taxanes BMS-188797 and BMS-184476, 47*t*
 MTM analogs, 49–50
 naturally derived agents, 5
 northern region modifications, 49–55
 southern region modifications, 44–48
 structures, 44
 structures of representative, 98*f*
 synthesis of 6-alpha fluoro paclitaxel, 52–54
 synthesis of 7-deoxy-6-thio methyl ether, 54
 synthesis of C-7 sulfur analogs, 51
Taxoids
 antibody-drug conjugates, 324, 327
 highly potent analogs of paclitaxel, docetaxel, and doxorubicin, 328*f*
 See also Antibody-drug conjugates
Taxoids and hybrids
 activity of 2-(3-azidobenzoyl)baccatin III, 74
 advanced second generation, 63
 discovery of IDN5109, 64–67
 dual function of second generation, 67–68

evolution from paclitaxel/docetaxel to second generation, 61–68
from 14β-OH-10 deacetylbaccatin III, 64–67
microtubule mutations, 74
multidrug resistance (MDR), 67
pharmacophore model for designing third generation, 70
pharmacophore model for microtubule-stabilizing anticancer agents, 68–71
photoaffinity labeling study identifying single amino acid residue in β-tubulin subunit, 74f
photoaffinity labeling taxoids for probing paclitaxel binding domains of microtubules and P-glycoprotein (Pgp), 73
photolabeling, for mapping Pgp, 75
photoreactive analogs of paclitaxel, 73f
schematic of evolution, 62
solid state ^{19}F NMR analyses of microtubule-bound fluoro-taxoids, 72
synthesis of SB-TE-1120, 70–71
synthesis of SB-TE-14801 and SB-TE-14811, 71
See also Microtubule-stabilizing anticancer agents
TAXOL® (paclitaxel)
activity against drug-resistant cancer cell lines, 64
advanced second-generation taxoids, 63
approval for use, 43–44, 60
comparing IDN5109 activity with paclitaxel and docetaxel, 66
cytotoxicity of antimitotic agents in drug-resistant cell line not expressing P-glycoprotein (Pgp), 89t
cytotoxicity of antimitotic agents in drug-resistant cell line overexpressing Pgp, 88t

discovery, 148
disrupting normal assembly/disassembly dynamics of microtubules, 82
dual function of second generation taxoids, 67–68
IDN5109 discovery, 64–67
initial isolation, 5
interaction sites between Taxol and β-tubulin, 82
in vivo antitumor activity assay against MX-1 mammary carcinona xenograft in mice, 66f
limitations in cancer treatment, 148–149
microtubule polymer as cellular target, 82
potency of various taxoids, 64
schematic of evolution of taxoids, 62
search for new natural products with Taxol-like activity, 82–83
stabilization of microtubules arresting cell division cycle, 60
structure, 44, 61
superior activity of IDN5109, 65
taxoids from 14β-OH-10-deacetylbaccatin III, 64–67
tubulin polymerization and microtubule stabilization properties, 150f
tumor growth delay for clinical candidate BMS-247550 versus paclitaxel in clinically derived tumor xenograft model, 110f
See also Discodermolide and Taxol®; IDN5109; Taxanes
TAXOTERE® (docetaxel)
discovery, 148
limitations in cancer treatment, 148–149
structure, 44
Taxus brevifolia, paclitaxel isolation from bark, 5
Tea-bag method, synthesizing new compounds, 4

Tetrahydrobenzodiazepine inhibitors
farnesyltransferase inhibitors, 206–211
intensive structure-activity relationship (SAR) investigation, 206, 208–210
rational drug design, 211
Tetrapeptide-based farnesyltransferase inhibitors, 202–205
Thiadiazoles
providing unprimed site binding interactions, 276
stromelysin inhibitors, 274
structure-activity relationship (SAR), 275f
See also Matrix metalloproteinase (MMP) inhibitors
Third generation taxoids. See Taxoids and hybrids
Tissue inhibitors of metalloproteinases (TIMPs), regulation of matrix metalloproteinase activity, 261
Total synthesis, epothilone derivatives, 99–100
Treatment of cancer, plants, 4–6
Trididemnum solidum, didemnin B, 5
Trocade
clinical trials for rheumatoid arthritis, 266–267
enzyme inhibitory profile, 267f
See also Matrix metalloproteinase (MMP) inhibitors
Tubulin polymerization, natural products with, 150f
Tumor angiogenesis, vascular endothelial growth factors (VEGF), 283–284
Tumor antigens. See Carbohydrate-based tumor antigens
Tumor growth delay, clinical candidate BMS-247550 versus paclitaxel, 110f
Tumor metastasis to bone, utility of matrix metalloproteinase (MMP) inhibitors, 276–277

U

UCN-01 (7-hydroxystaurosporine), cyclin-dependent kinase (CDK) modulator, 9-10
Ultra-tight binding inhibitors, reduction of cell mass, 12

V

Vascular endothelial growth factors (VEGFs)
indolinones SU5416 and SU6668, 291–293
inhibitory profiles of VEGF receptor (VEGFR)-kinase inhibitors, 288t
phthalazine PTK787/ZZK222584, 293, 295
portrayal of PTK787 in VEGF4-2-ATP binding site, 294
portrayal of SU6668 docked in VEGF4-2-ATP binding site, 292
portrayal of ZD4190 bound to VEGFR-2 ATP binding site, 290
prospects for designing second generation inhibitors as anti-angiogenic agents, 296
quinazolines ZD4190, 289
quinazolines ZD6474, 289, 290
receptors, 284
role in tumor antiogenesis, 283–284
schematic of ATP docked in binding site of VEGFR-2, 287
structure of VEGFR-2 kinase and its interaction with ATP, 285
structures of ATP and inhibitors of VEGFR-2 kinase, 286f
VEGFR-2 kinase inhibitors, 285–286, 288
VEGF. See Vascular endothelial growth factors (VEGFs)
Vinblastine
antitumor activity against human tumor xenografts, 186t

cytotoxicity of antimitotic agents in drug-resistant cell line not expressing P-glycoprotein (Pgp), 89*t*
cytotoxicity of antimitotic agents in drug-resistant cell line overexpressing Pgp, 88*t*
natural products, 4–5
Vinca alkaloids, natural products, 4–5
Vincristine
 antitumor activity against human tumor xenografts, 186*t*
 natural products, 4–5
 use with farnesyltransferase inhibitors (FTI), 192

Z

ZD4190
 inhibitory profiles, 288*t*
 portrayal of, bound to VEGFR-2 ATP-binding site, 290
 quinazoline, 289
 structure, 286*f*
 See also Vascular endothelial growth factors (VEGFs)
ZD6474
 quinazoline, 289, 291
 See also Vascular endothelial growth factors (VEGFs)